GRAPHICAL ANALYSIS

McGraw-Hill Book Co. Inc

PUBLISHERS OF BOOKS FOR

Electrical World ▽ Engineering News-Record
Power ▽ Engineering and Mining Journal-Press
Chemical and Metallurgical Engineering
Electric Railway Journal ▽ Coal Age
American Machinist ▽ Ingenieria Internacional
Electrical Merchandising ▽ BusTransportation
Journal of Electricity and Western Industry
Industrial Engineer

GRAPHICAL ANALYSIS

A TEXT BOOK ON

GRAPHIC STATICS

BY

WILLIAM S. WOLFE, M.S.

Head of Structural Department, Smith, Hinchman & Grylls. Formerly
Instructor in Architectural Engineering, University of Illinois.
Associate Member American Society of Civil Engineers.
Associate Member Society of Naval Architects
and Marine Engineers

FIRST EDITION

McGRAW-HILL BOOK COMPANY, INC.

NEW YORK: 370 SEVENTH AVENUE

LONDON: 6 & 8 BOUVERIE ST., E. C. 4

1921

PREFACE

THIS book has been developed from notes and blue prints prepared by the writer and used in his classes at the University of Illinois. Certain additional material has been added, a part of which has been briefly presented in a number of articles contributed by the writer to the technical press. The notes referred to above were modeled after a set prepared by Dr. N. C. Ricker, several years before, to whom the writer is greatly indebted.

Many additional problems might have been added and the discussion greatly extended. This is especially true of Chapters IV and VII to X inclusive. However, it has been thought best to keep the book to smaller proportions, and the writer believes that a thorough mastery of the constructions and solutions here presented will give the student an excellent grasp of the subject.

The object has been to deal with the analysis of stresses rather than with design or the computation of loads. Nevertheless it has seemed desirable to give some attention to the determination of loads, and Chapter IX has been devoted almost exclusively to design. Material and ideas have, of course, been drawn from many sources.

The writer is especially indebted to Prof. C. R. Clark for assistance in connection with the early part of the work and to Prof. L. H. Provine for encouragement and constructive criticism.

W. S. W.

DETROIT, MICH.
   October, 1921.

v

TABLE OF CONTENTS

PAGE

PREFACE.   v
NOTATION.   xiii

CHAPTER I

GENERAL METHODS

ART.
1. Introduction.   1
2. Definitions.   2
3. Graphical Construction and Measurement of an Angle.   3
4. Representation of Forces.   4
5. Composition of Forces.   6
6. Resolution of Forces.   8
7. Concurrent-Coplanar Forces, Conditions for Equilibrium.   10
8. Unknowns.   11
9. Non-Concurrent Coplanar Forces.   13
10. Funicular or Equilibrium Polygon.   14
11. Resultant a Couple.   18
12. Non-Concurrent Coplanar Forces, Conditions for Equilibrium.   19
13. Parallel Forces.   19
14. Resultant of a Number of Parallel Forces.   21
15. Reactions for a Beam.   23
16. Reactions for a Beam with a Distributed Load.   25
17. Reactions for a Beam with Inclined Loads.   26
18. Reactions for a Truss, Vertical Loads.   27
19. Reactions for a Truss with Vertical and Inclined Loads.   27
20. Horizontal Component Divided Equally between the Two Reactions.   31
21. Reactions of a Rafter Supported on Purlins.   32
22. Stresses in the Members of a Frame.   33
23. Stresses in a Bicycle Frame.   35
24. Passing a Funicular Polygon through Three Given Points.   37

CHAPTER II

CENTROIDS

25. Centroid of a Broken Line.   40
26. Centroid of an Arc.   42
27. Centroid of a Curve.   44
28. Centroids of Areas.   45

ART.   PAGE
29. Centroid of a Triangle.. 45
30. Centroid of a Quadrilateral.................................. 46
31. Centroid of a Trapezoid....................................... 47
32. Given the Centroid, Area and Distance between the Two Parallel
    Sides of a Trapezoid to Find the Lengths of the Parallel Sides..... 48
33. Centroid of a Sector.. 49
34. Centroid of a Circular Segment............................... 50
35. Centroids of Irregular Areas.................................. 51
36. Centroids of Volumes.. 53
37. Centroid of a Triangular Pyramid............................. 53
38. Complex Volumes which can be Divided into a Number of Regular
    Volumes... 54
39. Irregular Volumes, Division into Slices....................... 57
40. Centroid Located by the Use of Sections and Area Curves.......... 59
41. Stress Volumes... 61

CHAPTER III

MOMENTS

42. First Moment of a Force about a Point........................ 64
43. First Moment of a Number of Forces about a Point.............. 65
44. First Moment of an Area about a Given Axis................... 67
45. Moment Diagram for a Beam................................... 68
46. Second Moments.. 69
47. Second Moment of a Number of Parallel Forces................. 70
48. Moment of Inertia of an Area................................. 72
49. Mohr's and Culman's Methods................................. 73
50. Radius of Gyration... 74
51. Radius of Gyration of Rectangles, Parallelograms and Triangles..... 75
52. Moment of Inertia, Exact Method............................. 76
53. Higher Moments.. 77
54. Central Circle or Ellipse of Inertia.......................... 79
55. Complicated Problems... 80

CHAPTER IV

BEAMS

56. Construction of the Elastic Curve............................ 81
57. Simple Beams.. 85
58. Cantilever Beams.. 89
59. Beams with an Overhanging End.............................. 91
60. Beams with a Variable I...................................... 94
61. Beams with One End Fixed.................................... 96
62. Beams Fixed at Both Ends.................................... 99
63. Continuous Beams of Two Spans.............................. 102
64. Continuous Beams of Three Spans............................ 107

CHAPTER V

TRUSSES

ART. PAGE

65. Weight of Trusses.. 113
66. Other Weights... 114
67. Snow Load... 115
68. Wind Loads.. 115
69. Notation.. 116
70. Stresses Obtained Analytically............................ 116
71. Stresses by Analytical Moments............................ 117
72. Stresses by Graphical Moments............................. 117
73. Stresses Obtained Graphically by Joints................... 118
74. Stresses Obtained by Stress Diagram....................... 120
75. Stress Diagram, Upper and Lower Chord Loads............... 122
76. Wind Loads; Reactions and Stress Diagrams................. 124
77. Stress Diagram Combined Loads............................. 126
78. Maximum and Minimum Stresses, Reversals.................. 128
79. Cantilever Trusses.. 134
80. Mill Bent... 137
81. Three-Hinge Arch.. 141
82. A Large Three-Hinge Arch.................................. 145
83. A Large Mill Bent... 147
84. Cantilever Truss with Four Supports....................... 151
85. Combination Truss, Three-Hinge Arch and Mill Bent......... 154
86. Bridge Trusses of the K Type.............................. 156
87. Trussed Dome.. 158
88. Ring Dome, Dead Loads..................................... 159
89. Ring Dome, Wind Loads..................................... 163

CHAPTER VI

MOVING LOADS

90. Single Concentrated Moving Load........................... 168
91. Single Concentrated Moving Load and a Uniform Dead Load.... 171
92. Uniform Moving Load Longer than the Span.................. 175
93. Moving Uniform Load Shorter than the Span................. 177
94. Uniform Dead Load, and Uniform Moving Load Shorter than the
      Span... 179
95. Two Concentrated Moving Loads............................. 179
96. Three Concentrated Moving Loads........................... 184
97. Four Concentrated Moving Loads and a Uniform Dead Load.... 187
98. A Large Number of Concentrated Moving Loads............... 190
99. Maximum Shears and Moments in a Turntable................. 193
100. Moving Loads on Trusses................................... 196

CHAPTER VII

MASONRY

ART.   PAGE

101. Stresses in Rectangular Piers.................................... 201
102. Stress Volumes... 204
103. Problems, Rectangular Piers................................... 205
104. Problems, Irregular Piers..................................... 206
105. Kerns.. 209
106. Location of Points on the Edge of the Kern Analytically........... 214
107. Pressure on Wall Footings.................................... 216
108. Retaining Walls.. 218
109. Line of Pressure in a Pier.................................... 222
110. Masonry Chimneys.. 224
111. Line of Pressure in an Arch.................................. 228
112. Three-Hinged Arches.. 231
113. Three-Hinged Arch Symmetrically Loaded...................... 237
114. Two-Hinged Arches.. 238
115. Two-Hinged Arch, Method of Least Work...................... 238
116. Hingeless Arches, Method of Least Work...................... 242
117. Hingeless Arches, General Discussion.......................... 242
118. Solution of an Arch Using Theory of Least Crown Pressure......... 245
119. Investigation of a Gothic Vault............................... 245
120. A Study of Domes... 250

CHAPTER VIII

REINFORCED CONCRETE

121. Simple Rectangular Beams..................................... 254
122. I For Rectangular Beams...................................... 259
123. T-Beams... 260
124. I of T-Beams.. 264
125. Double Reinforced Concrete Beams............................ 264
126. Bending Stresses in Complex Sections......................... 267
127. Combined Stresses.. 270
128. Eccentrically Loaded Columns................................. 271
129. Reinforced Concrete Chimneys................................ 275
130. Deflection of Reinforced Concrete Beams...................... 278

CHAPTER IX

DESIGN

131. Design of Beams.. 281
132. Design of Plate Girders...................................... 297
133. Steel Trusses; Design and Details............................. 300
134. Wooden Trusses, Design and Details........................... 305
135. Diagram for Designing Wooden Compression Members.......... 313
136. Diagram for Designing Cast Iron O-Gee Washers.............. 314

ART.   PAGE
137. Diagram for Designing Steel Angle Struts..................... 317
138. Combined Stresses.. 317
139. Trussed Beams... 319
140. Diagram for Designing Eccentrically Loaded Steel Columns....... 322
141. Diagram for Designing Eccentrically Loaded Reinforced Concrete
        Columns.. 324
142. Problems.. 330

CHAPTER X

MISCELLANEOUS PROBLEMS

143. Steel Bases with Anchor Bolts................................ 339
144. Connection of Wind Bracing Girder to Columns................. 342
145. Steel Beams Reinforced with Concrete Fireproofing and Slabs..... 344
146. Continuous Frame of Two Fixed Columns and a Girder........... 347
147. Steel Towers... 354
148. Longitudinal Strength Calculations for a Ship................. 359
149. Moment of Inertia of a Large Reinforced Concrete Section......... 363
150. Stresses in a Reinforced Concrete Beam Foundation.............. 365

Index.. 369

NOTATION

Lengths have been referred to by letters or numbers with a dash between them, thus, A–B; meaning the length from A to B.

Forces or loads are often designated by F or P either with or without subscripts or primes, but any letter or pair of letters or numbers may be used.

In connection with moments the letter M is used to indicate the first moment, while I is used to indicate the second moment, often called the moment of inertia.

The pole from which the rays of a force polygon radiate is usually called p, and the pole distance H. In connection with the constructions for moment of inertia, and the deflection of beams, the second pole distance is called H' and the second pole p'.

The formula $H' = \dfrac{E \cdot I}{H \cdot n \cdot a}$ is used in the construction for finding the deflection of beams; E is the modulus of elasticity of the material, I is the moment of inertia of the section, H is the first pole distance, H' is the second pole distance, a is a value depending upon the scale, and n is any convenient number used to magnify the vertical dimension of the elastic curve.

In the formula $\dfrac{M}{S} = \dfrac{I}{c}$, used for beams, M is the bending moment in inch-pounds, S is the maximum unit stress, I is the moment of inertia of the section, and c is the distance from the neutral axis to the outermost fiber.

The left reaction of a beam or of a truss is usually called R_1 and the right R_2.

The two formulas $S_M = \dfrac{P}{A} + \dfrac{P \cdot e \cdot c}{I}$ and $S_m = \dfrac{P}{A} - \dfrac{P \cdot e \cdot c}{I}$ are used in connection with eccentric loads; S_M is the maximum unit stress on the section, S_m is the minimum unit stress on the section, P is the load, A is the area of the cross-section, I is the moment of inertia of the section using the axis about which bending occurs, e is the eccentricity, and c is the distance from the neutral

xiii

axis, to the fiber which has maximum stress when S_M is being found, and to the fiber which has minimum stress when S_m is being found.

In the formula $e_1 = \dfrac{I}{A \cdot c}$, e_1 is the distance from the centroid of the section to the edge of its kern measured at right angles to the neutral axis about which I is the moment of inertia, A is the area of the section and c is the distance to the outermost fiber on the opposite side of the neutral axis from that on which e_1 is being found.

In connection with reinforced concrete the letter n is used for the ratio of the modulus of elasticity of the steel to that of concrete.

The formulas $f_c = \dfrac{M \cdot c_c}{I}$ and $f_s = \dfrac{M \cdot c_s}{I} \cdot n$ are used in connection with reinforced concrete; f_c is the maximum unit stress in the concrete, f_s is the maximum unit stress in the steel, M is the bending moment in inch-pounds, c_c is the distance from the neutral axis to the outermost fiber of concrete, c_s is the distance from the neutral axis to the outermost fiber of steel, I is the moment of inertia, and n is the ratio of the moduls of elasticity of steel to that of concrete.

R_t is used for the resultant of all the tensile stresses, and R_c for the resultant of all the compressive stresses.

Other notation is explained in more or less detail where used.

GRAPHICAL ANALYSIS

CHAPTER I

GENERAL METHODS

1. Introduction.—There are numerous problems encountered in engineering and technical work that can be conveniently solved by means of graphical or geometrical constructions. There are of course, certain problems for which a graphical solution is undesirable, there are others for which either a graphical or an analytical solution is satisfactory and there are some for which the graphical solution has distinct advantages. In some cases a combination method, one using part analytical and part graphical methods is desirable.

The writer believes it would be to the advantage of many engineers and students if they were more familiar with graphical solutions. Often such solutions will save much time and labor, and an excellent check is obtained if the problem is solved graphically and certain parts checked analytically.

The graphical method often checks itself as will be seen later in connection with trusses.

Many students find it easy to follow through the graphical solution and see just what is being done. Also graphical constructions often make it easier to understand how the stresses act and how they may be analyzed. Therefore one of the best ways to teach analysis of stresses is by the use of graphics. In practice a great many problems are solved by what is called judgment. The better a man understands how the stresses follow through a member or structure, the better his judgment will be. Consequently it may sometimes be desirable to solve problems graphically in order to obtain a better understanding of the distribution of the stresses. This improves the individual judgment, so that the student will be better fitted to deal with similar problems in the future, when he may not have time to work out exact solutions.

Many problems which require for their analytical solution the development of complicated formulas can be solved by graphical constructions without the use of formulas. It is easier for the average student to understand why he goes through certain constructions than for him to understand the development of complicated formulas.

2. Definitions.—Brief definitions for a number of terms and expressions are given below. More complete definitions for some of them will be given later in certain parts of the text; also additional terms will be defined when used.

Concurrent Forces.—Forces whose action lines have one point in common.

Coplanar Forces.—Forces whose action lines all lie in the same plane.

Concurrent-coplanar Forces.—Forces whose action lines lie in one plane and have a common point.

Action Line.—The path in space along which a force acts.

Magnitude.—The size of a force as measured in pounds or some other unit.

Direction.—The direction of the action line of a force with respect to some reference line, usually shown by giving the angle which the action line makes with some reference line.

Sense.—The way a force acts along its action line, usually shown by an arrow or by the notation.

Vector.—A line used to represent a force in direction, magnitude, and sense, but not in action line.

Point of Application.—The point along the action line of a force at which it may be considered that the force is applied.

Equilibrium.—A system of forces may be said to be in equilibrium when the various forces composing the system balance each other; or in other words when their resultant is zero. Or we may say that any system of forces is in equilibrium when any one of them is the anti-resultant of all the others.

Composition of Forces.—Combining the forces of a system in order to obtain an equivalent system with a fewer number of forces, perhaps only one.

Resolution of Forces.—The reverse of the composition process. The process of finding for a given system of forces an equivalent system with a larger number of forces.

Component.—When a force is resolved into two or more forces, these new forces are called components of the original force.

Resultant.—The result of the composition process is called the resultant. The term is usually applied to one force which is the equivalent of a system.

Anti-resultant.—The anti-resultant is a force having the same magnitude, direction, and action line as the resultant but an opposite sense.

Force Polygon.—Is a polygon of vectors. When the polygon has only three vectors, which form the sides of a triangle, it is called a force triangle.

Funicular Polygon or Equilibrium Polygon.—A polygon of strings with each vertex at the action line of a force. The strings are parallel to the rays of a force polygon, and these rays are lines extending from the vertices of the force polygon to some common point called the pole.

The funicular polygon is one of the most important constructions used in connection with graphical analysis, and the student should not be content with the above brief and incomplete definition, but should give careful attention to the discussion and illustrations that will be given later.

3. Graphical Construction and Measurement of an Angle.— It may be well at this time to consider a graphical method of constructing an angle of a given number of degrees, without the use of a protractor, and also a graphical method of measuring the number of degrees in a given angle.

In Fig. 1 the line A–C is to be drawn making an angle of $31°\,20'$ with the line A–B which it is to intersect at any point A. Measure off the length A–e equal to 10 units. Now from a table of tangents, the tan of $31°\,20'$ is found to be .6088. The line e–d is drawn at right angles to A–e and of length equal to 6.088 of the units used for the line A–e. Connect A and d and the line C–A makes the desired angle with A–B, because A–e–d is a right angle and e–$d \div A$–e is the tan of $31°\,20'$.

Let it be required to measure the number of degrees in the angle C–A–B, shown in Fig. 2. From A measure off 10 units to the right, and locate point e. From e a line at right angles to

A–B is drawn and the intersection *d* obtained. The length of *e–d* is measured, using the same unit that was used for *A–e*. This length which was found to be 7.86, divided by 10 gives the

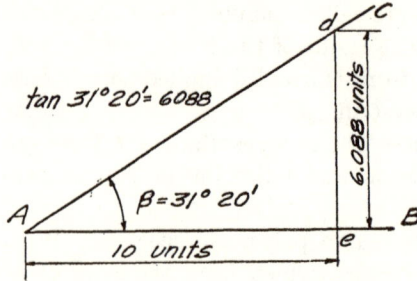

Fig. 1.

tan of β', and by referring to a table of tangents, β' is found to be 38° 10′.

4. Representation of Forces.—Forces will be represented by lines, areas, and volumes. A line is used for a concentrated force, an area for a force distributed along a line, and a volume for a force distributed over an area. In Fig. 3 the line *A–B* represents a

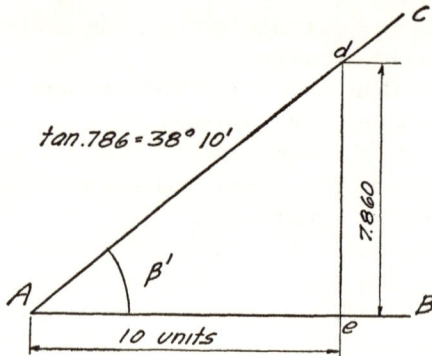

Fig. 2.

concentrated force, or a force acting along a line. A unit of *A–B* may represent a certain number of pounds, in which case the length of *A–B* measured to scale gives the magnitude of the force.

The area above the line *C–D* represents a force or load distributed along the line *C–D*. The ordinate *X* shows the intensity of the force in pounds per inch or in some other unit, and any length $Y-Z$ times the average intensity along it gives the total external force applied to it.

The area $K-L-M-N$ shows the top of a pier on which is applied a

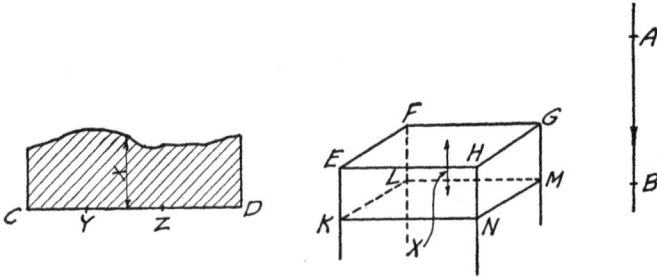

FIG. 3.

distributed force represented by the volume $E-F-G-H-K-L-M-N$, the ordinate X representing the intensity in pounds per unit of area, or in some other unit.

Fig. 4 shows a body with the concentrated force F applied at A. The line $B-C$ is *the action line* of the force F, and A is the *point of application*. The *direction* of the action line of F is indicated by the angle which it makes with the vertical or horizontal or some other reference line. *The sense* (indicated by an arrow or by the notation) shows the end of the action line towards which the force acts. Sometimes the term direction is used to indicate both sense and direction.

The line $D-E$ is a vector, or in other words it is a line which represents the force F in *direction, sense,* and *magnitude* to some *scale*, but not in *action line*. The distinction between vectors and action lines is very important, and

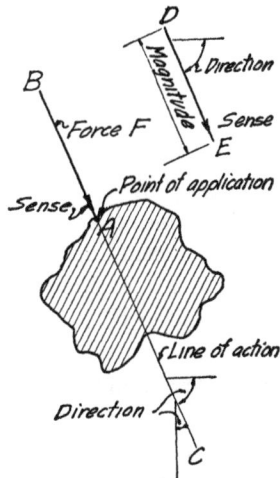

FIG. 4.

they should not be confused. The action line is the path in space along which the force acts, and it can not be moved without moving the force. On this action line we can show by a point where the force is applied and by an arrow its sense. Also by scaling off a certain length along the action line we can represent the magnitude of the force to some convenient scale. But the action line remains fixed with respect to the other forces of the system or with respect to the body on which it acts.

The vector, a line used to represent a force in direction, sense, and magnitude, may be moved any place as long as it is kept parallel to the action line of the force it represents. In Fig. 4, the line B–C is the action line of the force F, and D–E is a vector representing the force. The line B–C remains fixed with respect to the body, but the line D–E may be moved about. The length of D–E measured to the proper scale, gives the magnitude of the force F.

5. Composition of Forces.—In Fig. 5 two forces F_1 and

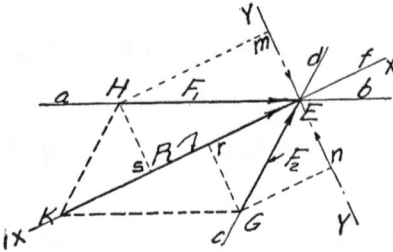 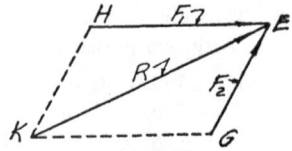

Fig. 5.   Fig. 6.

F_2 are shown with action lines a–b and c–d. A parallelogram of forces E–H–K–G is constructed and the resultant R obtained. A proof for this construction may be demonstrated in the laboratory, or geometrically in the following way. Let the line K–f in Fig. 5 be the x–x axis, then m–E is the y component of F_1 and s–E is its x component. Also n–E is the y component of F_2 and r–E is its x component. Since m–E and n–E are of equal magnitude and have the same action line but opposite sense, they neutralize each other. It follows that the resultant of the given forces must be the sum of the x components. But s–E plus r–E equals K–E, the diagonal of the parallelogram. Therefore the diagonal of the parallelogram gives the magnitude and direction of the resultant. The parallelogram E–H–K–G might have been drawn any place on the sheet, as shown in Fig. 6, in place of having E–H and E–G coincide with the action lines of the forces. In which case the action line of R would still pass through the intersection of a–b and c–d, and would be parallel to the diagonal of the parallelogram.

In Fig. 7 we have the same two forces F_1 and F_2 with action lines a–b and c–d intersecting at H. The vector K–G is laid off parallel to the action line of F_2, and equal to F_2 to some convenient scale. From G the vector G–E is laid off parallel to the action line of F_1 and equal to F_1, using the same scale that was used for F_2. The line K–E closes the force triangle and gives the magnitude and direction of the resultant R, but the action line passes through H. It should be noted that the triangle K–G–E in Fig. 7 is just half of a parallelogram similar to the one shown in Fig. 6. The side K–E of the triangle K–E–G is the same as the long diagonal of the parallelogram K–G–E–G'.

The anti-resultant, a force which would hold F_1 and F_2 in

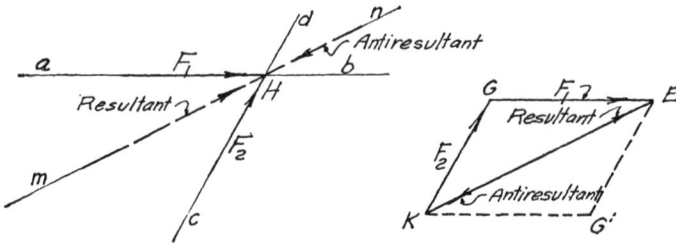

Fig. 7.

equilibrium, has the same magnitude, direction and action line as the resultant, but is of opposite sense as shown in Fig. 7.

In Fig. 8 the space diagram of five concurrent coplanar forces is shown. Their resultant is desired. The vectors are laid off in Fig. 9. Each vector must be drawn parallel to the action line of the force it represents, and when measured to the chosen scale it must equal the magnitude of that force. The force polygon was started at A, and the vectors laid off one after another. It really makes no difference in what order the vectors are laid off, because, starting at A, the end of the last vector will be at N no matter what the order. If F_4 is laid off after F_2 its end is at D', but the end of F_3 falls at E and the end of F_5 at N, the same as when the order was F_1, F_2, F_3, F_4, and F_5. The polygon must be drawn so that if one starts at A and follows along the polygon to B and then on around to N he is always moving in the direction of the arrows which indicate the sense. The line connecting A, the starting point of the polygon, and N gives the direction and magnitude of the resultant. An arrow pointing from A, the starting point of the polygon, towards N indicates the sense of the resultant. Of course the anti-resultant has an opposite sense.

The student may at first think that the force polygon of Fig. 9 is a long step in advance of the force triangle shown in Fig. 7. His attention is therefore called to the fact that the force polygon may be split into a number of force triangles, as shown by the lines A–C and A–D. The first force triangle is A–B–C, the second A–C–D, etc. The line A–C gives the resultant of F_1 and F_2, the

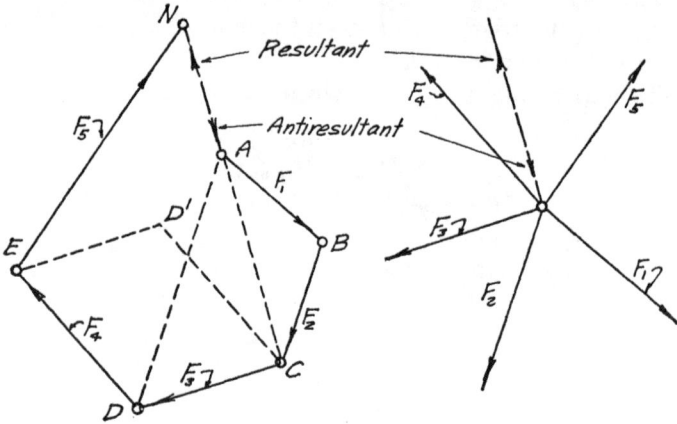
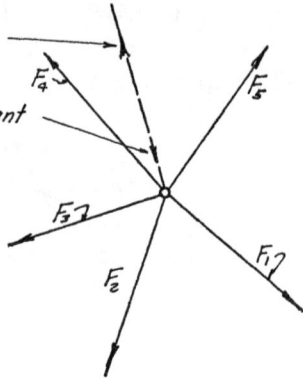

Fig. 9.   Fig. 8.

line A–D gives the resultant of this resultant and F_3, etc. In this connection see also Figs. 21 and 22.

6. Resolution of Forces.—In Fig. 10 consider the force F with action line a–b. Let it be required to resolve this force into two components at the point P. The line C–D, a vector for F, is drawn. Now if any line D–O_n is drawn from D, and any line C–O_n from C intersecting D–O_n at O_n, these two lines are vectors of components of F. The action lines of these components pass through P and are parallel to C–O_n and D–O_n. It is evident that there are an infinite number of pairs of such components. If the action lines of the components are given, then there is just one solution. Suppose g–h and m–n are given as the action lines of the two components F_2 and F_3. From C draw a line parallel to g–h and from D a line parallel to m–n intersecting the first line at E.

Then *C–E* gives the magnitude of F_3 and *E–D* that of F_2. The arrow showing the sense of *F* points from *C* to *D*. Go from *C* to *E* and then up to *D*; arrows showing the direction of the movement give the sense of F_2 and F_3.

In Fig. 11 a force *F* is shown with action line *a–b*. A vector representing *F* is shown as *C–D*, from *C* a horizontal line is drawn and from *D* a vertical intersecting the horizontal at *G*. It follows that *C–G* is the vector for the horizontal component and *G–D* the vector for the vertical component. The horizontal might have been drawn from *D* and the vertical from *C*, the same result being obtained; that is *V′* and *H′* would be equal to *V* and *H*. The

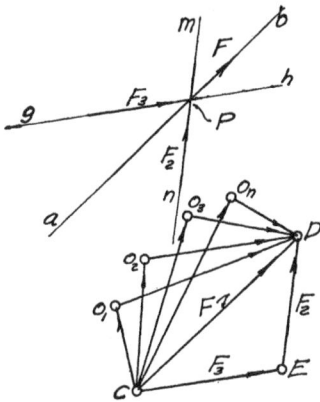

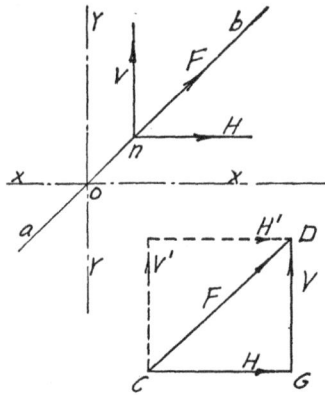

FIG. 10.　　　　　　FIG. 11.

action lines of *H* and *V* must intersect on the line *a–b*, the action line of *F*, but *n* may be any point along this line. The horizontal component is, of course, parallel to the *X* axis and at right angles to the *Y* axis. When considering a system of forces the summation of the horizontal components is often spoken of as summation F_x and the summation of the vertical components summation F_v. In many cases it is convenient to use the horizontal and vertical components of a force in place of the force itself.

In Fig. 12 resolution of the force *F* into many components is considered. The vector for *F* is shown as *C–D*. From *C* draw the line *C–L* of any length and with any direction; from *L* draw the line *L–E* with any length and with any direction, and continue the process until the desired number of vectors is obtained, the last one of course ending at D. Then F_1, F_2, F_3 . . . F_n are components of F. All of their action lines pass through O and are parallel to the corresponding vectors in the force polygon. These are, of course, an infinite number of groups of such components.

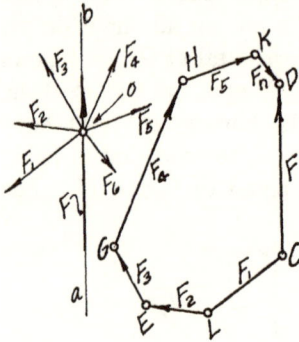

Fig. 12.

If the number of components required is given as well as their direction and sense, and the magnitude of all but two, we have a definite problem and there is just one answer.

7. Concurrent-coplanar Forces, Conditions for Equilibrium.—In Figs. 8 and 9 the resultant of a number of concurrent-coplanar forces was found, but nothing was said regarding conditions for equilibrium. Analytically such a system of forces is in equilibrium when summation F_x and summation F_y are each equal to zero; graphically the force polygon must close and the sense going around it must be continuous.

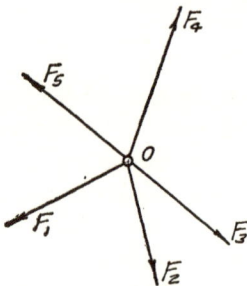

Fig. 13.

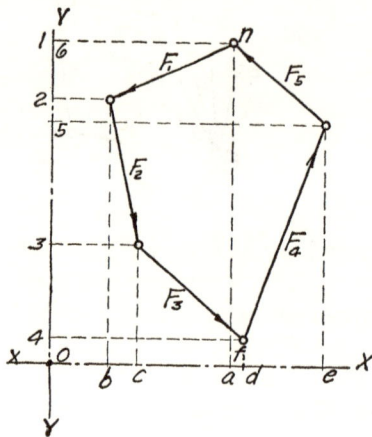

Fig. 14.

In Fig. 13 the space diagram for five forces in equilibrium is shown, and the force polygon has been drawn in Fig. 14.

The ends of the vector for F_1 are projected to the Y axis, giving the length 1–2 which represents the magnitude of the Y component of F_1. Similarly a–b represents the X component of F_1.

The X and Y components of the other forces are obtained in like manner. The Y component of F_1 is represented by 1–2, that of F_2 by 2–3, and that of F_3 by 3–4 all acting down. The Y component of F_4 is represented by 4–5 acting up, therefore its length is subtracted from 1–4. The Y component of F_5 is represented by 5–6, acting up, and if the polygon closes 6 will fall on 1 and summation F_y will be zero. The X components may be treated in a similar way.

Now if the force polygon were started at any point n and did not close, there would be a gap between n and the end of F_5, which would mean that either summation F_x or summation F_y or both, were not equal to zero. Also it is evident that if the sense of any one of the forces F_1, F_2 . . . F_5 were changed, the direction or sense of one of the Y components and of one of the X components would be changed and summation F_x or F_y would no longer be equal to zero even if the polygon did close. *Therefore, for equilibrium, the force polygon must close and the sense going around it must be continuous.*

8. Unknowns.—Given a system of concurrent-coplanar forces in equilibrium, it is possible to find two unknowns, sense and direction being considered together as one unknown. Two unknown magnitudes may be found, or two unknown directions, or one unknown direction and one unknown magnitude. The unknown direction of one force and the unknown magnitude of another or the unknown direction and magnitude of one force may be found. The last-named problem was solved in Figs. 8 and 9 by finding the anti-resultant.

(a) *Two Unknown Magnitudes.*—In Fig. 15 we have four forces in equilibrium, the magnitude of F_3 and F_4 being unknown. The force polygon in Fig. 16 is started at A and continued to C, using the known magnitudes of F_1 and F_2, and any convenient scale. From C a line is drawn parallel to the action line of F_3, and from A a line parallel to the action line of F_4. The intersection D marks off the vectors for F_3 and F_4, and these vectors, measured to the scale that was used for F_1 and F_2, give the unknown magnitudes. Attention should be called to the fact that the line parallel to F_4 might have been drawn from C, and that parallel to F_3 from A, and the same results obtained because C–D' equals D–A and D'–A equals C–D, A–D–C–D' being a parallelogram.

(*b*) *Two Unknown Directions.*—In Fig. 17, five forces in equilibrium are given, but the direction of two of them is unknown. In Fig. 18 the force polygon is started, using the forces with known magnitudes and directions. Then using D as a center an arc is drawn with a radius equal to the magnitude of F_4, mea-

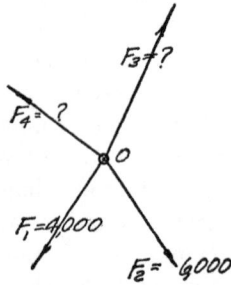

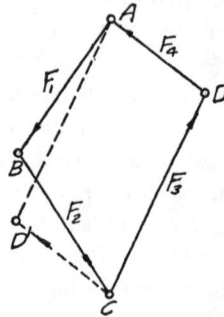

Fig. 15.   Fig. 16.

sured to the same scale that was used for F_1, F_2 and F_3. Now using A as a center an arc is drawn with a radius equal to F_5. The intersection of these two arcs locates the point E and the vectors for F_4 and F_5 are drawn as shown. Since the system of forces is in equilibrium, the sense going around the force polygon

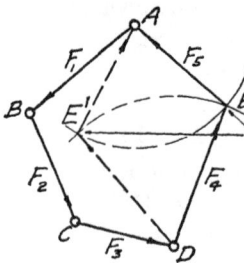

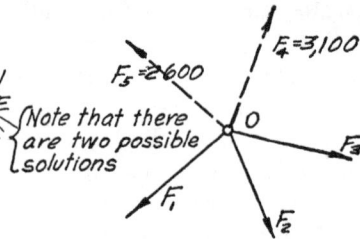

Fig. 18.   Fig. 17.

must be continuous, therefore the sense of F_4 and F_5 must be as indicated by the arrows. The action lines for F_4 and F_5 are drawn through O in Fig. 17 and made parallel to the corresponding vectors in Fig. 18. It is well to keep in mind that the arc having a radius equal to F_4, might have been drawn from A as a center, and that with F_5 as a radius from D as a center. The same results, that is the same directions for F_4 and F_5, would have been obtained, although the force polygon in Fig. 18 would have been of different shape.

(*c*) *One Force with an Unknown Magnitude and another with an Unknown Direction.*—Fig. 19 is the space diagram for four forces which are in equilibrium. The magnitude of F_3 is unknown, also the direction of F_4. Starting at A in Fig. 20, the vectors for F_1 and F_2 are laid off to some convenient scale, and from C a line is drawn parallel to the action line of F_3. With A as a center an arc is drawn with a radius equal to the magnitude of F_4 and the intersection D obtained. This point locates the end of the vector for F_3 and the line D–A is the vector for F_4. The action line of F_4 is parallel to D–A, and its sense must be as shown by the arrow

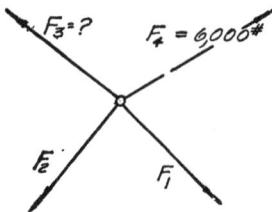

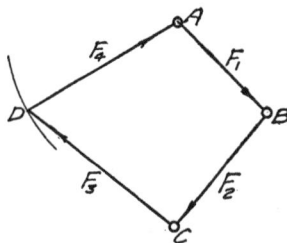

Fig. 19.   Fig. 20.

in order to make the sense going around the polygon continuous. The line for F_3 might have been drawn from A and the arc drawn from C, and the same results obtained. Note that in Fig. 20 the arc might be extended and another intersection obtained with the line C–D. This indicates that there are two possible solutions for the problem.

9. Non-concurrent Coplanar Forces.—The forces considered above have been concurrent, that is, the action lines for all the forces in any one system have had a common point. When such a system of forces is not in equilibrium the action line of their resultant passes through this common point. If the system under consideration is non-concurrent, no point on the action line of the resultant is known, therefore one must be located before the action line of the resultant can be drawn. Perhaps the first method that occurs to the student is to find the resultant of two of the forces by means of a parallelogram of forces, then the resultant of this resultant and a third force, by another parallelogram of forces, and then the resultant of this second resultant and the fourth force by the use of a third parallelogram of forces, etc.

The method shown in Figs. 21 and 22 is somewhat similar except that the force triangle is used. In Fig. 22 the vectors for F_1 and F_2 are laid off to some convenient scale, and the direction and magnitude of their resultant R_1 is given by the line A–C. The action line of R_1 passes through O, the intersection of the action lines of F_1 and F_2. The action line R_1 is now extended until it intersects that of F_3 at p. From C, the vector for F_3 is drawn and the point D obtained. The direction and magnitude of the

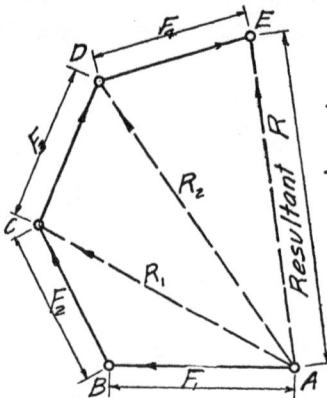

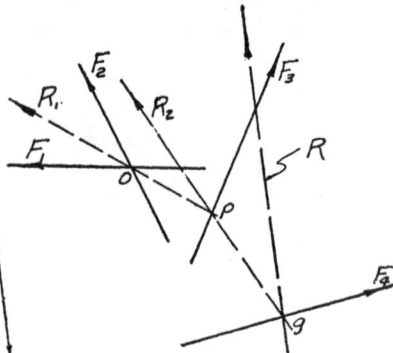

FIG. 22.   FIG. 21.

resultant of F_3 and R_1 is given by the line A–D, and the action line of this resultant R_2 passes through p. The action line of R_2 is extended until the action line of F_4 is intersected at g. The vector for F_4 is now drawn as shown, and the direction and magnitude of the resultant R is given by the line A–E. The action line of R passes through g and is parallel to A–E. This construction will be found less complicated than using parallelograms of forces, especially when there are a large number of forces in the system.

10. Funicular or Equilibrium Polygon.—There is another construction that may be used to solve a problem like the one just considered. While it may not be any more convenient for this particular problem, it has a very wide application, in fact it is one of the most important constructions with which we have to deal. It is called the funicular or equilibrium polygon and

will be presented here not only because it offers a good solution
of the above problem, but also because this is a good time for the
student to become acquainted with the construction.

In Fig. 23 four non-concurrent forces similar to those in Fig.
21 are shown. In Fig. 24 a force polygon is drawn and the magnitude and direction of R obtained. It now remains to locate the
action line of R which will, of course, be parallel to $A-E$. In
Fig. 24 choose any point for the pole p and draw the rays $A-p$,
$B-p \ldots E-p$. Now $A-p$ and $p-B$ are components of F_1, and
they may therefore be substituted for F_1. Also $B-p$ and $p-C$
are components of F_2, etc. At point 1, any point on the action

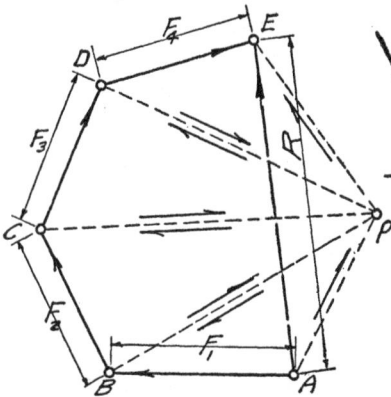 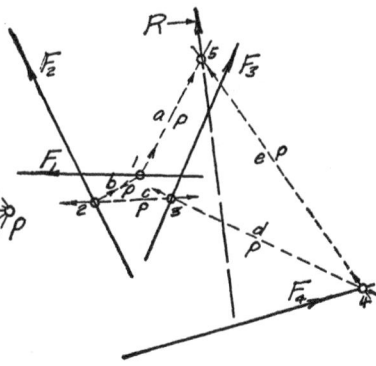

FIG. 24.　　　　　　　　FIG. 23.

line F_1, draw the strings $a-p$ and $p-b$ parallel to the rays $A-p$
and $p-B$ respectively, and extend $p-b$ until the action line of F_2
is intersected at point 2. At point 2 resolve F_2 into its two components $b-p$ and $p-c$; $p-b$, one of the components of F_1, has the
same action line as $b-p$, which is one of the components of F_2.
These two forces, $p-b$ and $b-p$, have the same direction, magnitude, and action lines, but opposite sense, therefore they neutralize
each other. The action line of the component $p-c$ is extended
until the action line of F_3 is intersected at 3. At this point F_3
is broken into its components $c-p$ and $p-d$ and $c-p$ neutralizes
$p-c$. The action line of the component $p-d$ is extended until the
action line of F_4 is intersected at 4, at which point F_4 is broken
into its two components $d-p$ and $p-e$, $d-p$ neutralizing $p-d$.

Now for each one of the four forces F_1, F_2, F_3, and F_4 two components have been substituted, and all of these components except a–p and p–e have neutralized each other. The resultant of a–p and p–e will therefore be the resultant of the system. Their action lines when extended intersect at 5, which is, therefore, a point on the action line of the resultant. And it follows that the action line of R is obtained by drawing a line through 5 parallel to A–E.

To sum up the above construction, each one of the vectors in Fig. 24 is broken into two components, each one of these components having one end at p. This means that one of the components of F_1 has the same direction and magnitude as one of the components of F_2, but an opposite sense. Also the other component of F_2 has the same direction and magnitude as one of the components of F_3, but an opposite sense, and so on.

Now in the space diagram the components shown in Fig. 24 are substituted for the various forces at points such that the action lines of the components that have the same magnitude and direction will coincide, and, since they have opposite sense, they will neutralize each other. At last there are only two components left and their resultant is the resultant of the system, its action line passing through the intersection of their action lines.

Since the funicular polygon is such an important construction it will be well to study it a little more carefully at this time. It should be kept in mind that the strings in the funicular polygon are the action lines of components of the various forces, and, since the action lines of the components of a force must intersect on the action line of the force, the strings of the funicular polygon must intersect on the action lines of the forces. Also the strings that intersect on the action line of any particular force must be parallel to the rays in the force polygon that are components of the vector which represents that force. It also follows that each string must connect the action lines of the forces whose vectors join at the vertex of the force polygon at which the ray parallel to the string in question ends.

It makes no difference where the pole p is chosen, or at what point along the action line of one of the forces the funicular polygon is started. Varying the location of these points simply changes the shape of the figures but the same action line will be obtained for the resultant. It makes no difference in what order the vectors are laid off in the force polygon, as long as each string in the funicular polygon connects the action lines of the forces whose vectors join at the vertex of the force polygon at which the ray parallel to the string ends.

In Fig. 25 the four forces of Fig. 23 are shown and in Fig. 26 a force polygon is drawn in which an order different from that of Fig. 24 is used. Any pole p is chosen and the funicular polygon started at any point 1 on the action line of F_1. The component a–p is a component of the resultant, as well as of F_1. Its action line is therefore extended indefinitely. The string b–p is the action line of a component of F_1 and also of F_4, therefore it is extended

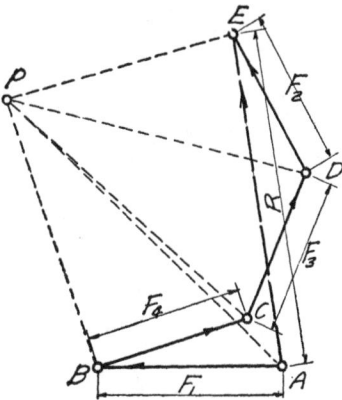

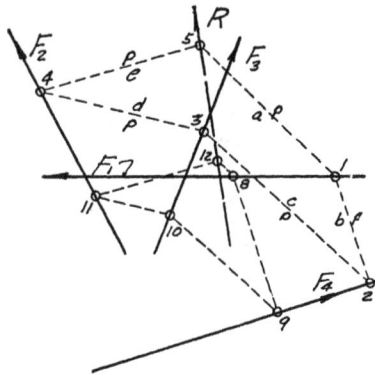

FIG. 26.   FIG. 25.

until the action line of F_4 is intersected at point 2. The string c–p is drawn from 2 parallel to the ray C–p and extended until the action line of F_3 is intersected. From 3 the string d–p is drawn intersecting the action line of F_2 at 4. The last string p–e is now drawn, and extended until the first string a–p is intersected, locating point 5, a point on the action line of the resultant R. It will be noted that the action line of R, in Fig. 25, has the same position with respect to the action lines of the forces in the system as it has in Fig. 23, showing that the different order used in the force polygon did not affect the result.

In order to show that it makes no difference at what point along the action line of a force the funicular polygon is started, a new polygon has been drawn starting at point 8. The point 12 located by the intersection of the first and last strings is found to fall on the action line of R already located.

11. Resultant a Couple.—A system of non-concurrent coplanar forces may have a force as its resultant, it may be in equilibrium, or it may have a couple for its resultant. When the force polygon does not close the resultant is a force, and the problem of locating its action line and determining its magnitude has already been considered. When the force polygon closes and the sense going around it is continuous, summation F_x and F_y are zero. Therefore the system must be in equilibrium or have a couple as its resultant.

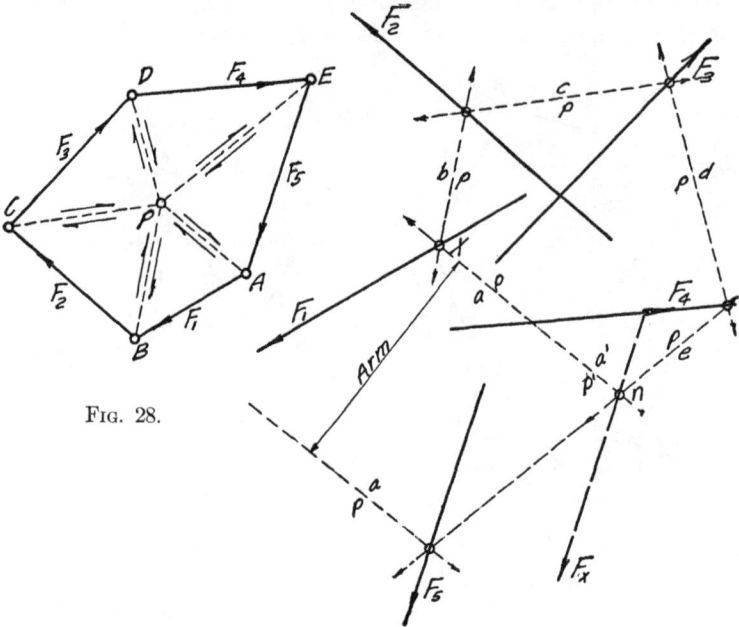

Fig. 28.

Fig. 27.

Consider a system of forces, Fig. 27, the forces of which have such magnitudes and directions that the force polygon Fig. 28 closes, and the sense going around it is continuous. The system must, therefore be in equilibrium or have a couple as its resultant. Choose any pole p in Fig. 28, and draw a funicular polygon starting at any point x on the action line of any one of the forces in Fig. 27. After drawing the first string a–p, the construction of the funicular polygon is continued until the last string is drawn. This last string is parallel to the first string a–p, but they do not coincide, therefore the funicular polygon does not

close.   Now in Fig. 27, two components have been substituted for
each of the given forces and all of these components neutralize
each other, except a-p and p-a, which are parallel to each other,
and have equal magnitudes but opposite sense.   They, therefore,
form a couple, or in other words, the resultant of the system is
a couple with arm, as shown.

The magnitude of the couple in inch pounds is given by the
product of the arm measured in inches to the scale at which the
space diagram, Fig, 27, was drawn, and the force p-A measured
in pounds to the scale that was used for the force polygon of
Fig. 28.

Now if the action line of F_5 were moved to the right, parallel
to its original position, the arm of the couple would be decreased,
and when the point n is reached it would be zero.   The funicular
polygon would close and the system would be in equilibrium
because the components a-p and p'-a' would neutralize each other.

12. Non-concurrent Coplanar Forces, Conditions for Equilibrium.—From mechanics we know that a system of non-concurrent coplanar forces in equilibrium has summation F_x and
summation F_y both equal to zero, and in addition, the moment
about any point must be equal to zero.   Graphically, when the
force polygon closes and the sense going around it is continuous,
summation F_x and summation F_y are equal to zero, and the system
is either in equilibrium or has a couple as its resultant.   If, in
addition, the funicular polygon closes, the system is in equilibrium
since summation M is equal to zero, but if it does not close the
resultant is a couple.   *In other words, for equilibrium* (1) *the force
polygon must close*, (2) *the sense going around it must be continuous*,
and (3) *the funicular polygon must close*.

13. Parallel Forces.—Parallel forces are a special case of non-concurrent coplanar forces, and the action line of their resultant
can be found by the use of the funicular polygon.   However, when
there are only two or three forces, it is often more convenient to
use a special construction.   The magnitude of the resultant is,
of course, given by the algebraic sum of the forces composing the
system.

(a) Resultant of Two Parallel Forces Having the Same Sense.—
In Fig. 29 two parallel forces having the same sense are shown,
and the action line and magnitude of their resultant is desired.
Any straight line c-b is drawn connecting the action lines of F,

and F_2. The length a–b is made equal to F_1 to some convenient scale, and c–d is made equal to F_2 to the same scale. The straight line a–d is now drawn and the intersection, o, located. This intersection is a point on the action line of the resultant, which is now drawn through o parallel to F_1 and F_2. The magnitude of the resultant is equal to the sum of the magnitudes of F_1 and F_2. A proof for this method of locating the action line of R is given below.

Triangle c–o–d is similar to triangle o–a–b

$$o\text{–}b : a\text{–}b :: o\text{–}c : c\text{–}d,$$

$$o\text{–}b \cdot c\text{–}d = a\text{–}b \cdot o\text{–}c,$$

$$o\text{–}b \cdot F_2 = F_1 \cdot o\text{–}c.$$

The moment of F_2 about o is o–$b \cdot F_2 \cdot \sin \phi$ and that of F_1 about o is o–$c \cdot F_1 \cdot \sin \phi$ and their sum is zero because they are of opposite

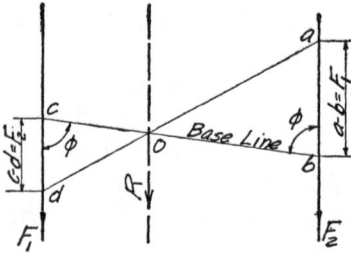

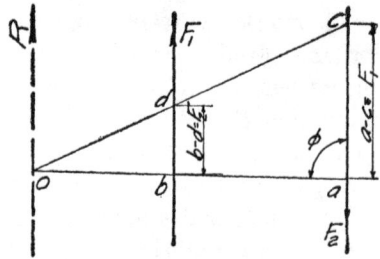

FIG. 29.   FIG. 30.

sign and of equal magnitude, since o–$b \cdot F_2 = F_1 \cdot o$–$c$. Now we know that the moment of a system of forces about any point is the same as the moment of their resultant about the same point. In this case the moment of the system about o is zero, therefore the moment of R about o must be zero, and to satisfy this requirement, the action line of R must pass through o.

(b) *Resultant of Two Parallel Forces with Opposite Sense.*—The resultant of two parallel forces of opposite sense may have its action line located by a construction similar to that given above. Consider the two forces F_1 and F_2 shown in Fig. 30. Draw any straight line a–b and extend it indefinitely in either direction. Make the line b–d equal to F_2, to some convenient scale, and a–c equal to F_1, to the same scale. Now draw the straight line c–d and extend it until the line a–b is intersected at o. The point o is a point on the action line of the resultant. The magnitude of the resultant is the algebraic sum of the two forces F_1 and F_2, which is their numerical difference. The proof for the construction just given follows:

Triangle $o\text{-}d\text{-}b$ is similar to triangle $o\text{-}c\text{-}a$

$$o\text{-}b : b\text{-}d :: o\text{-}a : a\text{-}c,$$

$$o\text{-}b \cdot a\text{-}c = b\text{-}d \cdot o\text{-}a,$$

$$o\text{-}b \cdot F_1 = F_2 \cdot o\text{-}a,$$

$$M = o\text{-}b \cdot F_1 \cdot \sin \phi - F_2 \cdot o\text{-}a \cdot \sin \phi = o.$$

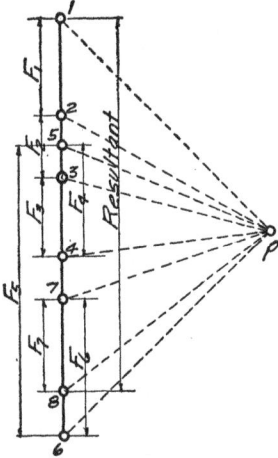

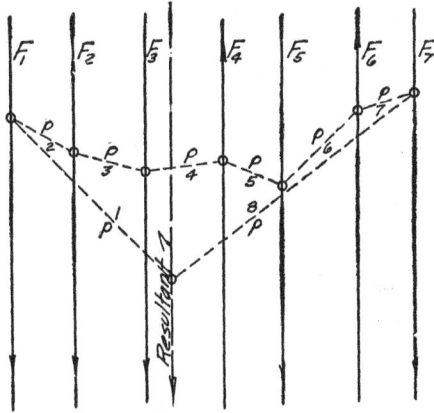

Fig. 32. Fig. 31.

The reader should note that the two constructions just given are very similar, but there is this marked difference. When the two forces have the same sense the lengths for F_1 and F_2 are laid off on opposite sides of the base line giving the intersection o between the action lines of the two forces. When the two forces have opposite sense the lengths representing their magnitudes are laid off on the same side of the base line, thus giving the intersection o on the outside near the larger force.

14. Resultant of a Number of Parallel Forces.—When the resultant of three or four parallel forces is desired, it is sometimes convenient to use the construction given above, applying it first to any two of the given forces, then to their resultant, and another, and so on until the system is reduced to one force. However, when there are a large number of forces the funicular polygon is preferable.

In Fig. 31 seven parallel forces are shown, and the magnitude, direction, and action line of their resultant is desired. The force polygon is drawn in Fig. 32 starting at point 1 with the vector for F_1. The sense of F_4 is up, so its vector is laid off extending up

FIG. 34.

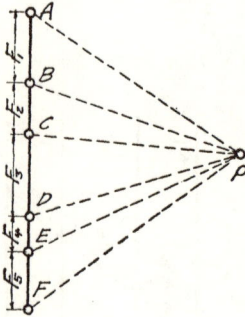

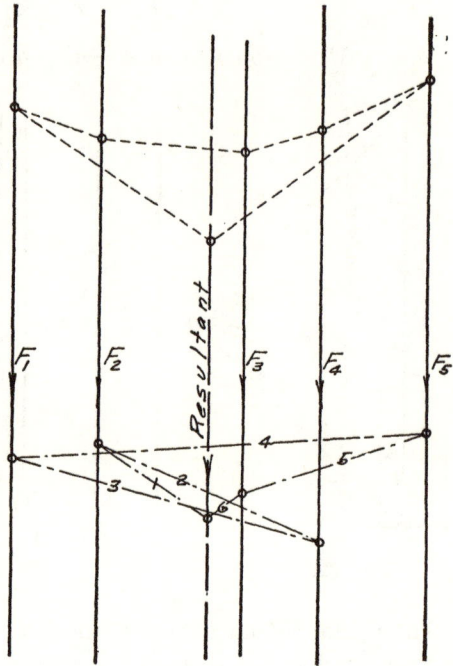

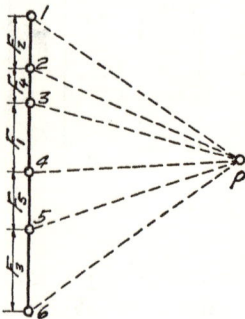

FIG. 35.   FIG. 33.

from point 4. The distance from point 1 to the end of the last vector, measured to scale, gives the magnitude of the resultant. Any point p is now chosen for a pole, and the funicular polygon drawn in Fig. 31. The intersection of the first and last strings, 1–p and 8–p, locates a point on the action line of the resultant.

The student should remember that it makes no difference in what order the forces are laid off in the force polygon, as long as the funicular polygon is correctly drawn.

In Fig. 33 five parallel forces are shown. In Fig. 34 a force polygon is drawn, taking the forces in order from left to right. From it the upper funicular polygon in Fig. 33 is constructed, and the action line of the resultant located.

In Fig. 35 another force polygon is drawn, taking the forces in a different order, and from it the lower funicular polygon in Fig. 33 is drawn. The intersection of the first and last strings is found to fall on the action line of the resultant already located. This shows that the different order used in Fig. 35 did not produce any error, but simply changed the shape of the funicular polygon.

15. Reactions for a Beam.—An ordinary beam supported at each end, and carrying a number of vertical concentrated loads is a

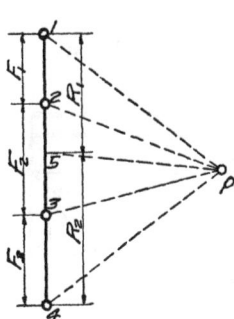

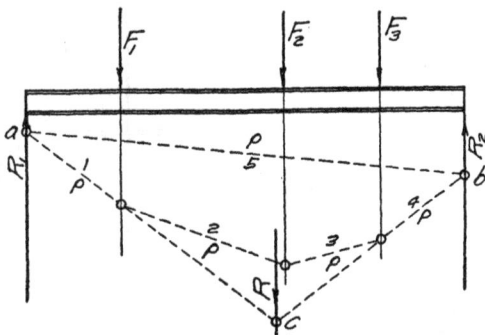

FIG. 37.   FIG. 36.

member in equilibrium acted upon by a system of parallel forces. Usually the loads are known and the reactions are to be found. The anti-resultant of a system of forces is one force which would hold the system in equilibrium. The reactions of an ordinary beam are two forces which hold the rest of the system in equilibrium. Consider the beam shown in Fig. 36 which carries the three concentrated loads F_1, F_2, and F_3. Draw the force polygon in Fig. 37 and choose any pole p. Then draw the strings p–1, p–2, p–3 and p–4 in the funicular polygon of Fig. 36. Extend the strings 1–p and 4–p until the action lines of R_1 and R_2 are intersected at a and b. Now the system of forces acting on the beam, including the reactions, is in equilibrium, therefore the funicular polygon must close. A straight line connecting a and b is the only line that will close the polygon, therefore, it must be the string p–5. Now the

strings p–1 and 5–p are the action lines of the two components of
R_1. From p, in Fig. 37, a line is drawn parallel to the string 5–p
and intersecting the line 1–4 at 5. The magnitude of the component p–1 is given by the length of the ray p–1, while the magnitude of 5–p, the other component of R_1, is given by the length
of the ray 5–p. The resultant of these two components is 5–1 or
R_1, and the line 5–1 measured to scale gives the magnitude of R_1.
The magnitude of R_2 is given by the length of the line 4–5, which is
the line that closes the force polygon.

It will be convenient to regard the vectors in Fig. 37 which
represent the loads and reactions as a force polygon even when they
all fall on a straight line. The reactions which hold the loads

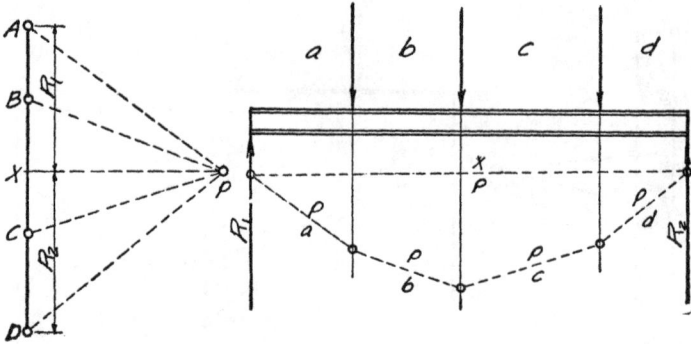

FIG. 39.   FIG. 38.

in equilibrium close the polygon, that is, connect the end with
the starting point. Attention should be called to the fact that the
action line of the resultant of the loads in Fig. 36 may be located
by extending the strings 1–p and 4–p until they intersect at c as
shown. But it is not necessary to locate this resultant in order to
find the reactions.

It is sometimes convenient to use the notation shown in Figs.
38 and 39. The force nearest the left end of the beam is called
a–b, the next b–c, etc. In the force polygon, Fig. 39, the vector
for the force a–b is laid off with one end marked A and the other B,
so that from A to B is the direction in which the force acts.
Then, starting at B, the vector for the force b–c is laid off so that
from B to C is the direction in which the force acts. This process
is continued until all of the loads are laid off. A pole p may

now be chosen, and the funicular polygon drawn in Fig. 38. Then parallel to the closing string p–X, a line is drawn in Fig. 39 from p, and the intersection X located. The string d–p is a component of the force c–d and also of R_2, therefore in Fig. 39 the vector for R_2 will have one end at D and the other at X. Its sense is up, the direction from D to X being up. The vector for R_1 is X–A, and its sense is also up, the direction from X to A being up.

16. Reactions for a Beam with a Distributed Load.—Let it be required to find the reactions of the beam shown in Fig. 40 which carries a distributed load, as indicated by the shaded area above. The vertical dimension of the shaded area at any section represents, to some scale, the intensity of the loading at that section in pounds per lineal foot or in some such unit. Also the resultant

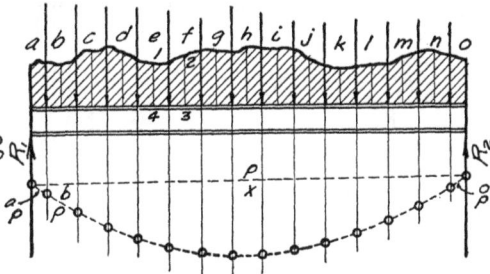

FIG. 41.   FIG. 40.

load on any length of beam 4–3 is given by the mean vertical dimension of the shaded area above, measured to scale, times the length 4–3; or in other words the area 1–2−3–4 measured to the proper scale.

The first step is to divide the distributed load into a number of slices and, for each slice, substitute an equivalent concentrated load. It will usually be found convenient to make all the slices of equal width, and obtain the magnitude of the equivalent concentrated loads by scaling the mean ordinates and multiplying by the width. The action line of the concentrated load should pass through the centroid of the slice in each case, but, if the width of the slices is small, the error in approximating the location of the centroid will be very small.

The concentrated loads that were substituted for the various slices of the distributed load are now laid off in Fig. 41. A pole

p is chosen and the funicular polygon in Fig. 40 drawn. From p a line is drawn parallel to p–X, locating X, which determines R_1 and R_2.

17. Reactions for a Beam with Inclined Loads.—The beams thus far considered have carried vertical loads only. A beam which carries inclined loads, as shown in Fig. 42, is a somewhat more difficult problem. In this particular case, R_2 is assumed to be vertical, thus simplifying things a little. The vectors for the loads are laid off in Fig. 43 in the usual way, and the pole p chosen.

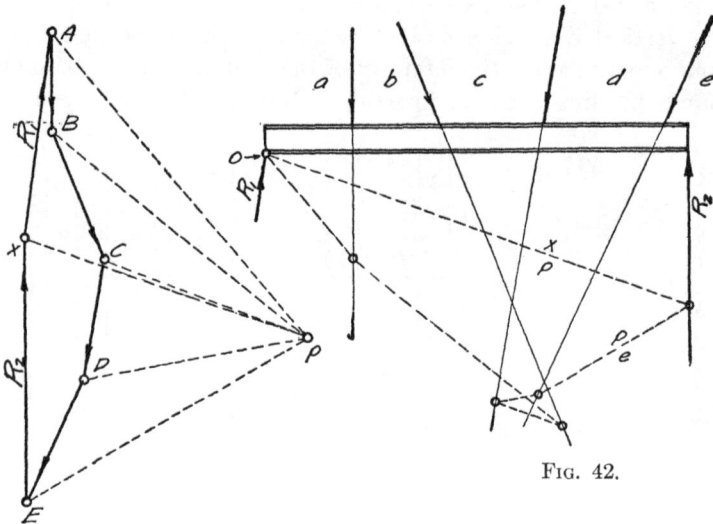

Fig. 42.

Fig. 43.

Since R_2 is vertical its action line is known, but the direction of the action line of R_1 is not known, All that is known about R_1 is that it passes through the point o. The funicular polygon is therefore started at o and constructed in the usual way. The closing string is drawn and parallel to it a line is drawn from p in Fig. 43.

The string p–e is a component of d–e and also of R_2. One end of the vector for R_2 is therefore at E in Fig. 43. A vertical line is drawn from E until the line from p parallel to the closing string is intersected, and the point X located. The length E–X gives the magnitude of R_2 and its sense is up, from E to X being up. The vector which closes the polygon is X–A. This vector gives the direction and magnitude of R_1. The sense of R_1 is up, from X to A being up, and its action line is drawn through o parallel to X–A.

It is seldom that beams have inclined loads, but trusses often do. A more detailed discussion of reactions under inclined loads will therefore be taken up in connection with trusses in this chapter and also in Chapter V.

18. Reactions for a Truss, Vertical Loads.—The construction used for finding the reactions of a beam may be used for finding the reactions of a truss. In fact a truss may sometimes be regarded as a deep beam having a large part of the web cut out. The truss

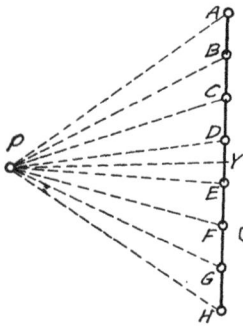
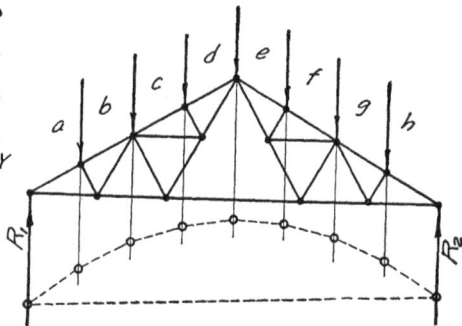

FIG. 45.   FIG. 44.

shown in Fig. 44 carries 7 loads at the panel points, and is supported at each end.

A force polygon is drawn as shown in Fig. 45, a pole p is chosen and the funicular polygon drawn in Fig. 44. Since the loads are all vertical, the reactions will both be vertical, and the funicular polygon may be started at any place along the action line of R_1 or R_2. Parallel to the closing string, a ray is drawn from p in Fig. 45, Y is located, and the magnitudes of the reactions determined,

It makes no difference where p is located, changing its location simply changes the shape of the funicular polygon in Fig. 44. If the pole p were taken on the other side of the load line A–H, the funicular polygon would be turned over.

19. Reactions for a Truss with Vertical and Inclined Loads.— When a truss is assumed to carry wind or inclined loads as well as vertical loads, the problem of determining the reactions is somewhat more difficult. In the first place, some assumption must be made regarding the way in which the horizontal component of the loading is balanced. When the resultant loading has a horizontal component one or both of the reactions must have a horizontal component. The method of supporting the truss determines to a large extent the way the horizontal component is divided between the two reactions. A roller or rocker bearing is sometimes used at one end so that the reaction there will be approximately vertical. Such a bearing also provides for expansion and contraction.

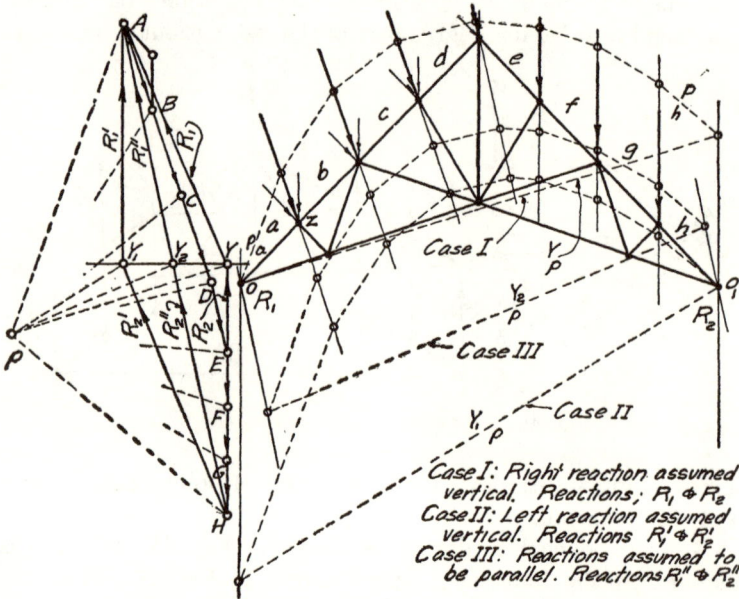

Case I: Right reaction assumed
vertical. Reactions; R_1 & R_2
Case II: Left reaction assumed
vertical. Reactions R_1' & R_2'
Case III: Reactions assumed to
be parallel. Reactions R_1'' & R_2''

FIG. 47.   FIG. 46.

In Fig. 46, a truss with vertical and inclined loads is shown. At joint Z there are two loads, a vertical dead load and a normal wind load. In Fig. 47 starting at A the vector for one of these is laid off to scale, and starting from its end the vector for the other, the end of which is marked B. The magnitude and direction of the resultant loading at joint Z is given by the vector $A–B$, and its action line passes through joint Z. This action line is now drawn passing through joint Z and parallel to $A–B$. The vectors for the loads at the other joints are laid off in order, and the action line for the resultant loads at the various joints drawn. The directions for these action lines are obtained from the resultant vectors in Fig. 47.

Now let it be assumed that the supports are so constructed that R_2 can take no horizontal component or, in other words, all of the horizontal component is taken by R_1, and R_2 is vertical. The action line of R_2 is known since it is vertical and passes through O_1, but the action line of R_1 is not known, though we do know that it passes through O. Therefore, the funicular polygon must pass through O, and the easiest way to have it pass through O is to start it there. Any convenient pole p is chosen in Fig. 47, and a funicular polygon drawn in Fig. 46, starting at O. The closing string for this polygon is drawn, and parallel to it a line from p in Fig. 47 which intersects the vertical from H at Y. The line H-Y gives the magnitude of R_2, and the line Y-A, which closes the force polygon, gives the direction and magnitude of R_1. The student is sometimes confused as to whether the end of R_1 is at the lower end of the force polygon or at the upper end. This point can easily be determined by using the idea of components. In Fig. 46 p-h is a component of the force g-h, and also of R_2. Therefore in Fig. 47, the ray p-H is a component of the vector G-H and also of R_2, and it follows that the vector of R_2 must have one end at H. The string p-a is a component of the force a-b and also of R_1, therefore the vector for R_1 must have one end at A.

So much for the case when R_2 is vertical.

Now let it be assumed that the supports are so constructed that R'_2 takes all of the horizontal component and R'_1 is vertical. For the solution of this condition the polygon must be started at O_1 which is the only known point on the action line of R'_2. The same pole has been used in Fig. 47 that was used for the first funicular polygon, but the same results could have been obtained by using any other pole, as long as the polygon was started at O_1. The second funicular polygon in Fig. 46 is now completed. The closing string is drawn, and parallel to it a line from p in Fig. 47 is extended until it intersects the vertical from A at Y_1. The line Y_1-A gives the magnitude of R'_1 for this second assumption. The line H-Y_1 which closes the force polygon gives the magnitude and direction of R'_2.

For a third assumption let it be assumed that the supports are of such a nature that the reactions are parallel to each other or, in other words, they are parallel to $A–H$, the resultant of all the loads. In this case the action line of both reactions may be drawn and the funicular polygon started any place along either of them. The funicular polygon for this case is the middle one shown in Fig. 46. Parallel to its closing string a line is drawn from p in Fig. 47 which intersects the straight line $A–H$ at Y_2. $H–Y_2$ gives the magnitude of R''_2, and $Y_2–A$ the magnitude of R''_1.

The reactions have now been obtained for three different assumptions, using a funicular polygon for the solution of each case. It will be found that the three points Y, Y_1 and Y_2 lie on

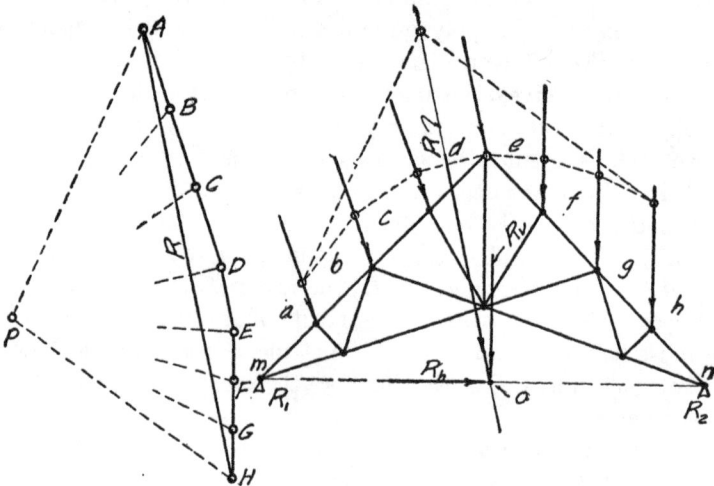

FIG. 49.   FIG. 48.

the same straight line, and that this line is parallel to the line connecting the two supports of the truss. When the two supports are at the same elevation the line is horizontal. When a truss has both supports at the same elevation we may therefore say that the vertical components of the reactions remain unchanged, no matter how the horizontal component is assumed to be divided between them. A proof for this is given in connection with Figs. 48 and 49. When any one of the points on the line $Y–Y_2$ is located, others may be found without drawing other funicular polygons. For example, suppose Y has been located by the use of a funicular polygon and we wish to find the reactions when R_1 is vertical. In place of drawing another funicular polygon in Fig.

46 in order to locate Y_1, we could have obtained Y_1 by drawing the vertical from A, and finding its intersection Y_1 with the horizontal from Y.

When the supports of a truss are at the same elevation the vertical components of the reactions are not affected by the way the horizontal component is divided between the two reactions. This has been shown in Figs. 46 and 47, but a more general proof will now be given. Let the truss shown in Fig. 48 be any truss with any loading. The force polygon is drawn in Fig. 49, and from it the funicular polygon of Fig. 48, thus locating a point on the action line of R, the resultant of all the loads on the truss. The direction and magnitude of this resultant are given by the line A–H in Fig. 49. The action line of this resultant of all the loads intersects m–n, the line connecting the two supports, at O.

At o let R be broken into horizontal and vertical components, R_h and R_v. R_h has no effect on the vertical component of R_1 or R_2 because it has no moment about either m or n. The vertical components of R_1 and R_2 are determined by R_v and the location of o. They must therefore remain constant, no matter how the assumption regarding the division of the horizontal component is changed.

When the supports are not at the same elevation it can be proved in a similar way that the components of the reactions normal to the line connecting the two supports are constant no matter how the component parallel to the line connecting the two supports is divided.

20. Horizontal Component Divided Equally between the Two Reactions.—Let it be required to find the reactions of the truss shown in Fig. 50 assuming that the supports are of equal lateral rigidity and that the horizontal component is, therefore, divided equally between the two reactions. The force polygon is drawn in Fig. 51 in the usual way, and from it a funicular polygon in Fig. 50. For convenience in constructing this polygon let it be assumed temporarily that R_2 is vertical. The funicular polygon is therefore started at m and continued until the vertical from n is intersected, then parallel to the closing string a line is drawn from p, and the intersection X is obtained. The point X marks off the reactions for the case when R_2 is vertical. Now, since a change in the division of the horizontal component does not affect the vertical components of the reactions, the vertical component of R_2

is always equal to H–X, and Y lies on a horizontal line through X. From X draw a horizontal line and from A a vertical line and obtain the intersection Z. The total horizontal component of the loading is therefore given by the line Z–X, which is now divided into two equal parts, the midpoint being Y. The line H–Y gives R_2 and Y–A gives R_1 for the case when the horizontal component is divided equally between the two reactions.

Now any other assumption might have been made regarding the proportion of the horizontal component going to each reaction,

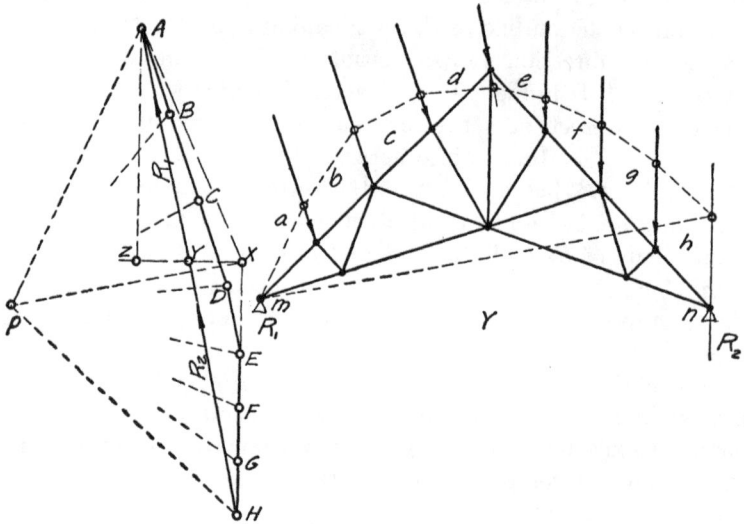

FIG. 51.   FIG. 50.

and the Y point determined by dividing the length X–Z into the proper proportions.

21. Reactions of a Rafter Supported on Purlins.—In Fig. 52, let m–n represent a rafter which is supported at m and n by purlins. We wish to find the reactions under the loading as shown. The force polygon is drawn in Fig. 53, in the usual way, and the pole p is chosen. For convenience in constructing the funicular polygon in Fig. 52, let it be assumed that R_2 is normal to the axis of the rafter. The polygon will therefore be started at O. A line is drawn from p parallel to the closing string until the line from G normal to the axis of the rafter is intersected at W. The point W marks off the reactions for the case when R_2 is normal to the axis of the rafter. Through W draw a line parallel to $m-n$ and from A a line at right angles to $m-n$ which gives the intersection Z. Now $Z-W$ is the total component of the loading parallel to $m-n$, and it may be divided in any way desired between the two reactions without affecting the normal components. Suppose the rafter is fastened to the purlins in such a way that it seems reasonable to assume that one-third of the parallel component is taken up by R_2. The reactions are obtained by locating Y on the line $Z-W$ so that $Z-Y$ is two-thirds of

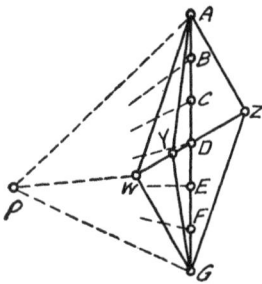
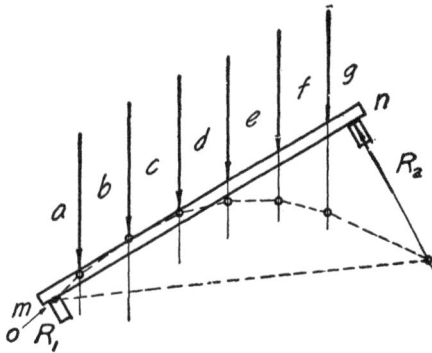

Fig. 53.   Fig. 52.

$Z-W$, and $Y-W$ one-third. Then $G-Y$ gives the direction and magnitude of R_2, and $Y-A$ of R_1.

The reactions of a truss with supports at different elevations may be found in a similar way.

22. Stresses in the Members of a Frame.—In Fig. 54 a simple truss is shown with three loads $a-b$, $b-c$, and $c-d$. The reactions may be found by graphical construction, as already explained, or by an analytical computation. At the left end of the truss, joint I, there are three forces acting, the reaction which is known, the stress in the lower chord, F_1, and the stress in the upper chord, F_2. The vector for the reaction R_1 is laid off in Fig. 55 and its ends marked Y and A. R_1 acts up, that is towards the joint, which fact we have here indicated by the use of an arrow.

The other two forces are known in direction, but not in magnitude. Through one end of the vector for R_1 a line parallel to the action line of F_1 is drawn, and through the other end a line parallel to the action line of F_2 is drawn. The intersection of these lines determines their length and therefore the magnitudes of the forces.

The sense going around the force triangle in Fig. 55 must be continuous because the forces are in equilibrium. The sense for F_1 and F_2 must therefore be as shown by the arrows in Fig. 55. F_2 acts down towards the joint which means compression in the upper chord, and F_1 acts towards the right, which means tension in the lower chord. The student should keep in mind that

Fig. 55.

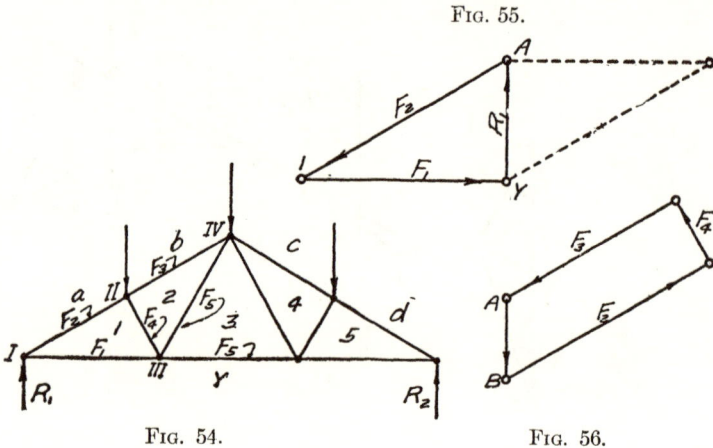

Fig. 54.   Fig. 56.

we are dealing with the forces that act at the joint, and that a force acting towards the joint means compression, while one acting away means tension. Therefore when F_2, the force which the upper chord delivers to joint I, acts towards the joint it means that there is compression in the upper chord, also since F_1, the force which the lower chord delivers to joint I, acts a way from the joint there is tension in the lower chord.

In Fig. 55 the full lines show how the force triangle was first drawn, and the dotted lines together with R_1 show how it might have been drawn.

At joint II there are 4 forces acting, F_2, F_3, F_4, and the load $a–b$. All of these forces are known in direction, but the magnitude of F_3 and of F_4 is unknown. The load $a–b$ acts down, and since we have found that there is compression in the lower section of the upper chord, F_2, when we consider joint II, will act up towards the joint. In Fig. 56 the vector for the force a–b is drawn, and its sense shown by an arrow pointing down. From B, the end of this vector towards which the arrow points, the vector for F_2 is drawn so that the arrow which indicates its sense points away from B. From A a line is now drawn parallel to F_3, and from the end of the vector for F_2 a line is drawn parallel to F_4. The intersection of these lines determines the magnitude of F_3 and also of F_4. The sense of F_3 and F_4 is shown by the arrows in Fig. 56 which are placed so that the sense going around the polygon is continuous. The sense of F_3 is found to be down, that is

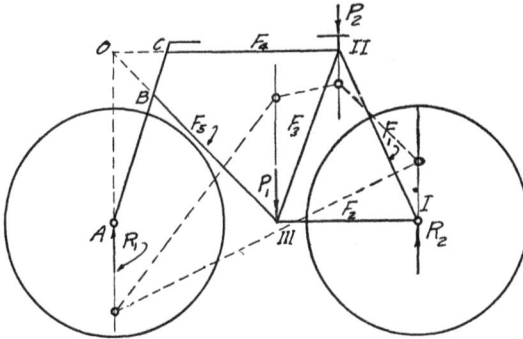

FIG. 57.

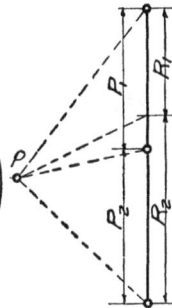

FIG. 58.

towards joint II, therefore there is compression in the member connecting joints II and IV. The sense of F_4 is found to be up, it therefore follows that there is compression in the member connecting joints II and III. This process may be continued, taking a joint at a time, until the stress in each of the different members is found.

The student will do well to check at least a part of his results analytically.

In Chapter V problems similar to the one just considered will be solved by the use of the stress diagram.

23. Stresses in a Bicycle Frame.—The bicycle frame makes a very interesting problem, and permits a good illustration of the way the stresses in a frame may be analyzed. Consider the frame shown in Fig. 57, carrying the two vertical loads P_1 and P_2.

Assume that the bicycle is at rest on a level surface, then the reactions must be vertical. They can easily be found by drawing a force polygon and funicular polygon the same as for trusses, or by an analytical computation. There are three forces acting at joint I, the reaction which has just been found in direction and magnitude, also F_1 and F_2. A force polygon is drawn for these forces in Fig. 59 and the magnitude and sense of F_1 and F_2 obtained. F_1 is found to be compression and F_2 tension. Fig. 60 shows the force polygon for joint II and Fig. 61 the polygon

FIG. 62. FIG. 61. FIG. 60. FIG. 59.

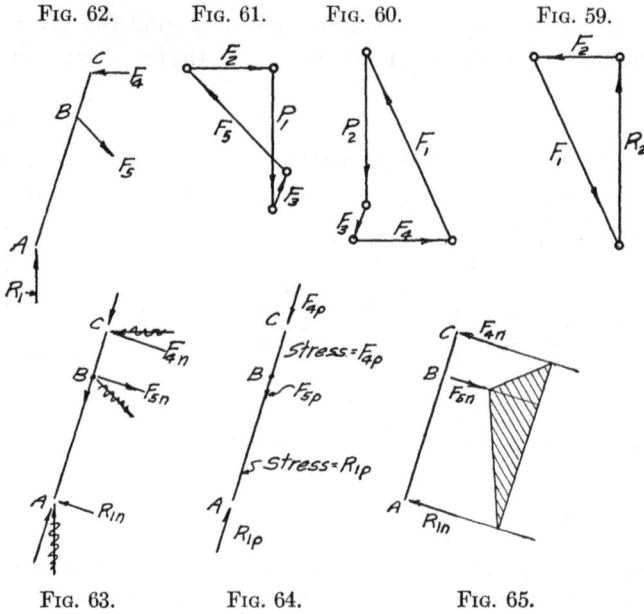

FIG. 63. FIG. 64. FIG. 65.

for joint III. All of the stresses have now been obtained except those in member A–C, and all the external forces acting on this member are known. This member and the forces acting on it are shown in Fig. 62. In Fig. 63 these forces are shown broken into their parallel and normal components. The parallel components produce direct stress, as shown in Fig. 64, and the normal components which produce bending, are shown together with the moment diagram in Fig. 65. The direct stress in A–B is just equal to R_{1p}, while the direct stress for C–B is equal to F_{4p}, and $F_{4p} + F_{5p}$ should equal R_{1p}. The maximum moment is at B

and is equal to R_{1n} times the length of A–B. The same result should be given by F_{4n} times C–B.

Attention should be called to the fact that the frame in Fig. 57 has been drawn so that the action lines of F_4, F_5 and R_1 intersect at a point O. In case these action lines did not intersect in a common point there would be a certain amount of moment thrown into all of the members of the frame.

24. Passing a Funicular Polygon through Three Given Points. Occasionally it is convenient to use the rather simple construction illustrated by Figs. 66 and 67. See Figs. 467, 468, 476, and 477. In Figs. 66 the space diagram for seven coplanar forces is shown, and a funicular polygon must be drawn which will pass through the three given points X, Y, and Z. In Fig. 67 the force polygon is drawn, taking the forces in order, and any pole p_1 is chosen. Then, starting at any point in Fig. 66, the preliminary polygon Ⓐ is drawn.

Suppose for the present we were to consider the forces between X and Y as supported on a beam coinciding with the line X–Y. Also that the beam is supported at X and Y by the reactions R_1 and R_2 which are parallel. The resultant of the four forces between X and Y is given by the line A–E in Fig. 67. Parallel to this line the action lines of R_1 and R_2 are drawn, as shown in Fig. 66, and the intersections X' and Y' obtained. The line X'–Y' may be considered a closing string, and R_1 and R_2 may be obtained by drawing p_1–U parallel to X'–Y' and locating point U. From U the line U–m is drawn parallel to X–Y. Now if any pole be taken along the line U–m and a polygon drawn starting at either X or Y, the reactions R_1 and R_2 are in no way affected because the loading has not been changed. This means that the location of U is not changed, and it follows that the closing string for this new polygon must have the same direction as m–U, which is parallel to X–Y. Therefore this closing string must coincide with X–Y because the polygon was started at one of these points. Since the closing string passes through the other point the polygon must also.

From the above discussion it is evident that if any point on the line U–m be taken as a pole and the polygon started at either of the two points X and Y, it will pass through the other. In a similar way it can be shown that if any pole be taken along the line V–n, and a polygon started at either of the two points Y and Z, it will

pass through the other.   Therefore, if p, the intersection of U–m
and V–n, be taken as the pole, and the polygon started at any one

Fig. 66.

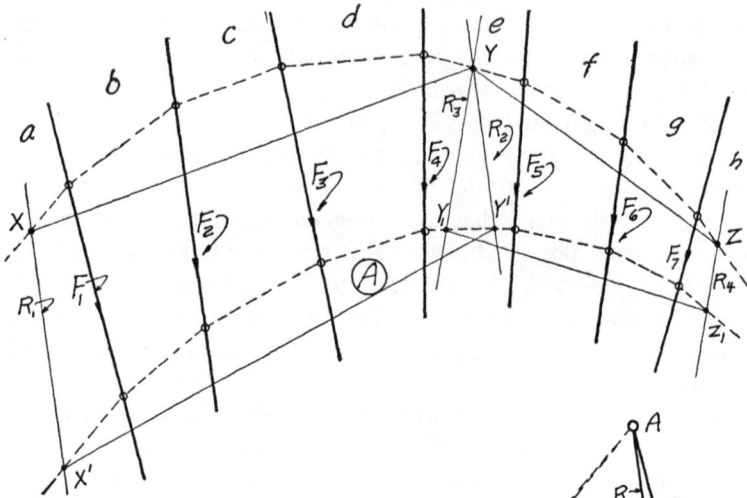

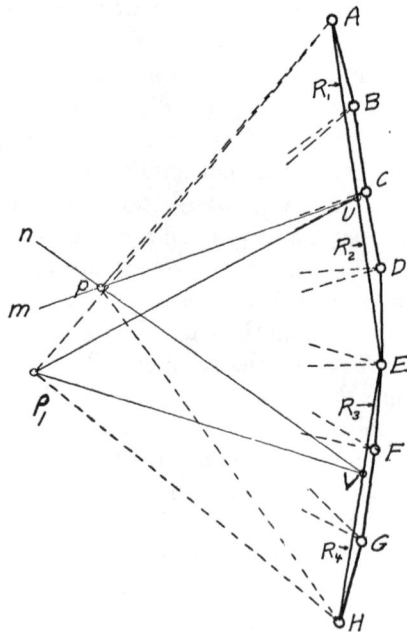

Fig. 67.

of the three points X, Y, and Z, it must pass through the other
two.

The construction explained above may be briefly outlined as follows:

Draw the force polygon of Fig. 67 and choose any point p_1. Using p_1 as a pole draw the polygon Ⓐ, starting any place in Fig. 66. From X and Y draw lines parallel to A–E, the resultant of the loads between X and Y, and locate the points X' and Y'. From p_1 draw the line p_1–U parallel to X'–Y' and locate U. Then from U draw the line U–m parallel to X–Y.

From points Y and Z draw lines parallel to E–H, the resultant of the forces between Y and Z, and locate Y_1 and Z_1. Draw p_1–V parallel to Y_1–Z_1 and from V draw V–n parallel to Z–Y. The intersection p locates the desired pole. If p is used as a pole and a polygon drawn in Fig. 66, starting at any one of the three points X, Y, and Z, it will pass through the other two.

The construction just illustrated is of special value in connection with the investigation of arches.

CHAPTER II

CENTROIDS

In many cases it is desirable to locate the center of gravity, or centroid, of a line, area, or solid, and it is often convenient to do this by the use of graphical construction.

Let it be considered that gravity, or any other attraction, acts upon a body, area, line, or group of lines. From the very nature of the center of gravity it is evident that the action line of the resultant of the attractive forces, applied to each and every element of the body, contains the centroid. Now if the attraction force be considered as acting in another direction, and the action line of the resultant obtained, the intersection of this resultant with the first will locate the center of gravity.

The volumes, areas, and lines here considered will be of uniform density unless otherwise noted.

25. Centroid of a Broken Line.—The centroid of a straight line is, of course its mid-point, but the centroid of a broken line may or may not fall on the line itself. Consider the broken line $A–B–C–D–E–F$ shown in Fig. 68. First assume that the attractive force acts in the direction shown by the action lines of F_1, F_2, F_3, etc. Lay off vectors in Fig. 69 which are proportioned to the lengths of the lines $A–B$, $B–C$, etc. Choose any pole p and draw the funicular polygon shown in the lower part of Fig. 68. This locates the action line of R. The location of the centroid is at some point along the length of this action line. Now assume that the attractive force acts in the direction shown by the action lines of F_1', F_2', etc.

In Fig. 70 the lengths of the vectors are made proportional to the lengths of the lines $A–B$, $B–C$, etc., and the pole p' chosen. From Fig. 70 the funicular polygon in the upper part of Fig. 68 is drawn, locating the action line of R'. The intersection of this action line with that of R locates the centroid g.

In Fig. 71, the centroid of the broken line is located by a some-

what different method. Assume that the attractive force acts
normal to the plane of the paper. The centroid of the line A–B

FIG. 70.   FIG. 68.

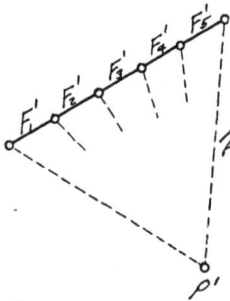

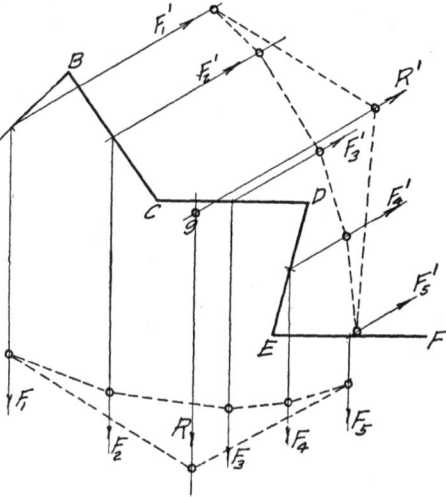

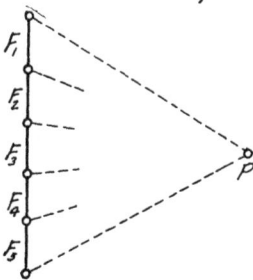

FIG. 69.

is of course at its mid-point. The same is true of the line B–C.
If these two mid-points are connected by a straight line, it is evident that some place along its length is the centroid of the broken

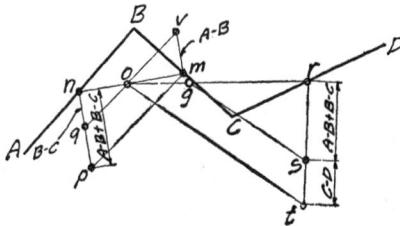

FIG. 71.

line A–B–C; also that this centroid o, is a point which divides
the connecting line inversely as the lengths A–B and B–C. The
construction for locating o is similar to that used in Fig. 29 of

Chapter I, except that the forces are normal to the plane of the paper and the construction has been revolved around the line *m–n* into the plane of the paper. The value of *A–B* is laid off to some convenient scale from *m*, and that of *B–C* from *n* as shown. The ends of these lengths are connected, and the centroid located by the intersection *o*. The lines *n–q* and *m–v* are of course parallel to each other, but not necessarily at right angles to *m–n*. The intersection *o* might have been located by drawing *p–q* to represent *A–B*, connecting *p* and *m*, and then drawing a line from *q* parallel to *p–m*.

The centroid of the broken line *A–B–C* is now connected with *r*, the mid-point of the line *C–D*. From *r* a line is drawn along which the value of *A–B* plus *B–C* is laid off to some convenient scale, and the point *s* located. From *s* the value of *C–D* is laid off, and the point *t* located. Now *o* and *t* are connected, and a line is drawn from *s* parallel to *o–t*. The intersection *g* locates the centroid. The construction just explained may be used when the broken line consists of more than three parts, but when there are a large number of parts the method given in Figs. 68 to 70 is often more convenient.

26. Centroid of an Arc.—Consider the arc *A–Z–F*, shown in Fig. 72. Through the center *O* draw the axis *Y–Y* bisecting the arc. Now the centroid is on this axis because it is an axis of symmetry, and it remains to locate the centroid along its length. Draw the regular circumscribing polygon *A–B–C–D–E–F*, and draw the axis *X–X* through *O* at right angles to *Y–Y*. Draw *A–w* and *B–s* parallel to *X–X* and *B–w* and *C–s* parallel to *Y–Y*. Now the moment of the line *C–B* about *X–X* is equal to *C–B* times *p–h*, which (triangles *h–p–O* and *B–C–s* being similar) is equal to *B–s* times *O–p*, (*O–p* being the radius of the arc and *B–s* being equal to *B'–s'*, the projection of *B–C* on the axis *X–X*). In like manner it can be shown that the moment of each line of the polygon is equal to the projection of the line on the axis *X–X* times *r*, and it follows that the moment of the entire polygon is equal to *m–n*, the projection of the polygon on the axis *X–X*, times *r*. This moment divided by the length of the ploygon will give the distance of the centroid *g* from the axis *X–X*. Extend the line *C–D* and make *D–t* equal to *D–E* plus *E–F*, or in other words, *Z–t* equals one-half the length of the polygon and is parallel to *X–X*. Connect *t* and *O* and from *F* draw *F–v* parallel to Y–Y. Then from v draw v–g parallel to X–X. The two triangles O–v–g and O–t–Z are similar, and the line v–g is equal to O–n or one-half the projection of the polygon on the axis X–X. The moment of the polygon about X–X is equal to m–n times r, which is equal to O–Z times $2 \cdot g$–v, and from similar triangles we have $M = O$–$Z \cdot 2 \cdot g$–$v = 2 \cdot t$–$Z \cdot g$–O. Now M divided by the length of the polygon gives the distance to the centroid

$$\frac{M}{\text{length of polygon}} = \frac{2 \cdot t\text{–}Z \cdot g\text{–}O}{\text{length of polygon}} = g\text{–}O$$

since $2 \cdot t$–Z is just equal to the length of the polygon. It therefore follows that g must be the centroid.

The discussion thus far has referred to the circumscribing

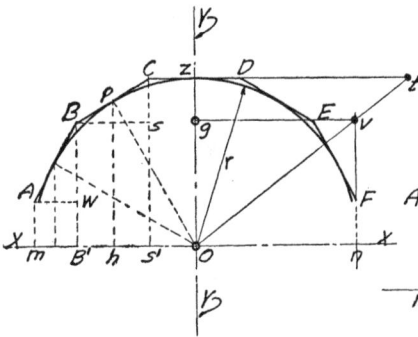
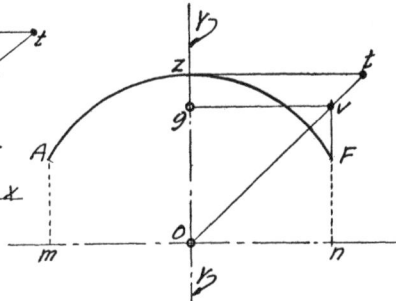

FIG. 72.   FIG. 73.

polygon and not to the arc, but when the number of sides of the polygon is increased indefinitely it approaches the arc as a limit. t–Z is then half the length of the arc and m–n is the projection of the arc on the axis X–X. Also the line Z–t is tangent to the arc at Z.

In Fig. 73 the construction necessary for locating the centroid is shown, the construction used for the proof being left off. The arc is bisected by the axis Y–Y, which is drawn through the center O. At right angles to this axis and passing through O, the axis X–X is drawn. The line Z–t is drawn parallel to X–X and tangent to the arc at Z, its length being made equal to half the length of the arc. The points t and O are connected, and from F the line F–v is drawn parallel to Y–Y and the intersection v obtained. From v a line is drawn parallel to X–X, and the intersection of this line with Y–Y locates g, the center of gravity of the arc.

27. Centroid of a Curve.—Let the heavy line in Fig. 74 be any irregular curve, the centroid of which is desired. The length is cut into a number of small divisions, and the centroid of each division approximated. In Fig. 75 a force polygon is drawn with vectors proportional to the lengths of the division of the curves,

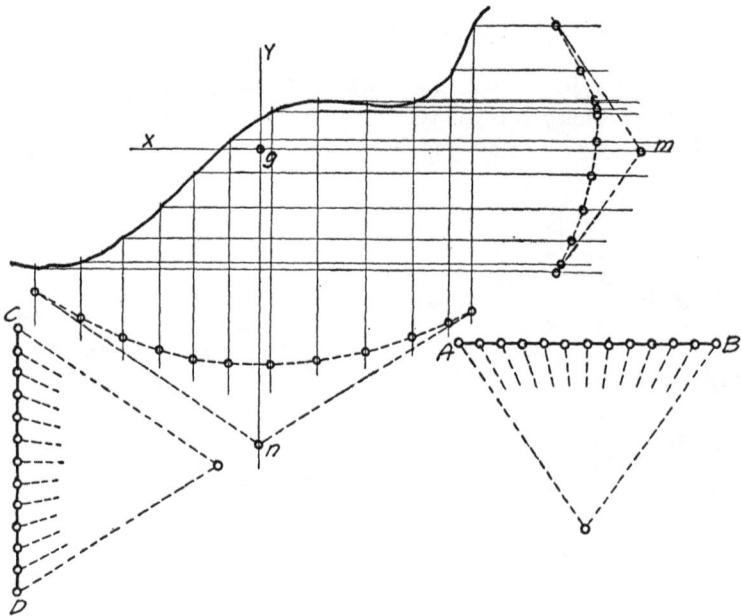

FIG. 76.   FIG. 74.   FIG. 75.

A–B having any desired direction. From the approximate centroid of each division of the curve a line is drawn parallel to A–B. Then from Fig. 75 the funicular polygon shown at the right of Fig. 74 is drawn, and the line m–X located by the intersection m. In like manner the intersection n is obtained by the use of Fig. 76 and a second funicular ploygon. The point n locates the line n–Y, which is parallel to C–D, and the intersection of Y–n and X–m locates g, the centroid of the curve.

There was a slight approximation made in the location of the centroid of each division of the curve, but if the divisions are small these small errors will produce no appreciable error in the final result.

A convenient way of measuring the length of a curve is to step off its length with dividers counting the number of steps, then step off the same number of steps along a straight line and scale the length. The length of each step should, of course, be reasonably small so that the length of each corresponding portion of curve will approximate very closely that of its chord.

28. Centroids of Areas.—Thus far attention has been given to lines, but it is often necessary to locate the centroids of areas. For some areas this is very easy, while for others it is more or less laborious. When an area has two or more axes of symmetry the problem is easy, for each axis of symmetry contains the centroid, which is therefore located by the intersection of any two such axes. The square, rectangle, and circle are such areas. There are other areas, such as the triangle, trapezoid, and quadrilateral for which the centroid may be located by a rather simple construction. But when considering the more complicated areas it is often necessary to divide them up into a number of simple divisions, locate the centroid of each division and then by the use of some construction such as the funicular polygon, locate the centroid of the entire area.

29. Centroid of a Triangle.—Consider the triangle, $A-B-C$, shown in Fig. 77. Connect the vertex A with d the mid-point of the opposite side. Now if $m-n$ be any slice parallel to $B-C$, it is evident that the line $A-d$ contains its mid-point. The area of the triangle may be considered made up of many such slices parallel to $B-C$, all of which have their mid-points on the line $A-d$; therefore the centroid of the triangle must lie some place on the line $A-d$. A line is now drawn from the vertex B to e, the mid-point of the side $A-C$. The centroid of the triangle must be on this line for the same reason that it was found to be on $A-d$. Therefore the intersection g must be the centroid of the triangle. It can easily be proved that g is the third point of the lines $A-d$, $B-e$, and $C-f$; that is $d-g$ is one-third of $d-A$, and $e-g$ is one-third of $B-e$. In other words, the centroid lies at the third point of any median line. It follows that if a line is drawn through a triangle parallel to any side and distant from that side one-third of the altitude measured from that side it will contain the centroid. Fig. 78 shows the centroid located in this way.

30. Centroid of a Quadrilateral.—Consider the quadrilateral
A–B–C–D shown in Fig. 79. Draw the lines A–C and B–D and
locate e, the mid-point of B–D. Draw e–A and C–e. The centroid g_1, of triangle A–B–D lies at the third point of e–A, and g_2,
the centroid of triangle B–C–D is at the third point of C–e. It follows that g, the centroid of the quadrilateral, must lie at some point
on the line g_1–g_2, and that this point g will divide the line g_1–g_2 into
segments inversely proportional to the areas of the triangles. These

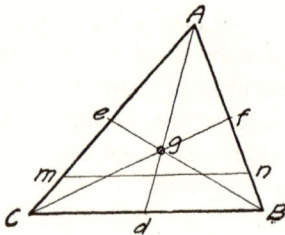

Fig. 77.

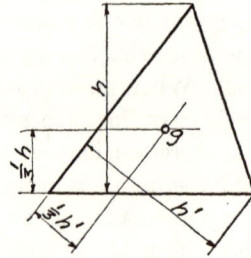

Fig. 78.

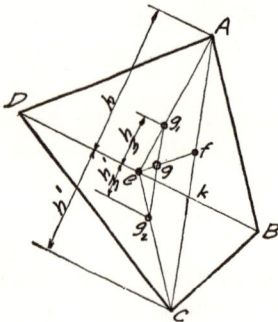

Fig. 79.

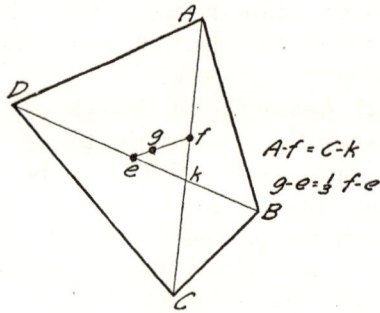

Fig. 80.

two triangles have B–D as their common base. Their areas then
are proportional to their altitudes, which in turn are proportional to
the lines k–A and C–k. That is C–k : A–k :: area of triangle
B–C–D : area of triangle A–B–D. The point f is now located so
that A–f is equal to C–k. Then A–f : C–f :: the area of triangle
C–B–D : the area of triangle A–B–D. Draw the line e–f. Triangles A–e–f and g_1–e–g are similar, also e–f–C and e–g_2–g.
Then g_1–g : g–g_2 :: A–f : f–C :: area of triangle C–B–D : area of
triangle A–B–D. Therefore g, the intersection of f–e and g_1–g_2,

is the centroid of the quadrilateral, because it divides g_1-g_2 into segments inversely proportional to the areas of triangles A-B-D and C-B-D. It should be noted that e-g is one-third e-f because e-g_1 is one-third e-A and e-g_2 is one-third C-e.

In Fig. 79 there are a number of lines which were used in connection with the development of the proof for this method and the student may get the idea that the construction is more complicated than it really is. In Fig. 80 just the construction necessary to locate the centroid is shown. The diagonals A-C and B-D are drawn, e is located as the mid-point of B-D, and A-f is made equal to C-k. Points f and e are connected. The centroid g is now located by making e-g equal to one-third e-f.

31. Centroid of a Trapezoid.—The centroid of a trapezoid may be located in several different ways, one of which is illustrated in Fig. 81. Let it be required to find the centroid of the trapezoid A-B-C-D. The trapezoid is divided into two triangles by the line B-D and the centroid of each triangle is located. This was done by drawing a line from a vertex to the mid-point of the opposite side and locating its third point. It is evident that the centroid of the trapezoid will be at some

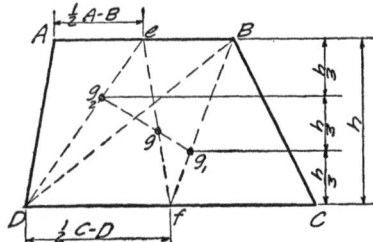

FIG. 81.

point on the line e-f, a line connecting the mid points of A-B and C-D, because if the area is divided into thin slices parallel to C-D, the line e-f will contain the center of each and every one of them. Also the centroid of the trapezoid must be on the line g_1-g_2, a line connecting the centroids of the two triangles. g, the centroid of the trapezoid, is therefore located by the intersection of the lines e-f and g_1-g_2.

Another construction for locating the center of gravity of a trapezoid is shown in Fig. 82. The side A-B is extended to the left and A-n made equal to C-D; also the side C-D is extended to the right and C-m made equal to A-B. The line e-f is drawn connecting the mid-points of the parallel sides; also the points m and n are connected. The intersection g locates the centroid of the trapezoid. A proof for the above construction may be developed as follows: Draw the diagonals A-C and B-D. The triangles $B-n-p$ and $m-p-D$ are similar and $n-B$ is equal to $D-m$. It follows that $B-p$ is equal to $p-D$ or p is the mid-point of $B-D$. Triangles $n-A-h$ and $m-C-h$ are similar, therefore $n-A : m-C :: A-h : C-h$ and $C-D : A-B :: A-h : C-h$. Triangles $A-q-B$ and $C-q-D$ are similar and $C-D : A-B :: C-q : A-q$. Therefore $A-h : C-h :: C-q : A-q$ or $\dfrac{A-h}{C-h} = \dfrac{C-q}{A-q}$ and by adding to the numerator on each side its denominator we have $\dfrac{A-h + C-h}{C-h} = \dfrac{C-q + A-q}{A-q}$ which may be written $\dfrac{A-C}{C-h} = \dfrac{A-C}{A-q}$. From this it is found that $A-q = C-h$. Now this trapezoid is a special case of the quadrilateral, the construction for the location of the centroid of which was given in Figs. 79 and 80. Point h in Fig. 82 corresponds

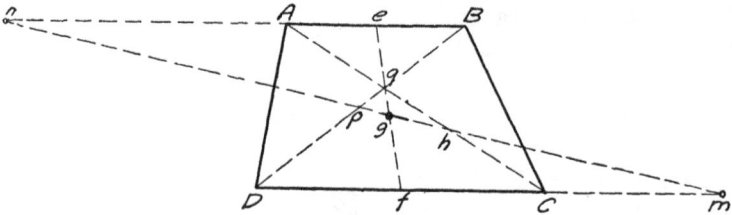

Fig. 82.

to f in Fig. 80, and p to e. Therefore the centroid of the trapezoid shown in Fig. 82 must be on the line $h-p$. Since the centroid g lies on the line $e-f$, connecting the mid-points of $A-B$ and $C-D$, it must be located by the intersection of $e-f$ and $m-n$.

Attention should be called to the fact that the lines $A-C$ and $B-D$ were drawn in connection with the proof and are not necessary for locating the centroid.

Other methods may be used for locating the centroid of a trapezoid, but they are not of enough importance to make it desirable to give them here.

32. Given the Centroid, Area, and Distance between the Two Parallel Sides of a Trapezoid to Find the Lengths of the Parallel Sides.—A problem of this nature is sometimes encountered in connection with footings. In Fig. 83 column 2 has a larger load than column 1, the resultant of the two loads falling at g. In case the footing is allowed to extend only a short distance beyond each column, a trapezoid footing with its centroid at g may be desirable. Our problem is to find the length of $A-D$ and of $B-C$. The total loading divided by the allowable soil pressure gives the area of the footing, and this area divided by $e-f$ gives the mean width, or one-half $A-D$ plus one-half $B-C$. From g draw the line $g-h$ parallel to $A-D$. Also draw $m-n$ and $m'-n'$ parallel to $A-D$ and passing through the third points of the line $e-f$. This means that $m-n$ passes through the centroid of triangle $A-C-D$, and that $m'-n'$ passes through the centroid of triangle $A-C-B$. Along $m-n$ lay off $o-s$ equal to the value already obtained for one-half $A-D$ plus one-half $C-B$, draw the line $o-p$ and connect s and p. The line $s-p$ gives the intersection r. Through r draw $t-v$ parallel to $o-p$. Then $s-t$ measured to the proper scale gives one-half $C-B$, and $v-p$ measured to the same scale one-half $A-D$. This must be true because the areas of triangles $A-B-C$ and $A-C-D$ are proportional to $B-C$ and $A-D$ respectively. If $s-t$ and $v-p$ were not proportional to the areas of these triangles, $t-v$ and $s-p$ would not intersect on $g-h$. The student should note the similarity between the lower part of Fig. 83 and Fig. 29 in Chapter I. In Fig. 29 of Chapter I the two forces are given and from them the resultant located. In Fig. 83 the resultant is known and it is necessary to break it into two components. One is, in a sense, the reverse of the other. If the trapezoid had been given and the location of its centroid required, a construction similar to that shown in Fig. 83 might have been used. After drawing $m-n$ and $m'-n'$ make $s-t$ equal to one-half $B-C$ and $p-v$ equal one-half $A-D$, Then draw $v-t$ and $s-p$. Their intersection locates r which in turn locates the line $h-g$. The intersection of this line and $e-f$ locates g, the centroid of the trapezoid because r divides the distance between $m-n$ and $m'-n'$ into parts inversely proportional to the areas of the two triangles.

33. Centroid of a Sector.—Consider the sector shown in Fig. 84. If the sector were divided into small triangles somewhat as shown at the left of Fig. 84, the centroid of each would be found at a distance two-thirds r from O. Therefore, as far as the location of the centroid is concerned, the area of the sector may be considered as concentrated along the arc $m-s-n$, which is a line of uniform density. The centroid of this arc is located by the method illustrated in Fig. 73. The line $Y-Y$ is an axis bisecting the sector. The length $s-t$ is made equal to the length of the arc

s–n, t and *O* are connected and from *n* a line is drawn parallel to
O–Y. From the intersection of this line with *t–O* a line is drawn
parallel to *X–X*. The intersection *g* locates the centroid of the arc
m–s–n and also of the sector.

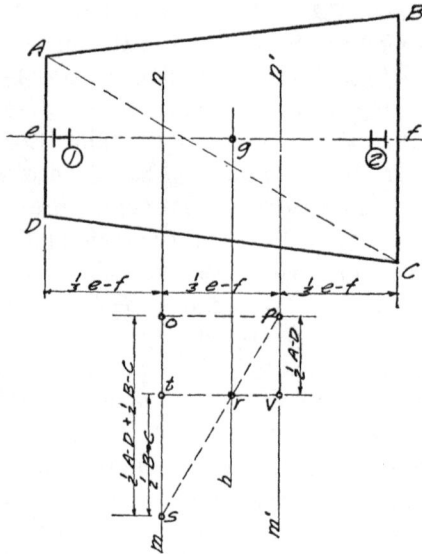

FIG. 83.

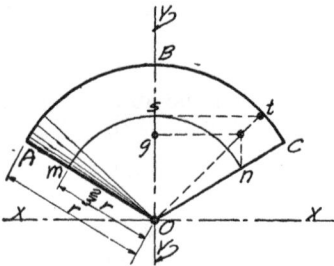

FIG. 84.

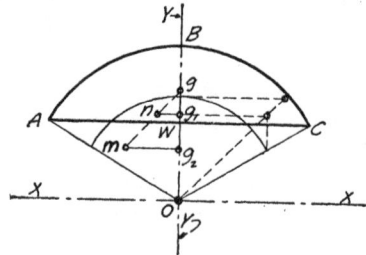

FIG. 85.

34. Centroid of a Circular Segment.—Let it be required to
find the centroid of the segment shown in Fig. 85. The bisecting
axis *Y–Y* is drawn, and through *O* the axis *X–X* at right angles
to *Y–Y*. Then, using the construction of Fig. 84, the centroid
of the sector *A–B–C–O* is located at *g₁*. This sector may be

divided into two parts, the given segment and the triangle O–A–C. The centroid of the sector has been located, and g_2 the centroid of the triangle lies on w–O, one-third of the distance from w to O. Now, if a force is assumed to act at g_1 with a magnitude proportional to the area of the sector, and another one of opposite sense at g_2 proportional to the area of triangle A–C–O, and the resultant found, the action line of this resultant will pass through the centroid of the segment. The resultant of these two forces, which are

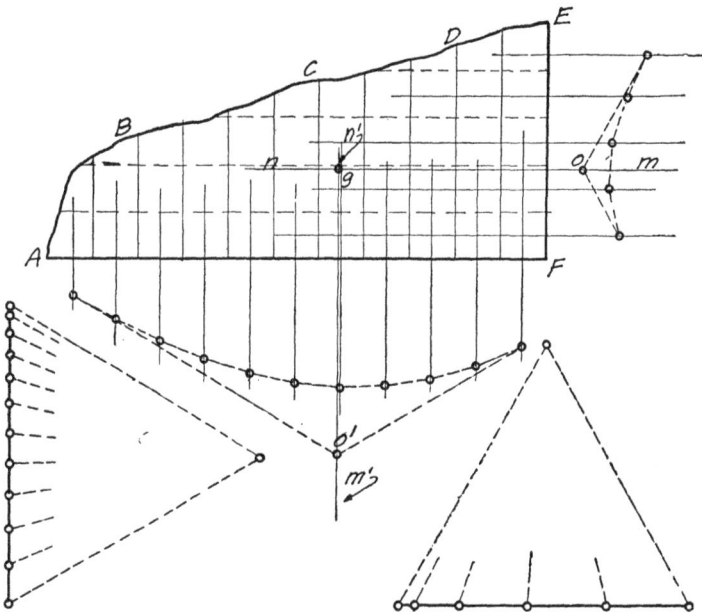

FIG. 87.   FIG. 86.   FIG. 88.

parallel but of opposite sense, is located by the construction shown in Fig. 30, Chapter I. The length of g_2–m is made proportional to the area of the sector, and g_1–n is drawn parallel to it on the same side of Y–Y and proportional to the area of the triangle A–C–O. Now m and n are connected and the line extended until the axis of symmetry Y–Y is intersected at g. The nt g is the centroid of the segment.

35. Centroids of Irregular Areas.—Fig. 86 shows an irregular area the centroid of which is desired. In such a case the most convenient method is to divide the area into small parallel slices, shown by the full vertical lines. The centroid of each of these slices is now approximated, and from their approximate centroids lines are drawn parallel to the slices. The area of each one of these slices is determined approximately by scaling the mean length and multiplying it by the width. In Fig. 87 the vectors are laid off proportional to these areas, and from Fig. 87 the lower funicular polygon of Fig. 86 is drawn.

Attention should be called to the fact that, in order to obtain satisfactory results, the slices should be reasonably small, and yet not so small that their number becomes excessive and thus the chance of accumulation of small errors dangerous. The vertical

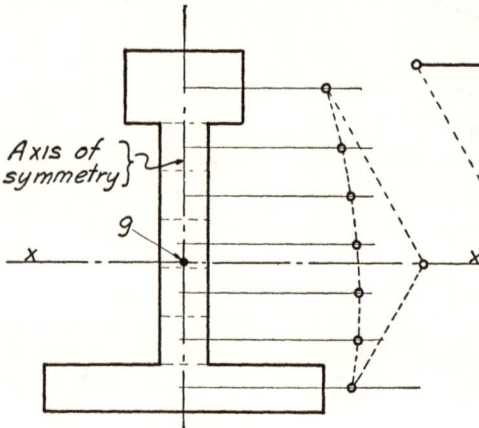
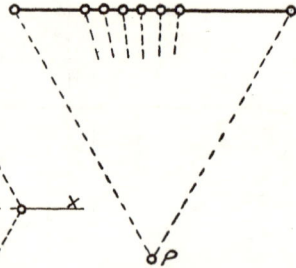

Fig. 89.   Fig. 90.

line, or more strictly speaking, the line parallel to the slices, located by the intersection of the first and last strings of the funicular polygon, contains the centroid of the area, but as yet the location of the centroid along its length is unknown. The area is now divided into another set of slices, in this case horizontal slices as shown by the dotted lines. Now the centroid of each slice is approximated, the areas found, and the force polygon of Fig. 88 drawn, using vectors proportional to the areas of the slices. From this force polygon, the funicular polygon at the right of Fig. 86 is drawn and the horizontal line *m–n* located by the intersection *o*. The intersection *g* locates the centroid of the area.

Figs. 89 and 90 illustrate the construction for locating the centroid of a somewhat different shaped area. Since this area has an axis of symmetry, only one force polygon and one funicular polygon are required to locate the centroid g.

36. Centroids of Volumes.—The centroid of some volumes can be located easily by means of planes of symmetry. There are others that have their centroids on some axis at a certain proportional distance from one end. For still others the centroid can be most conveniently located by cutting the volume into small slices and approximating the centroid and content of each slice. Then the centroid of the whole may be located by the use of force and funicular polygons. There are some volumes for which it is convenient to find the areas for various sections, plot these areas, draw an area curve, and then locate the centroid of the area under the curve. Then again it may be possible to split a given volume into a few smaller volumes, the centroid of each of which can be easily located, and the centroid of the entire volume located by the use of force and funicular polygons or by some other convenient method.

It is evident that the centroid of a sphere is the common point of three or more planes of symmetry, the centroid of a cube is also the common point of three or more planes of symmetry or the mid-point of a line connecting the centroids of two opposite sides. The centroid of a rectangular parallelopiped is the common point of three planes of symmetry, or the mid-point of a line connecting the centroids of two opposite sides. The centroid of a prism with parallel bases is the mid-point of a line connecting the centroids of the two bases.

37. Centroid of a Triangular Pyramid.—Consider the triangular pyramid shown in Fig. 91. First locate the centroid d of the triangular base A–B–C by drawing lines from the vertices A and B to the mid-points of the opposite sides. If the pyramid were divided into very thin slices parallel to the triangular base, the line connecting d and V would contain the centroid of each of them. Therefore the centroid of the pyramid must be on the line V–d. Now locate the centroid e of the side A–C–V by drawing lines from the vertices A and V to the mid-points of the opposite sides. Connect e and B. The centroid of the pyramid must lie on this line B–e for the same reason that it is on the line d–V, since any side of a triangular pyramid may be considered the base. Therefore the intersection of B–e and d–V must be the centroid of the pyramid. Now it can be proved that this centroid g is at the fourth point of d–V or of e–B, that is d–g equals one-fourth d–V. The triangle V–f–B with the lines e–B and d–V is shown in Fig. 92. In this figure e–h is drawn parallel to f–B. Since e–f equals one-third f–V, k–d equals one-third d–V. Now triangles e–k–g and g–B–d are similar and e–k equals two-thirds f–d also f–d equals one-half B–d. Then e–k equals one-third B–d. It follows that k–g equals one-third d–g and k–d equals one and one-third d–g equals four-thirds d–g equals one-third d–V. Therefore d–g equals one-fourth d–V.

It is also true that the centroid of any pyramid is on the line connecting the centroid of the base with the vertex, and one-fourth the distance up from the base. For if the pyramid is divided into thin slices parallel to the base the line connecting the centroid of the base with the vertex will contain the centroid of each of them,

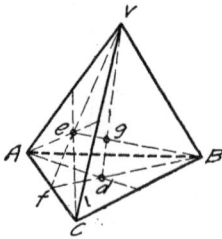

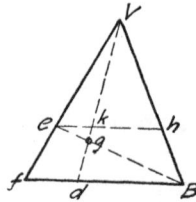

Fig. 91.   Fig. 92.

and therefore the centroid of the pyramid. Now if the base is divided into triangles, and the pyramid into triangular pyramids, the centroid of each one of these pyramids will be up from the base one-fourth the distance to the vertex. Therefore the centroid of the large pyramid must be up one-fourth the distance from the base to the vertex and on the line connecting the centroid of the base with the vertex.

38. Complex Volumes which can be divided into a Number of Regular Volumes.—The elevation of a rather simple complex volume is shown in Fig. 93, and a plan of the same volume in Fig. 94. It is convenient to divide this volume into two parts Ⓐ and Ⓑ, in plan Q–H–O–T and K–M–N–O. Now the portion Ⓐ, shown as A–B–E–F in elevation and Q–H–O–T in plan, has the dimension normal to the side A–B–E–F constant. A plane parallel to A–B–E–F and half the way back will be a plane of symmetry and will therefore contain the centroid. This plane of symmetry is shown in plan by the line r–s. The side A–B–E–F is a trapezoid and its centroid g_1 is located by the construction already explained for the trapezoid. The volume Ⓐ may be cut into thin slices all parallel to this side and directly back of it. A line projected directly back from g_1 would contain the centroid of each of these slices, and therefore the centroid of volume Ⓐ. The point g_1 in Fig. 93 is the elevation of the centroid of volume Ⓐ. By projecting down from g_1 until r–s in Fig. 94 is intersected, g_1', the centroid of volume Ⓐ in plan, is located. In a similar way the centroid of volume Ⓑ is located in elevation as g_2 and in plan as g_2'. Now the centroid of the entire volume must lie on the line connecting the centroids of the two volumes Ⓐ and Ⓑ, which line is shown in elevation as g_1–g_2 and in plan as g_1'–g_2'. In Fig. 95 vectors are drawn proportional to the two volumes Ⓐ and Ⓑ, a pole is chosen, and the funicular polygon in the lower part of Fig. 94 is drawn. The intersection v locates a plane which contains the centroid of the large volume. The trace of this plane, u–v, intersects g_1'–g_2' and g_1–g_2 at g' and g respectively, therefore the centroid of the entire volume is shown in elevation as g and in plan as g'.

Figs. 96 and 97 show an interesting volume, the centroid of which is desired. The first step is to divide the given volume into smaller volumes, the centroid of each of which can be easily determined or approximated. A vertical plane passing through B–D will cut off a pyramid at the left. The base of this pyramid is a rectangle whose edge is shown in plan as B–D, and in elevation as f–k. Now if a horizontal plane is passed through f, a triangular pyramid is cut off, the base of which is shown in plan by B–C–D and in elevation by f–i. There remains a triangular prism shown in plan by B–C–D and in elevation by f–i–J–k. The centroid of each one of these small volumes can be easily located. Also it is evident that a vertical plane passed through A–C is a plane of symmetry for the given volume, as well as for each one of the divisions. The base of the pyramid at the left is a rectangle, the centroid of which is shown in plan as o, and in elevation as s. From s a line is drawn to the vertex E, and g_1 located at the quarter point. Projecting up, g_1' is obtained on the line A–C. The point g_1 shows the elevation, and g_1' the plan of the centroid of this pyramid. Now consider the triangular prism, shown in plan by the triangle B–C–D, which has its centroid at g_3'. The prism

Fig. 93.

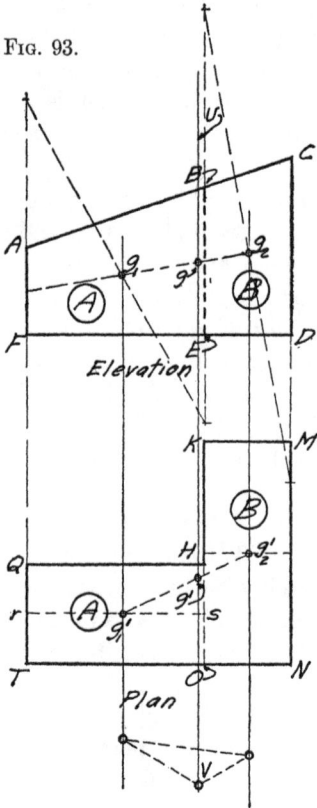

Elevation

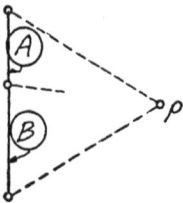

Plan

Fig. 94.

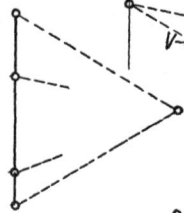

Fig. 95.

Fig. 96.

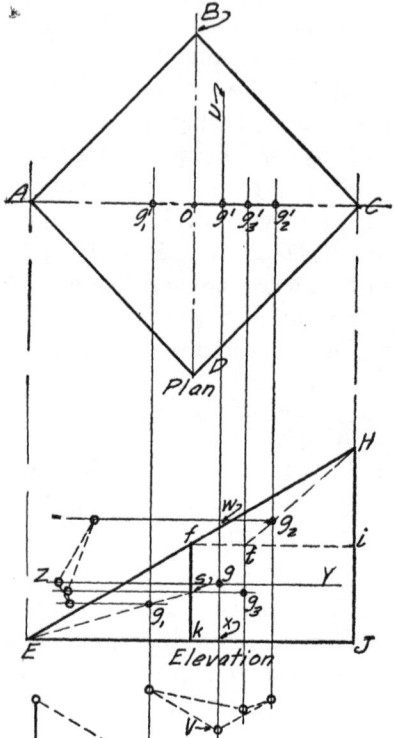

Plan

Elevation

Fig. 97.

Fig. 98.

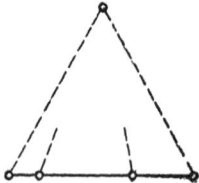

Fig. 99.

may be considered as made up of a lot of thin slices, all parallel
to the base and of the same shape, the centroid of each being on a
vertical line through the centroid of the base. Therefore the
centroid of the prism must lie on this vertical line and half-way up.
Now project down from g_3' and locate g_3 mid-way between f–i and
k–J; g_3 is the elevation, and g_3' is the plan of the centroid of the
prism.

The triangular pyramid, above the prism just considered, has
the centroid of its base shown in elevation by t, and in plan by
g_3'. Connect t with the vertex H and locate g_2 at the quarter point.
The point g_2 is the elevation of the centroid of the pyramid, and
g_2', obtained by projecting up from g_2, shows it in plan.

Vertical lines are now drawn from g_1, g_2, and g_3, and in Fig. 98
vectors are laid off proportional to the different volumes. A pole
is chosen and the lower funicular polygon in Fig. 97 drawn.
The intersection v locates a vertical plane which contains the
centroid of the given volume. The trace of this plane is u–v
which intersects the trace of the plane of symmetry at g'. Therefore g' shows the centroid of the given volume in plan and in
elevation the centroid is at some point along the line w–x. Horizontal lines are now drawn from g_1, g_2, and g_3. In Fig. 99 vectors are laid off proportional to the various volume divisions, and
after choosing a pole, the funicular polygon shown at the left of
Fig. 97 is drawn. The intersection z locates a horizontal plane
which contains the centroid of the entire volume. The trace of
this plane is z–y, and it intersects w–x at g. The centroid of the
entire volume is therefore shown in elevation as g and in plan as g'.

39. Irregular Volumes. Division into Slices.—Let it be
required to find the centroid of the volume shown in Figs. 100 and
101, which is a portion of a hollow circular cylinder. The line A–B
is the trace of a vertical plane of symmetry. The centroid will
therefore lie above this line. The volume cannot be divided into
small regular volumes the centroid of each of which is known, but
it may be cut into a number of thin parallel slices at right angles to
the plane A–B. If the slices are thin the centroid of each may be
assumed to be mid-way between its sides. Through these approximate centroids vertical lines are drawn as shown in Fig. 101. The
volume of each slice may be obtained with very small error by
multiplying the thickness by the mean width times the mean
height. The mean widths are obtained from Fig. 100, and the

mean heights from Fig. 101, each being approximated. The mean width of the fifth slice is *e–d* plus *c–f* and the mean height is *r–s*. It is convenient to make the thickness of all the slices the

Fig. 100.

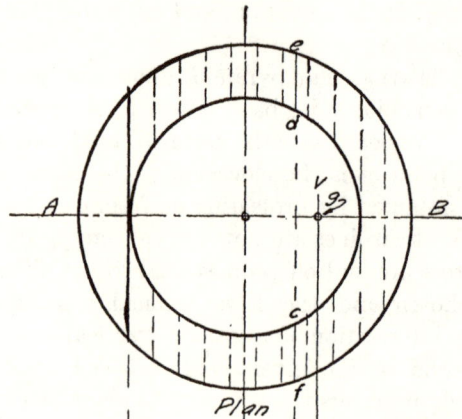

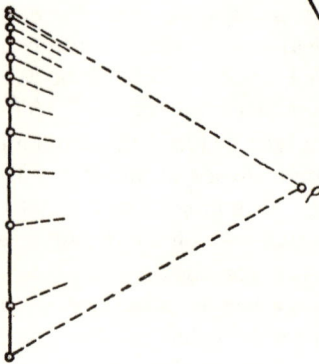

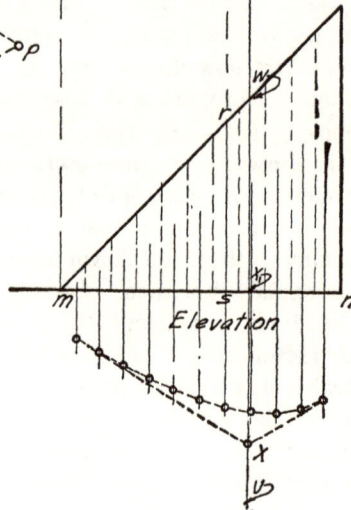

Fig. 101.

Fig. 102.   Fig. 103.

same, thus simplifying the computations for the volumes. In Fig. 102, vectors are laid off proportional to the volumes of the slices, the pole *p* is chosen, and the funicular polygon, Fig. 103, is drawn. The intersection x locates a vertical plane at right angles to A–B which contains the centroid. The trace of this plane is u–v, which intersects A–B at g. The point g therefore shows the centroid in plan and in elevation it lies at some point along the line w–x_1. The distance of the centroid above the base m–n may be determined by dividing the volume into thin horizontal slices, and drawing a force and a funicular polygon similar to Figs. 102 and 103. The intersection of the first and last strings of the funicular polygon would locate a horizontal plane containing the centroid. The intersection of its trace with w–x would locate the centroid in elevation.

40. Centroid Located by the Use of Sections and Area Curves. —Figs. 104, 105, and 106 show a volume or body which has no plane of symmetry and cannot be cut into simple volumes. The centroid might be located by using the method of slices, but it is sometimes more convenient to use sections and areas, especially when a planimeter is available for measuring the areas. Vertical sections are taken at intervals along the length of the volume, the outline of each section being shown in Fig. 104. The area of each one of these sections is found from Fig. 104 by the use of a planimeter, and the values plotted above the base line in Fig. 107. A smooth curve is now drawn through the points thus located, and an area curve is obtained. This area curve is a curve such that the ordinate to it at any point measured to the proper scale gives the area of the corresponding section of the volume. It should be noted that the area under this curve measured to the proper scale gives the volume or content of the volume. The vertical scale is, of course, the same as was used in laying off the areas of the various sections, that is, 1 inch equals a certain number of square inches or square feet. The horizontal scale is the same as that used in Fig. 105, that is, 1 inch equals a certain number of inches or feet. Therefore, for the units of the area under the area curve, we have square feet times feet equals cubic feet, or square inches times inches equals cubic inches.

Since the ordinates in Fig. 107 represent the distribution of volume in Fig. 105, the vertical line which passes through the centroid of the area, shown in Fig. 107, will locate a vertical plane normal to the paper which will contain the centroid of the given volume. The area shown in Fig. 107 is, therefore, divided into small vertical slices, and vectors are laid off in Fig. 108 proportional to the areas of these slices. Vertical lines are now drawn down from the approximate centroids of these slices and the funicular polygon of Fig. 109 drawn from Fig. 108. The intersection z

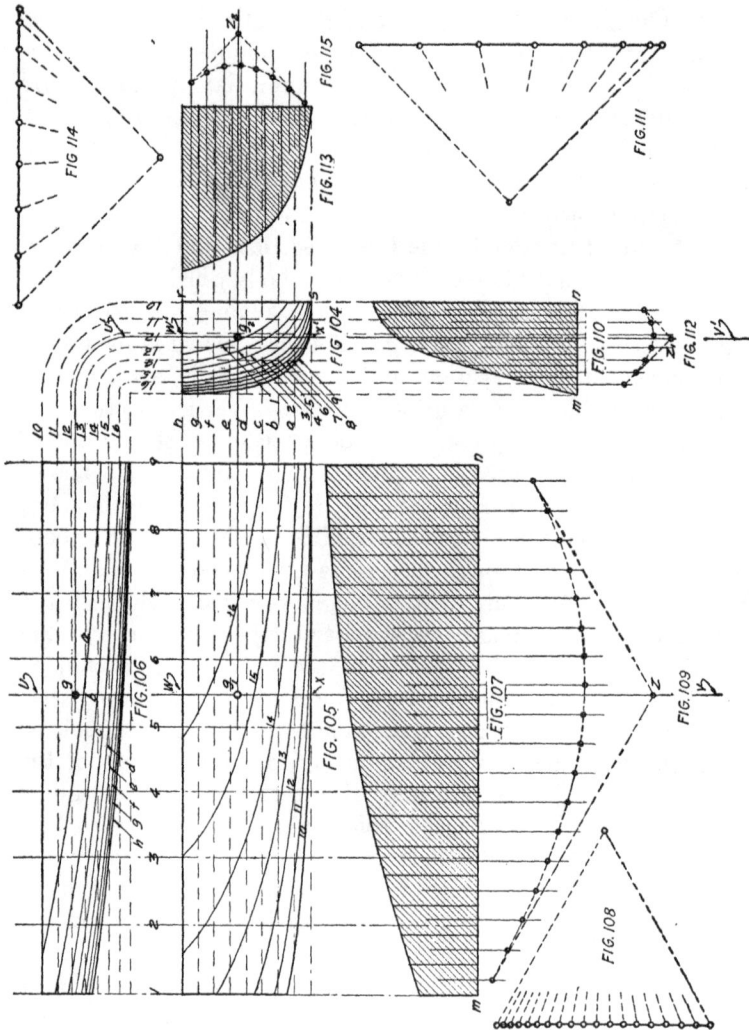

Figs. 104–115.

locates the vertical line u–v which contains the centroid of the area shown in Fig. 107. The line u–v is therefore the trace of a plane which contains the centroid of the given volume.

One plane has now been found which contains the centroid, but two others must be found before the centroid can be definitely located. Let the volume be cut by a number of vertical planes at right angles to the sections which were taken for the area curve of Fig. 107. These planes are shown by the vertical lines in Fig. 104, by the horizontal lines in Fig. 106, and by the curves in Fig. 105. The areas of the various sections cut by these planes may be obtained from Fig. 105 by the use of a planimeter. The vertical lines of Fig. 104 are extended down to the base line m–n in Fig. 110, and above m–n the areas of the various sections obtained from Fig. 105 are plotted to some convenient scale. A smooth curve drawn through these points gives the curve of areas. The area under this curve is now divided into small vertical slices, and by the use of Figs. 111 and 112, the intersection z_1, is obtained, which locates a vertical line containing the centroid of the area shown in Fig. 110. The line u'–v' is therefore the trace of a vertical plane which contains the centroid of the given volume.

Two vertical planes have now been located which contain the centroid of the volume. The centroid therefore lies somewhere on their line of intersection, which line is shown in Fig. 105 as w–x, in Fig. 104 as w'–x', and in Fig. 106 as g. It now remains to locate a horizontal plane which contains the centroid. This horizontal plane is located in a way similar to that used in locating the vertical planes. Horizontal sections are taken, shown by the horizontal lines in Figs. 104 and 105, and by the curved lines in Fig. 106. The areas of these sections are measured in Fig. 106, and the area curve is drawn in Fig. 113. Then by the use of Figs. 114 and 115 the intersection z_2 is located. The horizontal line from z_2 intersects w'–x' at g_2, and w–x at g_1. The centroid of the given volume is therefore shown in plan by g, in elevation by g_1, and in end view by g_2.

In many cases it is necessary only to locate the distance of the centroid from a given plane, in which case it is not necessary to locate the centroid itself, but only a plane which contains it and is parallel to the given plane.

41. Stress Volumes.—Stress volume is the term sometimes applied to a volume or solid used to represent the distribution or intensity of stress or pressure over the section of some member or structural element.

Fig. 116 shows the elevation of a pier or column, and in Fig.

117 a section is shown at a larger scale. The stress volume for this section is a volume whose base is $A-B-C-D$, and whose dimension, normal to the base at any point, represents to some convenient scale the pressure or stress at that point. It should be noted that the base of the stress volume is an area, and is measured in square inches or square feet, while the vertical dimension represents pressure, and will therefore be measured in pounds per square inch or per square foot. Now square inches, times pounds per square inch, gives pounds. Therefore the content of a stress volume is a certain number of pounds, and this number of pounds

Fig. 118.   Fig. 119.

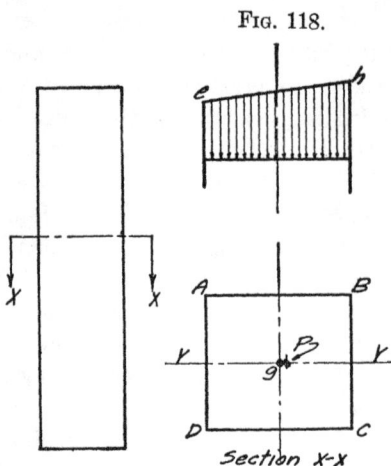
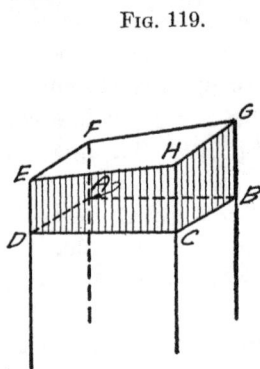

Fig. 116.   Fig. 117.

is numerically equal to the resultant of all the applied forces normal to the section considered. Also the action line of this resultant will pass through the centroid of the stress volume. In case the stress volume is given and the location of the resultant is unknown, it may be found by locating the normal line which contains the centroid of the stress volume. The point where the resultant cuts the section is called the center of pressure.

Fig. 118 shows the elevation of the stress volume when the center of pressure is at P, a little to the right of the centroid g, but on the axis of symmetry $Y-Y$. A portion of the pier, together with the stress volume, is shown in Fig. 119. $A-B-C-D-E-F-G-H$ is the stress volume.

The centroid of a stress volume, or a line containing its centroid, may be located by the same constructions that are used for ordinary volumes.   Stress volumes will be considered in more detail in following chapters, especially in the chapter on Masonry.

CHAPTER III

MOMENTS

First and second moments and higher moments, if desired, can be found conveniently by graphical construction. The first moment of a force about a point is the force times the perpendicular distance between its action line and the point; the second moment is the force times the square or second power of this perpendicular distance; and a higher moment is the force times a higher power of this perpendicular distance. The first moment of an area about an axis is the summation of all the little elementary areas times their respective perpendicular distances from the axis. The second moment of an area about an axis is the summation of all the little elementary areas times the square of their perpendicular distances from the axis. A higher moment is the summation of all the little elementary areas times a higher power of their perpendicular distances from the axis.

The letter M is often used as a symbol to indicate the first moment; while I, and moment of inertia, are used as expressions for the second moment.

The simpler problems can often be solved more quickly by analytical methods, but for the longer and more complicated problems, the graphical solution is of special value. The simpler problems will be considered first, because by their use it is easier to explain and illustrate the method.

42. First Moment of a Force about a Point.—In Fig. 120 consider the force F, the moment of which about the point o is desired. Through o draw a line parallel to the action line of F. The first moment $M = F \cdot d$, d being the perpendicular between the action line of F, and the line parallel to it passing through o. In Fig. 121 draw the vector for F, using any convenient scale, and choose any convenient pole p. Then from 3, any point on the action line of F, draw the line 3–1 parallel to B–p, and 3–2 parallel to A–p. Now triangle 1–2–3 and triangle A–p–B are similar,

therefore $F : H :: 1\text{-}2 : d$ or $F \cdot d = 1\text{-}2 \cdot H = M$.   That is, M
is equal to the intercept 1-2, measured to the scale at which Fig.
120 was drawn, times H, measured to the scale used for the vector
$A\text{-}B$.

Moments which tend to produce rotation in a clock-wise direction
may be called $+$ and those which tend to produce rotation in a
counter clock-wise direction $-$.   The important thing is to keep
in mind that moments which tend to produce rotation in the same
direction have the same sign and those which tend to produce
rotation in opposite directions have opposite signs.

43. First Moment of a Number of Forces about a Point.—
(a) *Parallel Forces.*—In Fig. 122 four parallel forces are shown

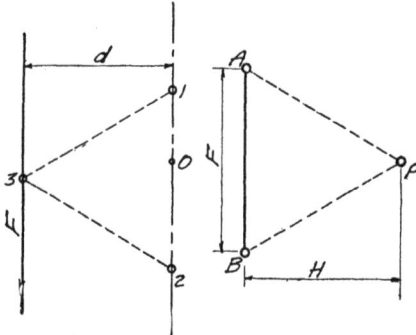

FIG. 120.   FIG. 121.

and their moment about the point o is desired.   Through o a line
is drawn parallel to the action lines of the forces, and in Fig. 123
vectors are laid off proportional to the magnitudes of the forces,
taking them in order from left to right.   Any pole p is chosen in
Fig. 123, and from 1, any point on the action line of F_1, two strings
are drawn parallel to the two rays in Fig. 123, which are components of the vector F_1.   These strings are extended until the
line through o is intersected.   The string $b\text{-}p$ is a component of
F_2 as well as of F_1, therefore from point 2, the intersection of $b\text{-}p$
and F_2 a line is drawn parallel to $C\text{-}p$, which is the other component for the vector F_2; and so on until the last string has been
drawn.   In each case the string has been extended to the right
until the line $Y\text{-}Y$ through o is intersected, giving the intersections
5, 6, 7, 8, and 9.

It follows from what has been said in connection with Figs. 120 and 121, that the moment of F_1 about o is $H \cdot 8$–9; the moment of F_2 about o is $H \cdot 8$–7; of F_3, $H \cdot 6$–7, and of F_4, $H \cdot 5$–6. That is,

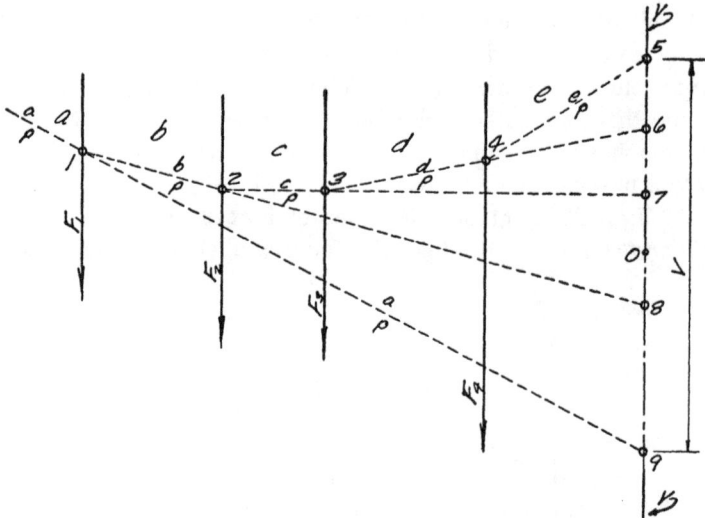

Fig. 122.

the moment of each force about o is H times a certain intercept along Y–Y. The resultant moment of all four forces about o will, therefore, be H times the sum of all of these intercepts, and this sum is the length 5–9 or the intercept V. In other words, the total first moment of all four forces about o is the pole distance H times the intercept V. $M = H \cdot V$.

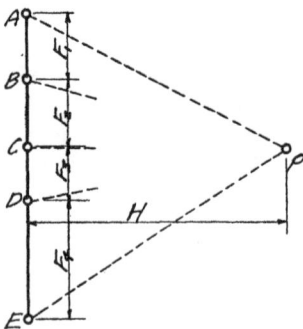

Fig. 123.

Attention should be called to the fact that the polygon 1–2–3–4–5 is a funicular polygon for the given forces, and the first and last strings have been projected over to the line Y–Y, their intersections giving the intercept V.

(*b*) *Non-parallel, Non-concurrent Forces.*—Any straight line may be drawn intersecting the action lines of the given forces and passing through o, the point about which moments are to be taken.

Resolve each one of the given forces into two components at the point where its action line cuts the line just drawn through o. One of these components should make an angle of 90° with this line through o, and the other should coincide with it. The given system has now been resolved into double the original number of forces, half of which are parallel to each other and the other half have their action lines all passing through o. The moment of the

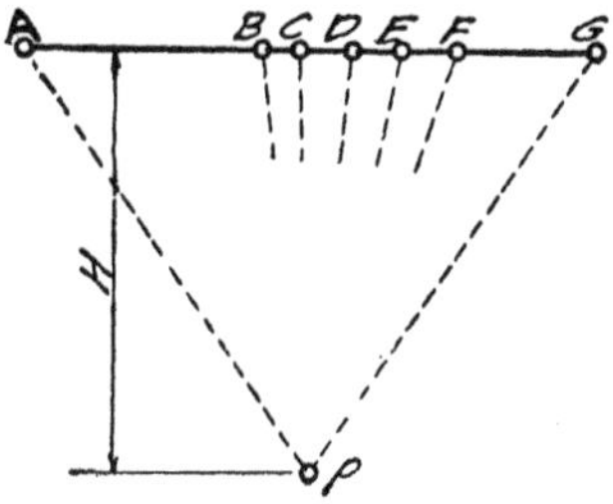

FIG. 125.

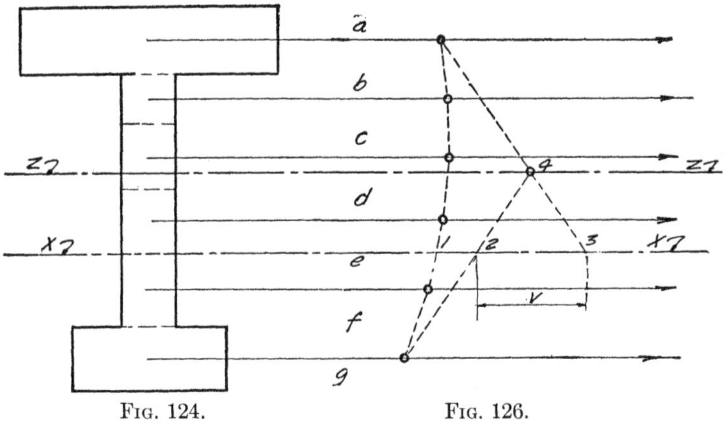

FIG. 124.   FIG. 126.

given system about o will, therefore, be given by the group of parallel forces. The moment of this group can be found by the use of constructions similar to those given in Figs 122 and 123.

Another method of determining the moment produced by a system of non-parallel, non-concurrent forces, is to first locate the resultant of the entire system and then find its moment about the given point.

44. First Moment of an Area about a Given Axis.—Let it be required to find the first moment of the area shown in Fig. 124

about the axis x–x. Divide the area into slices, and from the approximate centroid of each slice, draw lines parallel to the axis x–x. In Fig. 125 lay off vectors proportional to the areas of the various slices and choose any pole p. Then draw the funicular polygon shown in Fig. 126. Now the length 1–3 times H gives the moment of that part of the area above x–x about x–x, also 1–2 times H gives the moment of that part of the area below x–x about x–x. Now these two moments are of different sign, therefore, the total moment of the area about x–x is equal to their difference, which is $H \cdot V$. The pole distance H is measured to the scale used for the vectors in Fig. 125, that is, 1 inch equals a certain number of square inches or square feet, and V is measured in feet or inches to the scale that was used in Fig. 124. Inches times inches squared gives inches to the third power for the units of the moment. It should be noted that point 4, the intersection of the first and last strings, locates the neutral axis, or the axis about which the first moment is zero, or in other words, an axis which contains the centroid.

45. Moment Diagram for a Beam.—In Fig. 127 a beam is shown carrying a number of concentrated loads. Draw the force polygon in Fig. 128, choose any pole p, and from Fig. 128 draw the funicular polygon in Fig. 127, obtaining the closing string 1–7. From p a line is drawn parallel to 1–7, and the point Y located. The length F–Y gives the magnitude of R_2, and Y–A that of R_1. Draw x–x, a vertical line at any section of the beam and extend the various strings until they intersect x–x. Now the moment of R_1 about x–x is H times the intercept 6–2; that of the force a–b is H times the intercept 2–3; that of b–c is H times the intercept 3–4, etc. The moment from the loads tends to produce rotation in a direction opposite to that in which the moment from R_1 tends to produce rotation. Therefore to obtain the resultant moment at x–x, the moment of the loads must be subtracted from the moment of the reaction. The moment of R_1 and also of each load is the product of H and a certain intercept. Therefore the resultant moment is equal to H times the intercept 2–6 less the intercepts 2–3, 3–4, and 4–5, and equals $H \cdot$ 5–6 or $H \cdot V$. Note that V is the intercept between the funicular polygon and the closing string. It follows that the moment at any section of a beam is given by the product of the intercept between the funicular polygon and its closing string, times the pole distance H. The pole distance H is measured in pounds to the scale that was used for the vectors in Fig. 128 and the intercept V in feet or inches to the scale at which Fig. 127 was drawn.

The area enclosed between the funicular polygon 1–12–11–10–9–8–7 and its closing string 1–7 may, therefore, be called the moment diagram. Attention should be called to the fact that it makes no difference whether or not the closing string is horizontal, because the vertical intercepts are not affected by the inclination of the closing string, as long as the pole distance H remains unchanged. Also, it makes no difference whether the

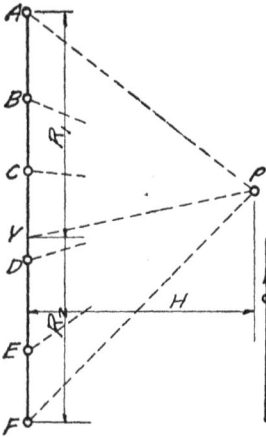
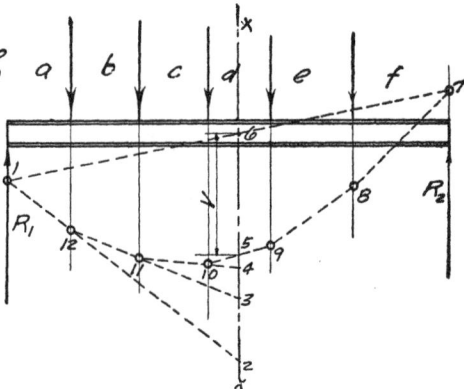

Fig. 128.   Fig. 127.

area of the moment diagram is above or below the closing string, since the vertical intercept times the pole distance always gives the moment, and a glance at the beam will tell which side has tension and which side has compression.

For further discussion of moment diagrams for beams see Chapter IV.

46. Second Moments.—Let it be required to find the second moment I of the force F_1 about the point o, see Fig. 129. I is equal to $F_1 \cdot d^2$. In Fig. 130 a vector is drawn representing F_1 to some convenient scale, and a pole p is chosen with any convenient pole distance H. Then starting from any point 1 on the action line of F_1, the strings a–p and p–b are drawn, and the intersections 2 and 3 obtained. Now take any pole p', with any convenient pole distance H', and draw the lines p'–2 and p'–3. Then in Fig. 131, from any point 4 on the action line of F_1, draw 4–5 parallel to 3–p' and 4–6 parallel to 2–p', thus obtaining the intersections 5 and 6.

Now $I = F_1 \cdot d^2 = F_1 \cdot d \cdot d = M \cdot d = H \cdot 2$–$3 \cdot d = H \cdot V \cdot d$, because in connection with first moments M was shown to be equal to $H \cdot V$. It is evident that the two triangles 2–p'–3 and 4–5–6 are similar, since their sides are respectively parallel. Therefore, $H' : V :: d : 5$–6, and it follows that $d \cdot V = H' \cdot 5$–6 $= H' \cdot V'$. Therefore $I = H \cdot V \cdot d = H \cdot H' \cdot V'$. That is, I is equal to the

Fig. 129.

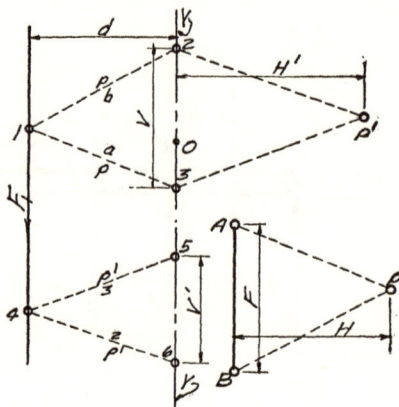

Fig. 131.   Fig. 130.

product of the first pole distance H, the second pole distance H', and the intercept V'. The pole distance H is to be measured to the scale which was used for the vector in Fig. 130, while H' and V' are to be measured to the scale at which Fig. 129 was drawn.

This construction may seem rather complicated, but it should be kept in mind that it is of special value for complicated problems, and that the simple problems are used first for the purpose of teaching the method.

47. Second Moment of a Number of Parallel Forces.—In Fig. 132, three forces F_1, F_2, and F_3, are shown, and their second moment about the point o is desired. In Fig. 133, vectors are laid off proportional to the forces, to some convenient scale, and the pole p located with some reasonable pole distance H. From

Fig. 133 a funicular polygon is drawn in Fig. 132, and the various strings projected over until the line $Y-Y$ is intersected, and the points 4, 5, 6, and 7 are obtained. Now any pole p' is chosen,

FIG. 132.

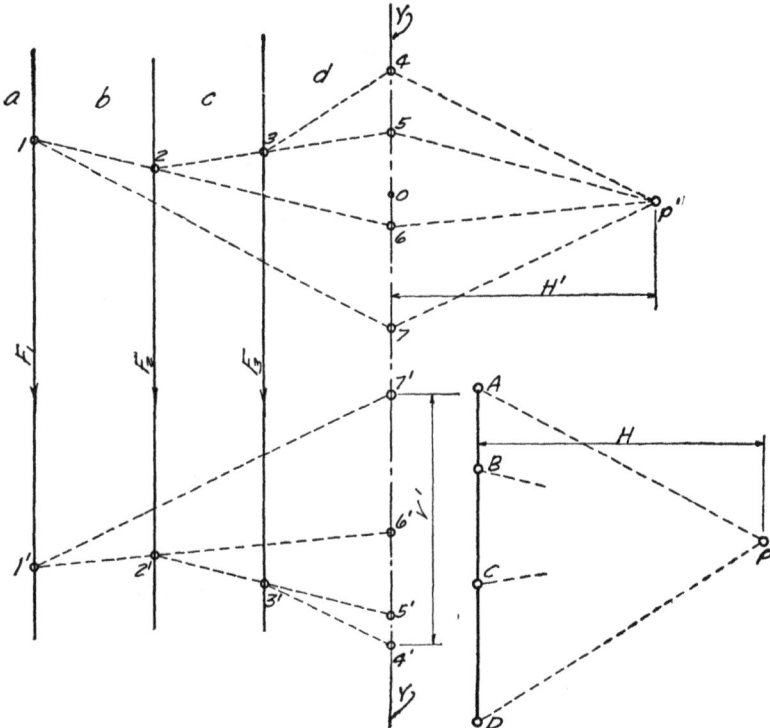

FIG. 134. FIG. 133.

with any convenient pole distance H', and the lines p'-4, p'-5, etc., are drawn.

In Fig. 134, 1'-7' is drawn parallel to 7-p', 1'-6' parallel to 6-p', etc. Now, the second moment of F_1 about o is

$$I_1 = H \cdot H' \cdot 7'\text{-}6'$$

of F_2 $\qquad I_2 = H \cdot H' \cdot 6'\text{-}5'$

of F_3 $\qquad I_3 = H \cdot H' \cdot 5'\text{-}4'$

$$I = I_1 + I_2 + I_3 = H \cdot H' \cdot (7'\text{-}6' + 6'\text{-}5' + 5'\text{-}4') = H \cdot H' \cdot V'.$$

I being the total moment of all the forces about o.

Attention should be called to the fact that in Fig. 134 the two lines that intersect on the action line of F_2 are parallel to the two lines from p' which intersect Y–Y at the points where the two strings which intersect on the action line of F_2 in Fig. 132, intersect Y–Y. The same sort of thing is true in connection with the other forces.

48. Moment of Inertia of an Area.—Figs. 136, 137, and 138

Fig. 136.

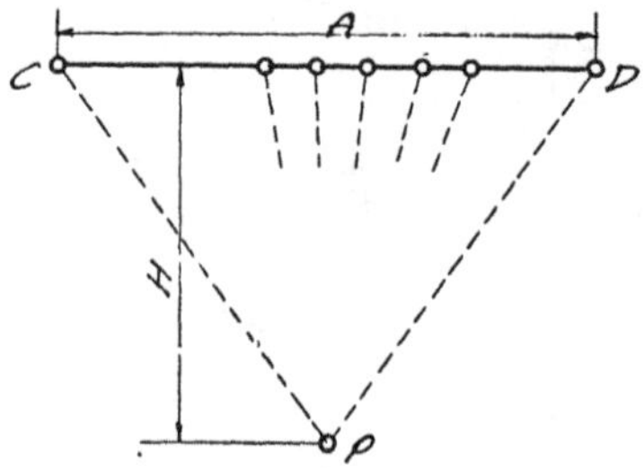

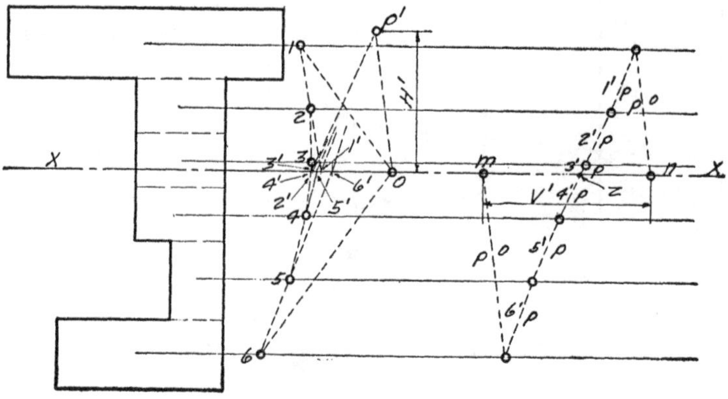

Fig. 135.   Fig. 137.   Fig. 138.

show the construction for finding the moment of inertia of the area shown in Fig. 135 about the neutral axis x–x. The area is divided into a number of slices and from the centroid of each slice a straight line is drawn parallel to the axis x–x. In Fig. 136 vectors are laid off proportional to the areas of the various area divisions, and a pole p is chosen, with pole distance H. From Fig. 136 the funicular polygon of Fig. 137 is drawn and the intersection o locates the neutral axis x–x. The various strings are now extended until the axis x–x is intersected, giving the points $1'$, $2'$, $3'$, etc. The pole p' is chosen with any convenient pole distance H', and the rays p'–$1'$, p'–$2'$, etc., are drawn. Parallel to these rays the lines of Fig. 138 are drawn and the intercept V' obtained. Attention should be called to the fact that the string in Fig. 137 which connects any two area lines, if extended, locates by its intersection with the axis x–x the end of the ray which is parallel to the line in Fig. 138, which in turn connects these same two lines. Now the length n–z times H times H' gives the approximate moment of inertia of that part of the area above x–x about x–x; also the length z–m times H times H' gives the approximate moment of inertia of that portion of the area below x–x about x–x. Both these values have the same sign, therefore the moment of inertia of the entire

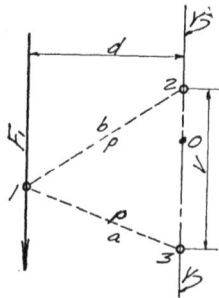 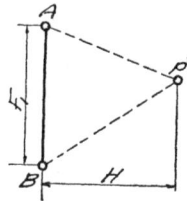

FIG. 139.　　　　　　FIG. 140.

area about its neutral axis x–x is given approximately by the product $H \cdot H' \cdot V'$.

49. Mohr's and Culmann's Methods.—The construction just illustrated is sometimes called Culmann's method, while a slightly different construction is called Mohr's method. In order to get a clear idea of the difference between these two methods, it will be found convenient to go back to the problem of finding the second moment of a single force about a point. Consider the force F_1 shown in Fig. 139, and let it be required to find its moment of inertia about the point o. A vector representing F_1 is drawn in Fig. 140, and any pole p chosen with any convenient pole distance H. From 1, any point on the action line of F_1, the two strings a–p and p–b are drawn, and the intersections 2 and 3 obtained. Now $I = F_1 \cdot d^2 = A{-}B \cdot d \cdot d = H \cdot V \cdot d$, and since $d \cdot V$ equals two times the area of triangle 1–2–3, I is equal to $H \cdot 2$ times the area of triangle 1–2–3. The second moment is therefore obtained without any additional construction, the second pole and polygon of Culmann's method being unnecessary. However, it is necessary to measure or compute an area.

This construction may be extended to give the moment of inertia of a number of forces, or of an area. Considering the forces shown in Fig. 132, their moment of inertia about o could be obtained by Mohr's method without using p' or drawing Fig. 134, I being given by H times 2 times the area 1–2–3–4–7.

Suppose it were desired to find the moment of inertia of the area shown in Fig. 135 about its neutral axis x–x by Mohr's method. Figs. 136 and 137 would be drawn as shown, except that in Fig. 137 the various strings would not have to be extended to the axis x–x, nor would p' have to be located or its rays drawn. The area enclosed by the funicular polygon 1–2–3–4–5–6–0 would be found and multiplied by 2 times the pole distance H, in order to obtain the moment of inertia. The area enclosed by the funicular polygon of Fig. 137 may be measured by the use of a planimeter. If a planimeter is not available the area may be divided into narrow horizontal slices all of equal width and the sum of their measured mean lengths multiplied by the common width to give the area. This area in Fig. 137 is measured to the scale that was used in Fig. 135.

50. Radius of Gyration.—The radius of gyration of an area is equal to the square root of $(I \div A)$, A being the area for which I is the moment of inertia. Let it be required to find the radius of gyration of the area shown in Fig. 135 about the axis x–x. Fig. 141 shows Fig. 136 redrawn, and the construction necessary for finding the radius of gyration graphically, added. The line e–f is drawn parallel to C–D, its ends falling on the strings p–C and p–D, and its length being equal to H' of Fig. 137. Through p the line o–u is drawn at right angles to e–f, and p–o made equal to V' of Fig. 138. Now with u–o as a diameter draw the semi-circumference o–$s \cdot u$, and from p draw the line p–s at right angles to o–u. $I = H \cdot H' . V'$, and in Fig. 141 C–$D : H :: e$–$f : u$–p. But C–D represents the area shown in Fig. 135, and e–f was made equal to H', therefore $A : H :: H' : u$–p or $A \cdot u$–$p = H \cdot H'$, and it follows that $I = A \cdot V' \cdot u$–p, also $r = \sqrt{I \div A} = \sqrt{V' \cdot u$–$p}$. Now from geometry we know that s–p is the mean proportional between u–p and

p–o, that is $(s$–$p)^2 = u$–$p \cdot p$–o, and therefore $r = \sqrt{V^1 \cdot u$–$p}$ $= \sqrt{u$–$p \cdot p$–$o} = s$–p.

51. Radius of Gyration of Rectangles, Parallelograms and Triangles.—Let it be required to find the radius of gyration of the rectangle shown in Fig. 142 about its neutral axis x–x, which is parallel to one of its sides. We know that $I = \frac{1}{12} bd^3$ and $r = \sqrt{I \div A} = \sqrt{\frac{1}{12} d^2}$, since $A = b \cdot d$, or we may say $r = \sqrt{\frac{1}{6} d \cdot \frac{1}{2} d}$. Now in Fig. 142 e–$f = \frac{1}{2} d$, and g–f is made equal to $\frac{1}{6} d$. With g–e as a diameter, draw the semi-circumference g–h–e. Then

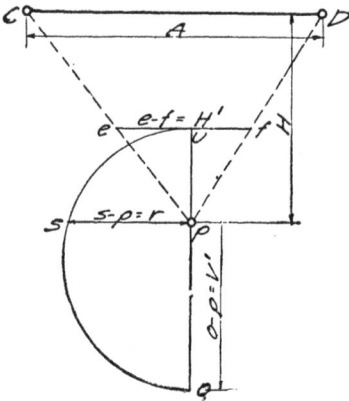

FIG. 141.

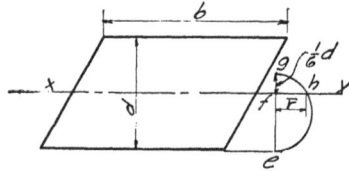

FIG. 143.

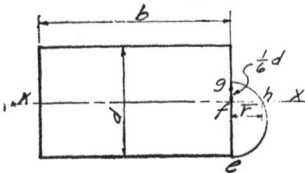

FIG. 142.

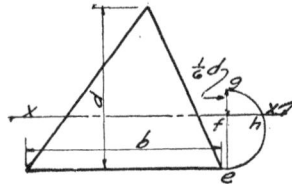

FIG. 144.

f–h is the mean proportional between e–f and f–g, or f–$h = \sqrt{e$–$f \cdot f$–$g} = \sqrt{\frac{1}{6} d \cdot \frac{1}{2} d} = r$.

The construction for obtaining the radius of gyration r, of a parellelogram, about a neutral axis parallel to one of its sides is very similar, see Fig. 143. The I of this parallelogram is equal to $\frac{1}{12} bd^3$, and therefore $r = \sqrt{\frac{1}{12} d^2} = \sqrt{\frac{1}{6} d \cdot \frac{1}{2} d}$. In Fig. 143 e–$f = \frac{1}{2} d$, and f–g is made equal to $\frac{1}{6} d$, so that $\sqrt{e$–$f \cdot f$–$g} = \sqrt{\frac{1}{6} d \cdot \frac{1}{2} d} = h$–$f = r$.

In Fig. 144 a triangle is shown, and its radius of gyration is desired about the neutral axis x–x, which is an axis parallel to one of its sides. The I for the triangle about such an axis, is $\frac{1}{36} \cdot d \cdot d^3$ and $r = \sqrt{I \div A} = \sqrt{\frac{1}{18} d^2} = \sqrt{\frac{1}{6} d \cdot \frac{1}{3} d}$ since $A = \frac{1}{2} b \cdot d$. In Fig. 144 e–$f = \frac{1}{6} d$, and f–g is made equal to $\frac{1}{3} d$, then $\sqrt{\frac{1}{6} d \cdot \frac{1}{3} d} = \sqrt{e\text{–}f \cdot f\text{–}g} = f$–$h = r$, since f–h is the mean proportional between e–f and f–g.

52. Moment of Inertia. Exact Method.—The construction shown for the moment of inertia in Figs. 135 to 138 is approximate in that the horizontal lines are drawn from the centroids of the various area divisions, in place of from a point at distance r from the axis. The I of an area division about an axis, other than its neutral axis, is $I = \bar{I} + A \cdot d_1^2$, where d_1 is the perpendicular distance from the axis about which I is desired, to a parallel axis through the centroid of the area division, while $\bar{I}$ is the moment of inertia about this parallel neutral axis. Now in the approximate method shown in Figs. 135 to 138, the values $A \cdot d_1^2$ have been summed up but $\bar{I}$ has been neglected. This error is small when the area divisions are small, but it may become quite large when these divisions are larger.

Let it be required to find the moment of inertia of the area shown in Fig. 145 about any axis x–x, using the exact method. Let the area be divided into a few convenient area divisions as shown. When using the exact method there is no objection to having large area divisions, but for convenience, they should be either rectangles, parallelograms, or triangles. The I of each one of these divisions about x–x is equal to $\bar{I} + A \cdot d_1^2$, or the sum of the moment of inertia of the division about its own neutral axis parallel to x–x, and its area times the square of the perpendicular distance between the neutral axis of the division and x–x. In other words $I = I + A \cdot d_1^2 = A \cdot r^2 = A \cdot \bar{r}^2 + A \cdot d_1^2$. Dividing through by A, we have $r^2 = \bar{r}^2 + d_1^2$. Consider the area division 3, shown separately in Fig. 146; o–f is equal to d_1 and f–h is $\bar{r}$. Since the triangle o–f–h is a right triangle, $\bar{r}^2 + d_1^2 = (o\text{–}h)^2 = r^2$. When the lines were drawn from the centroids of the area divisions, we obtained by our moment of inertia construction, the summation of all the areas times the square of the distance of their centroid from the axis; that is, we obtained summation $A \cdot d_1^2$. What we want is summation $A \cdot r^2$. Therefore if the line from each area division is moved out until it is distant r from x–x, and the construction already given for I employed, we obtain summation $A \cdot r^2$.

The student should note that the exact method differs from the approximate method only in the location of the lines from the various area divisions.   In the approximate method these lines are drawn from the centroids of the various divisions while in the exact method, they are moved out from x-x until they are distant from it a length equal to the radius of gyration about x-x of each respective area division.   Either Mohr's or Culmann's construction may be used with the exact method.   In Figs. 145 to 148,

Fig. 147.

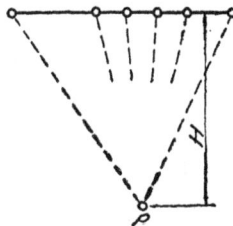

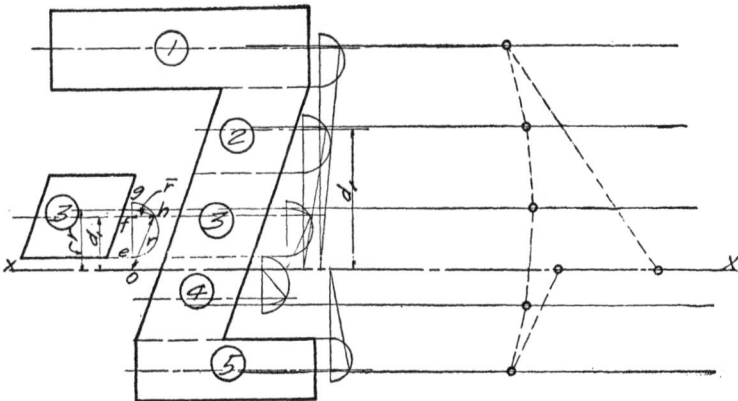

Fig. 146.   Fig. 145.   Fig. 148.

Mohr's construction has been used, and I is obtained by multiplying the area enclosed by the funicular polygon of Fig. 148 by two times the pole distance H.

53. Higher Moments.—It is seldom necessary to obtain higher moments than the second moment, yet it is interesting to investigate how higher moments may be obtained by graphical construction.

Let it be required to find the third and fourth moments of F_1

about the point o in Fig. 149. The third moment is $F_1 \cdot d^3$, and the fourth moment $F_1 \cdot d^4$. Draw Fig. 150 and the two lines from point 4 in Fig. 151 as though the second moment were to be obtained. The second moment I equals $H \cdot H' \cdot V'$ and the third

Fig. 149.　　　　Fig. 150.

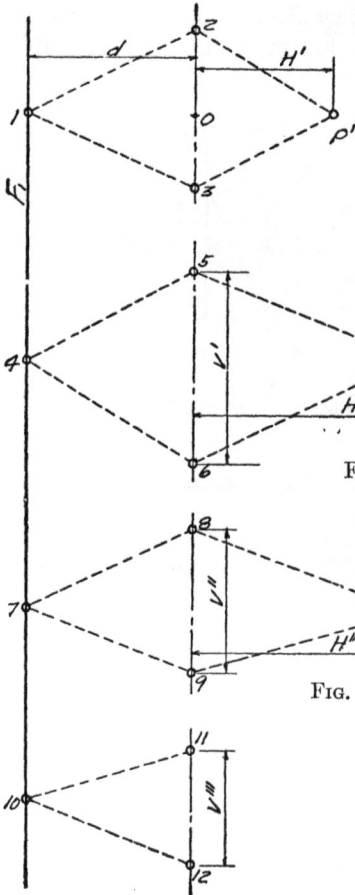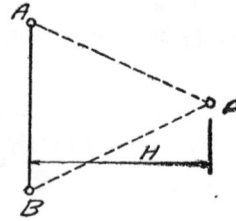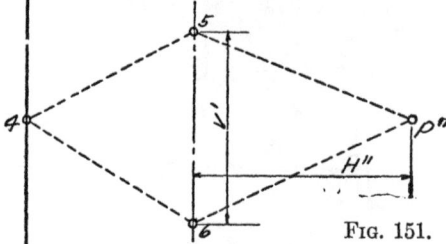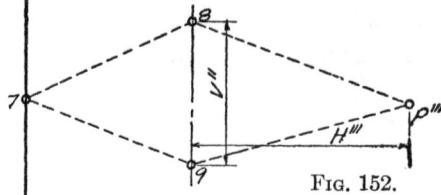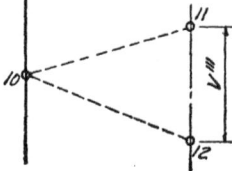

Fig. 151.

Fig. 152.

Fig. 153.

moment $F_1 \cdot d^3$ equals $I \cdot d$ or $H \cdot H' \cdot V' \cdot d$. Choose any pole p'' in Fig. 151, and draw strings 7–8 and 7–9 in Fig. 152. Now triangle 5–p''–6 is similar to triangle 7–8–9 and $V' : H'' :: V'' : d$ or $H'' \cdot V'' = V' \cdot d$. Then the third moment $= H \cdot H' \cdot V' \cdot d =$

$H \cdot H' \cdot H'' \cdot V''$. The fourth moment equals the third moment times d or $H \cdot H' \cdot H'' \cdot V'' \cdot d$. In Fig. 152 choose any pole p''', and draw the lines 10–11 and 10–12 in Fig. 153. Now triangle 8–9–p''' and triangle 10–11–12 are similar, therefore $V'' : H''' :: V''' : d$ or $H''' \cdot V''' = V'' \cdot d$, and it follows that the fourth moment equals $H \cdot H' \cdot H'' \cdot V'' \cdot d = H \cdot H' \cdot H'' \cdot H''' \cdot V'''$. The fifth or sixth or any other higher moment may be obtained by continuing the above process.

The construction here given for higher moments may be

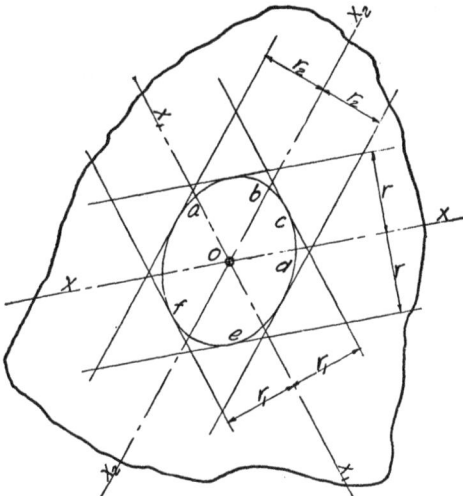

Fig. 154.

extended to deal with complicated areas or systems of forces just as the construction for I was extended.

54. Central Circle or Ellipse of Inertia.—Let the heavy irregular line in Fig. 154 represent the outline of an area whose centroid is at o. The axis x–x is any axis passing through the centroid. Now let it be assumed that the moment of inertia and the radius of gyration of the area about x–x are found by graphical construction. After obtaining the value of r, draw the two lines parallel to x–x, one on either side, and distant from x–x the radius of gyration r, measured to the scale at which the area was drawn. Now take any other axis passing through the centroid, such as x_1–x_1, and by graphical construction obtain the radius of

gyration about this new axis. Also draw the two lines parallel
to x_1–x_1 and distant from it a length corresponding to the radius
of gyration.

Several other axes may now be drawn, and the r about them
found as for x–x and x_1–x_1. Then two lines may be drawn parallel
to each axis and distant from it a length corresponding to the radius
of gyration about it. Now draw a smooth curve tangent to the
lines which are parallel to and distant r from the various axes.
This curve a–b–c–d–e–f is of such a nature that the distance from
any of its tangents to the corresponding parallel axis through the
centroid gives the radius of gyration about that axis. It is usually
a circle or an ellipse, and is called the central ellipse or central
circle of inertia. When this curve has been drawn for an area, the
moment of inertia may be obtained for any neutral axis in the
following way:

Parallel to the given axis draw a line tangent to the central
circle or ellipse of inertia and measure the perpendicular distance
between the axis and this tangent. Then the square of this distance times the area gives the desired moment of inertia.

When the curve is a circle, I about every axis is the same, when
the curve is an ellipse, an axis normal to the major axis of the
ellipse is the axis giving the maximum moment of inertia. The
axis normal to the minor axis of the ellipse gives the minimum
moment of inertia.

When the area has an axis of symmetry its central ellipse of
inertia will have either the minor or major axis coinciding with the
axis of symmetry.

Very little space is here devoted to the central ellipse or central
circle of inertia, because there are many other things of more
importance and interest to the average engineer.

55. Complicated Problems.—The problems considered in
this chapter have been rather elementary, but the solution of
more complicated problems will be found in Chapters VIII and X.

CHAPTER IV

BEAMS

Graphical constructions are especially well adapted to the solution of complicated problems dealing with beams. Continuous beams with a complicated loading, fixed beams, and beams having a variable moment of inertia can be conveniently solved.

Simple beams with a simple loading will be considered first, after which the more complicated problems will follow.

56. Construction of the Elastic Curve.—Let the simple beam shown in Fig. 155, carrying a single concentrated load, be considered. The force polygon Fig. 156 is drawn, and from it the moment diagram, Fig. 157. The reader is referred to Chapter III, for a discussion of the construction and use of moment diagrams. The reactions R_1 and R_2 can be obtained from Fig. 156, as has been explained in Chapter I. It may be well to recall at this time that the bending moment at any section of the beam is given by measuring the intercept in the moment diagram, at the section, and multiplying it by the pole distance H. The pole distance H is to be measured to the scale used in laying off the force A–B, while the intercepts in the moment diagram are to be measured to the scale at which the beam was drawn. The maximum moment may be obtained by using the maximum intercept in the moment diagram.

In order to proceed with the construction of the elastic curve, the moment diagram is divided into small vertical slices of width dx, and from the centroid of each, a vertical line is drawn extending down.

Let the first area division at the left be called a_1–b_1, the second b_1–c_1, etc., and let each slice be measured to the scale at which the beam was drawn and its area found. The areas thus computed are laid off along the load line in Fig. 158, to a scale of 1 inch equals a certain number of square inches or square feet. Now let the pole distance H' in Fig. 158, be made equal to $\dfrac{E \cdot I}{H}$, in which I is

81

Fig. 155.

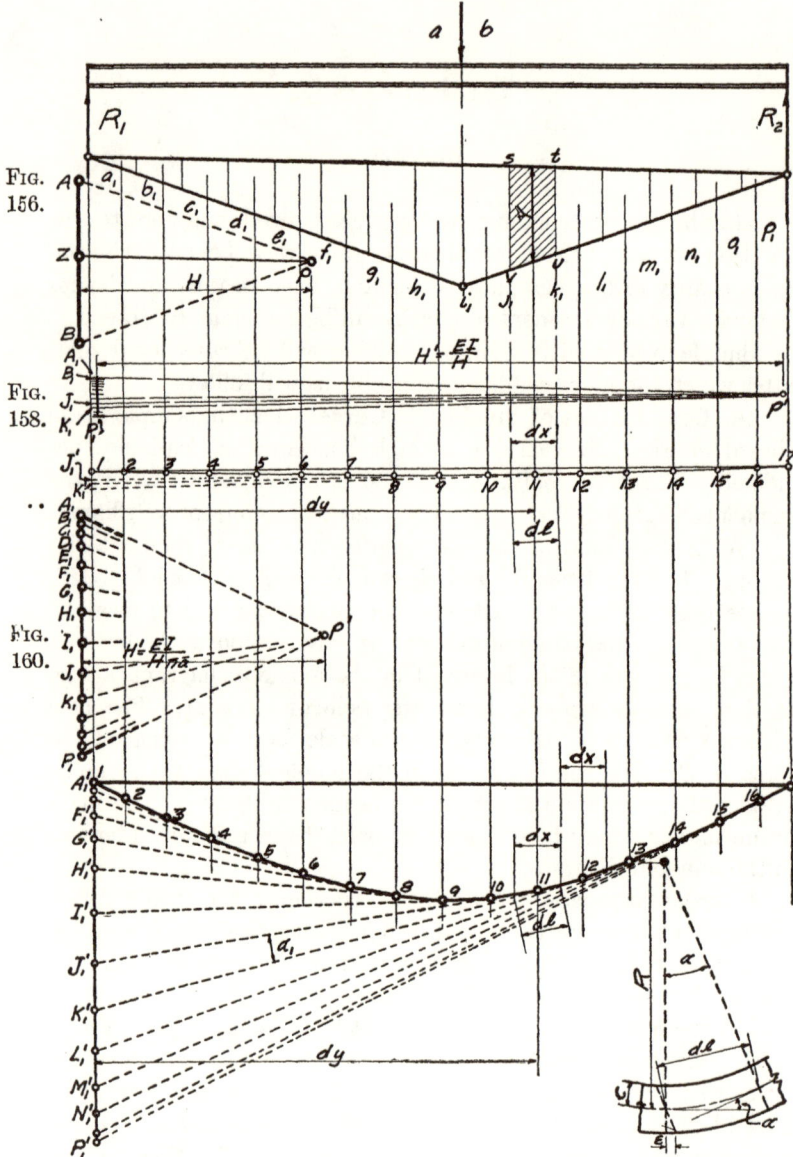

Fig. 156.

Fig. 157.

Fig. 158.

Fig. 159.

Fig. 160.

Fig. 161.

Fig. 162.

the moment of inertia of the beam, E its modulus of elasticity, and H the pole distance of Fig. 156.

In the above expression E is in pounds per square inch, I is in inches to the fourth power, and H is in pounds; therefore H' will be in square inches. Now the value of H' just obtained, is measured to the same scale that was used in laying off the areas of the moment diagram. If the areas of the moment diagram were measured in square feet and laid off along the load line in Fig. 158 to a scale of 1 inch equals a certain number of square feet, the value of H', obtained from the above formula, should be divided by 144 in order to reduce it to square feet.

From Fig. 158 the funicular polygon 1-2-3-4 . . . 17 in Fig. 159 is drawn and the strings extended back until they intersect the vertical line from the reaction R_1, giving the points A_1', B_1', C_1', etc. The vertical dimensions in Fig. 158 are small and the pole distance is rather large. Fig. 160 is the same as Fig. 158, except that the pole distance has been shortened and the vertical dimensions have been magnified; also Fig. 161 is the same as Fig. 159, except that the vertical dimension has been magnified. The student should refer to Figs. 160 and 161 when the notation is not given in Figs. 158 and 159, or when the construction is not clear.

Now any one of the vectors along the load line in Fig. 158 represents the area of the corresponding slice in the moment diagram. For example, J_1-K_1 represents the area of the slice s-t-u-v of the moment diagram, or in other words J_1-$K_1 = Y \cdot dx$, and since $M = Y \cdot H$, $H = \dfrac{M \cdot dx}{J_1 - K_1}$. Now H' was made equal to $\dfrac{E \cdot I}{H}$, then $H = \dfrac{E \cdot I}{H'}$. From these two values of H, we obtain

$$\frac{E \cdot I}{H'} = \frac{M \cdot dx}{J_1 - K_1} \text{ or } J_1 - K_1 = \frac{M \cdot dx \cdot H'}{E \cdot I}.$$

The same is true for any other vector in Fig. 158, provided M is the average moment for the corresponding dx of beam. Triangles J_1-K_1-p' and J_1'-K_1'-11 are similar, therefore,

$$J_1 - K_1 = \frac{J_1' - K_1' \cdot H'}{dy} = \frac{M \cdot dx \cdot H'}{E \cdot I},$$

since $J_1 - K_1 : H' :: J_1' - K_1' : dy.$

In Fig. 159 the length $J_1'-K_1'$ is approximately equal to dy tan α_1. Using this valve in the above equation:

tan $\alpha_1 = \dfrac{M \cdot dx}{E \cdot I}$, in which dx is any portion of the beam, M is the average moment along that portion, and α_1 is the angle of rotation for the length dx. Remember that α_1 is the angle of rotation which is obtained by the constructions in Figs. 156 to 159. An expression will now be derived giving the value of tan α, when α is the angle of rotation in the actual beam for the length dx. Draw Fig. 162, which is a portion of the beam drawn to an exaggerated scale. From mechanics we know that $\epsilon = \dfrac{S \cdot dl}{E}$, and from similar triangles $R \cdot \epsilon = dl \cdot c$. From these two equations, it follows that $S = \dfrac{E \cdot c}{R}$, and, since $S = \dfrac{M \cdot c}{I}$, we derive the equation $R = \dfrac{E \cdot I}{M}$. The length R is usually very large and dl is usually comparatively small; therefore tan α is approximately equal to dl divided by R. Using the above value of R, tan $\alpha = \dfrac{M \cdot dl}{E \cdot I}$, or since dl and the corresponding dx are almost equal in all actual beams,

tan $\alpha = \dfrac{M \cdot dx}{E \cdot I}$. This is exactly the same as the expression which gives the tan of the angle of rotation obtained in Fig. 159; therefore these two angles, α_1 and α, must be the same. It therefore follows that the broken line in Fig. 159 has the same curvature as the elastic curve. When dx becomes very very small this broken line approaches the true elastic curve as a limit, thus the true elastic curve is a smooth curve inside the broken line curve of Fig. 159, and is tangent to each one of the short lines.

The line 1–17 represents the original position of the neutral axis of the beam, and the deflection at any section is given by the vertical intercept in Fig. 159 between the broken line and the line 1–17, measured at the given section and to the scale at which the beam was drawn.

These vertical intercepts in Fig. 159 are very small, and they usually will be for all ordinary beams, when H' is made equal to $\dfrac{E \cdot I}{H}$. If the pole distance H' is shortened, it can be proved by similar triangles that the intercepts in Fig. 159 will be increased in direct proportion. Suppose the beam is drawn to a scale of 1 inch equals a inches, that is, to $\frac{1}{a}$ actual size, then if the pole distance H' is divided by a the deflections in Fig. 159 will come out full size. That is, if the pole distance $H' = \frac{E \cdot I}{H \cdot a}$, the deflections may be obtained by measuring the intercepts to a scale of 1 inch equals 1 inch. Usually it is desirable to shorten the pole distance even more by some convenient number n, usually taken as 2, 3, 4, or 5. Therefore the common formula for the pole distance H' is $H' = \frac{E \cdot I}{H \cdot n \cdot a}$, and the deflection may be obtained by measuring the intercepts to a scale of 1 inch equals 1 inch and dividing by n. In Fig. 155, the beam was drawn to a scale of 1 inch = 4 feet, which is $\frac{1}{48}$ actual size; therefore a in the formula $H' = \frac{E \cdot I}{H \cdot n \cdot a}$ is taken as 48 in computing the pole distance for Fig. 160. From Fig. 160, Fig. 161 is drawn, and the deflection at any section may be obtained by measuring the vertical intercept between the line 1–17 and the broken line 1–2–3 . . . 17, at the given section, to a scale of 1 inch equals 1 inch, and dividing by n.

57. Simple Beams.—It may be well to work out an actual problem and obtain numerical results in order to make clear just how the graphical method is applied. A beam 20 feet long will be considered as carrying a uniform load of 1000 pounds per foot, and a concentrated load of 10,000 pounds at the center. See Fig. 163. The uniform load is divided into a number of divisions, and an equivalent concentrated load substituted for each division. Unless these divisions of the uniform or distributed load are reasonably small, a considerable error will be produced. In the present problem they are taken 2 feet long to give good results. The loads are now laid off in order along the load line in Fig. 164, and 30,000 pounds is chosen as the pole distance H. From Fig. 164, the moment diagram of Fig. 165 is drawn. The maximum vertical intercept in the moment diagram, measured to the scale at which the beam was drawn, is found to be about 3.35 feet. This intercept times the pole distance, $H = 30,000$ pounds, gives the bending moment in foot pounds. $M = 3.35 \times 30,000 = 100,500$ foot-pounds, or 1,206,000 inch-pounds. Computing the moment analytically, 1,200,000 inch-pounds is obtained for the maximum moment, which shows that the graphical result was in error by about $\frac{1}{2}$ of 1 per cent.

It is well for the beginner to start with simple problems and check his results analytically, in order to be sure that he understands the graphical method and knows how to use it. Also he should remember that it is in the long complicated problems, and

Fig. 164   Fig. 163.

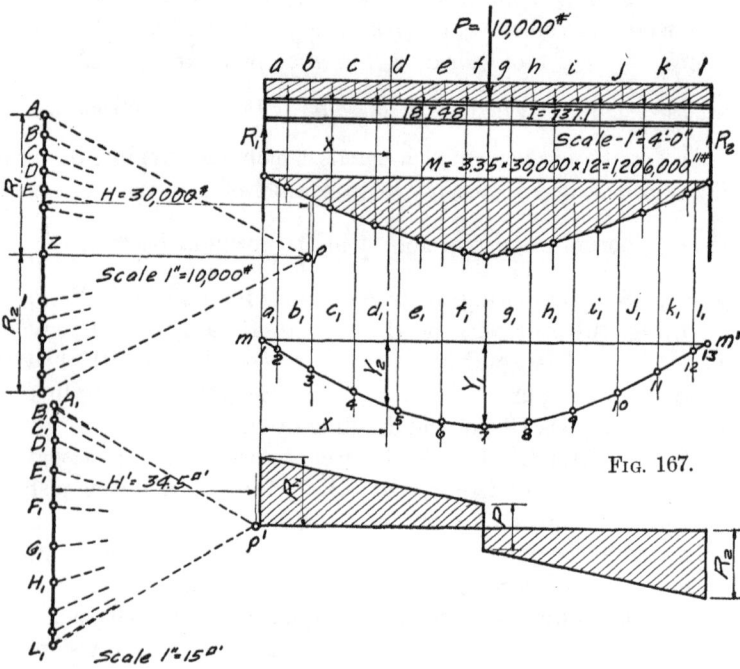

Fig. 165.

Fig. 167.

Fig. 166.   Fig. 168.

not in the short simple problems, that graphical methods are of special value.

Having found the maximum bending moment, the beam may be designed by referring to a hand-book. In this case an 18-inch *I* 48 pounds is used having a moment of inertia of 737.1. The moment diagram, Fig. 165, is now divided into a number of small vertical slices, the areas of which are found, using the scale at which the beam was drawn. The smaller the slices are made, the closer will the broken line curve 1–2–3 . . . 13 approach the smooth one of the elastic curve, and when the slices are very small their centroids and centers will be practically the same point. When the slices are large, the centroid may be at some little distance from the center, in which case it is important that the centroid be used. In Fig. 169 the full lines show what is obtained when large divisions are used, the vertical lines being drawn from their centroids, while the dotted lines show what is obtained when small slices are taken and the vertical lines drawn from their centers.

The areas of the slices of the moment diagram are laid off along the load line in Fig. 166, using any convenient scale which will make the over-all-dimension A_1–L_1 of desirable length. In this problem these areas were found in square feet and laid off to the scale of 1 inch = 15 square feet. The reader should remember that the scales here given are those for the original drawings, which have been reduced.

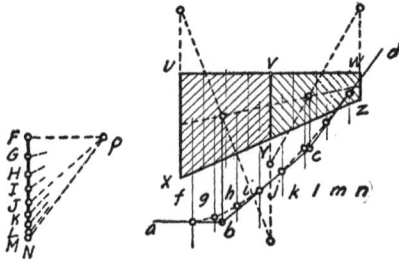

FIG. 170.   FIG. 169.

The next thing to determine is the pole distance H'. In the formula $H' = \dfrac{E \cdot I}{H \cdot n \cdot a}$, E, I, and H are known. The beam was drawn to the scale of 1 inch = 4 feet, that is, $\frac{1}{48}$ actual size, therefore, a will be taken equal to 48. When n is assumed equal to 3 and the result obtained from the above formula is divided by 144 to reduce it to square feet, H' is found to be 34.5 square feet. If a larger value were assumed for n the pole distance H' would be correspondingly smaller. The value obtained for H' is measured off in Fig. 166, using the same scale that was used in laying off the areas of the moment diagram. From Fig. 166 the funicular polygon of Fig. 167 is drawn. The line m–m' represents the position of the neutral axis of the beam before bending, while the polygon below represents the position of the neutral axis after bending, the vertical dimension being magnified. The maximum deflection is obtained by measuring the largest vertical intercept Y_1 in Fig. 167 to the scale of 1 inch = 1 inch and dividing by n, which is in this case equal to 3. In this problem the maximum deflection is found to be approximately .32 of an inch. The deflection at any section distant X from the left support is equal to $Y_2 \div n$, which in this case is equal to $Y_2 \div 3$.

Attention should be called to the fact that it does not make any difference whether the line m–m_1 comes out horizontal or not, because the vertical intercepts will be just the same, as long as the pole distance H_1 is not changed.

The shear diagram which is shown in Fig. 168 has been drawn, not because it was needed for the construction of the other figures, but because a shear diagram is sometimes of value in designing a beam, and in studying the stresses produced. The shear at any section is equal to the summation of all the forces to the left, or to the right, of the section.

Beginning at the left end of Fig. 168, the reaction R_1 is laid off to some convenient scale above the base line. Now moving toward the right, the uniform load is passed over at the rate of 1000 pounds per foot, that is, the shear decreases at the rate of 1000 pounds per foot, until P is reached, when it drops 10,000 pounds. This drop of 10,000 pounds goes below the base line, showing that the shear changes sign. Continuing the movement to the right, the uniform load, which is now the same sign as the shear, causes the shear to increase at the rate of 1000 pounds per foot. Just before the right reaction is reached, the shear is, of course, equal to R_2.

The sign of the moment or of the shear as indicated by position above or below its base line, will be considered of little importance here. The moment diagram, Fig. 165, which is below its base line, has just the same area and the same vertical intercepts that it would have if it were drawn above the base line by locating the pole in Fig. 164 on the left side of the load line. The important thing is to find when the moment and shear change sign, and to give due consideration to this change. By reference to the beam and the system of loading, it is easy to tell which side of the beam has tension and which side has compression.

Simple beams with a more complicated or a more irregular loading than that shown in Fig. 163, can have their moment and shear diagrams, and deflection curves drawn as just explained and illustrated. The only difference will be that the figures will have a different shape than Figs. 164 to 168.

58. Cantilever Beams.—Consider the cantilever beam shown in Fig. 171, which has its support at the left end, and carries both uniform and concentrated loads. The uniform load is divided into small divisions, and an equivalent concentrated load substituted for each division. These loads, together with the large concentrated loads, are laid off in order along the load line of Fig. 172. For convenience, the pole distance H is made equal to 30,000 pounds and p is taken on the left side of the load line. If p had been taken on the right side of the load line, the moment diagram, Fig. 173, would have been turned over. However, the vertical intercepts and the areas would not have been changed if the same pole distance had been used. In constructing the moment diagram, Fig. 173, the funicular polygon is drawn in the usual way. The string l is extended back to the left, and by our theory of moments, any vertical intercept, Y', times H gives the moment of all the forces to the right of the section. The maximum moment is given by the maximum vertical intercept, Y, times the pole distance H. It makes no difference whether the line o–u is horizontal or not, but the pole distance H must always be measured perpendicular to the load line. The moment diagram is now divided into slices, the vertical lines drawn, and the areas laid off along the load line in Fig. 174. The pole distance H' is computed from the formula $H' = \dfrac{E \cdot I}{H \cdot n \cdot a}$, and p' is located. From Fig. 174 the broken line in Fig. 175, representing the elastic curve, is drawn. Now in a cantilever beam, the direction of the neutral axis right at the support is supposed to be the same after bending as it was before, so from m draw the line m–m' tangent to the elastic curve. This means extending the first length of the broken line. The line m–m' will represent the position of the neutral axis of the beam before bending, and the deflection at any section along the beam may be obtained by measuring the vertical intercept between the line m–m' and the broken line, to the scale of 1 inch = 1 inch, and dividing by n. The maximum deflection is found by dividing the maximum vertical intercept Y_1 by n. It makes no difference whether or not the line m–m' comes out horizontal, as long as vertical intercepts are taken.

If p' in Fig. 174 had been taken on the left side in place of on

FIG. 171.

FIG. 173.

FIG. 172.

FIG. 175.

FIG. 174.

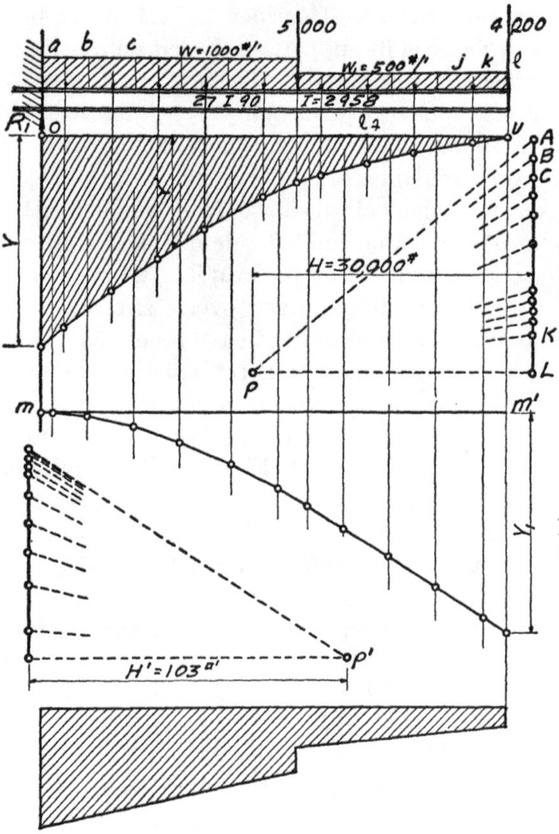

FIG. 176.

the right side of the load line, the deflection curve in Fig. 175 would have been turned bottom side up; that is, it would have appeared as if the beam were deflecting up. Nevertheless, the numerical results could have been obtained just as well as from the figure that is given. The shear diagram is shown in Fig. 176.

The constructions shown in Figs. 171 to 176 are typical for cantilever beams; different loadings, scales, pole distances, and loca-

tions of poles, simply change the shape of the figures and turn
them over.

FIG. 177.

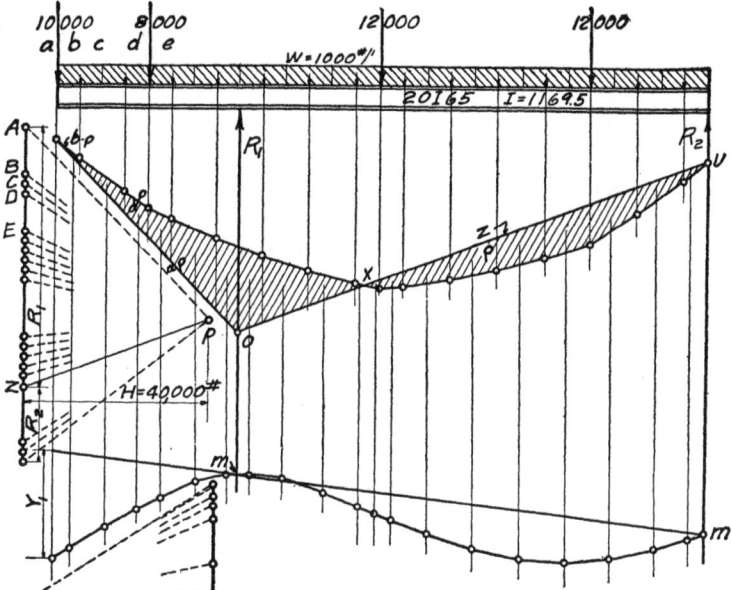

FIG. 178.

FIG. 179.

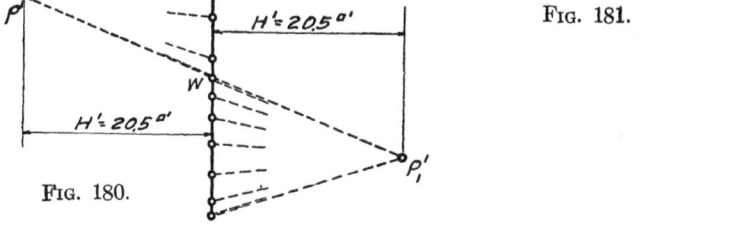

FIG. 181.

FIG. 180.

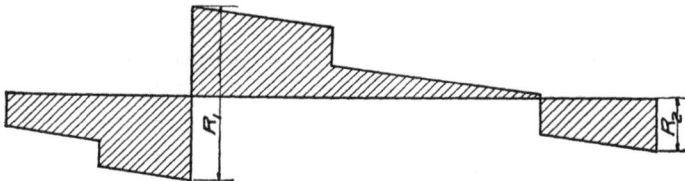

FIG. 182.

59. Beams with an Overhanging End.—The beam shown in
Fig. 177 illustrates a typical overhanging beam. In addition to a
uniform load which includes its own weight, the beam carries a

number of unequal concentrated loads placed at irregular intervals. A more complicated loading might have been considered,
but the loading here assumed is just as instructive.

As usual, the uniform load is divided into small divisions.
The loads carried by the beam are laid off in order along the load
line in Fig. 178, and the pole p with the pole distance H, chosen.
The moment diagram, Fig. 179, is drawn in the usual way. The
string $a–p$ is a component of the force $a–b$ and also of R_1, it therefore connects the action line of $a–b$ with that of R_1. The closing
string $z–p$ is drawn, and parallel to it the closing ray from p, which
gives the point Z, marking off the reactions R_1 and R_2. Now the
vertical intercepts in Fig. 179 times the pole distance H give the
moments produced at various sections along the beam.

It is noted that the moment diagram is part on one side of the
closing string and part on the other; this shows that the moment
passes through zero and changes sign at X.

When working with an overhanging beam, it is usually desirable
to make the pole distance H rather short in comparison with the
load line, in order to keep the vertical intercepts in the moment
diagram from being too small for convenience. The student
should not be worried over the fact that the moment diagram,
Fig. 179, does not have just the same shape as the diagram he had
in mechanics for overhanging beams. The moments obtained
from Figs. 178 and 179 will check with those computed from the
formulas of mechanics.

In order to construct the elastic curve, the moment diagram is
divided into slices, the vertical lines are drawn, and the areas of
the slices are laid off to some convenient scale along the load line in
Fig. 180. From the left end over to the point X, the moment is
of one sign and the areas of the slices are laid off along the load line
in Fig. 180 going down. The pole distance H' is computed as
usual and the pole p' located. The areas to the right of X, which
are of different sign, may be laid off, in Fig. 180, coming up from
W and the pole p' used; or these areas may be laid off going down
from W and the pole p_1' used. The pole p_1' is distance H' from
the load line, and the line $p'–W–p_1'$ must be a straight line. From
Fig. 180, the funicular polygon in Fig. 181, representing the
elastic curve, is drawn. The line $m–m'$ represents the position
of the neutral axis before bending, and is drawn through the
intersection of the elastic curve with the reactions. Therefore,

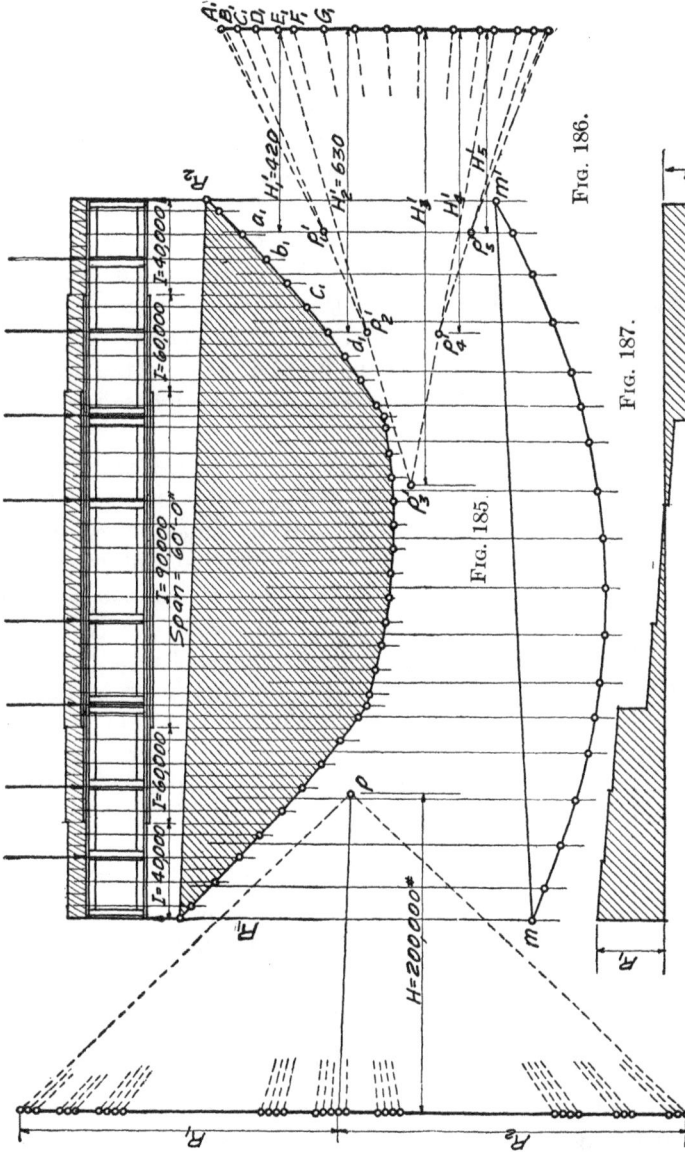

FIG. 183.

FIG. 184.

FIG. 185.

FIG. 186.

FIG. 187.

FIG. 188.

the vertical intercepts between the line m–m' and the broken line give n times the deflection. The maximum deflection appears to be at the left end of the beam, and its numerical value is obtained by measuring the intercept Y_1 to the scale of 1 inch = 1 inch and dividing by n.

The shear diagram is shown in Fig. 182.

60. Beams with a Variable I.—While a variation in the moment of inertia along the length of a simple beam or girder has no effect upon the moment produced by a given loading, it has a decided effect upon the amount of deflection. Consider the plate girder of 60 feet span shown in Fig. 183 which carries, in addition to a uniform load, a number of concentrated loads. These loads are laid out along the load line in Fig. 184 in the usual way, starting with the load at the left end of the beam and taking the others in order. From Fig. 184, the moment diagram Fig. 185 is drawn, and the reactions R_1 and R_2 are found.

The moment diagram is now divided into slices and the vertical lines drawn from their centroids. It is important that a division line between areas comes directly under every section where there is an important change in the moment of inertia. The areas of the various slices of the moment diagram are now computed and laid off to some convenient scale along the load line of Fig. 186. In this problem it was found desirable to start with the areas at the right end of the moment diagram, so that the deflection curve in Fig. 187 would be right side up when the poles were taken on the left side of the load line. Now the first two areas at the right end are under a portion of the beam which has an I of 40,000 as its moment of inertia. This I of 40,000 is substituted in the formula $H' = \dfrac{E \cdot I}{H \cdot n \cdot a}$ in order to obtain H'_1, which is used to locate the pole p'_1. The third and fourth area divisions are under a portion of the beam that has 60,000 for its I, which value is used in the above formula in order to obtain H'_2. It will be observed that I is the only variable in the formula $H' = \dfrac{E \cdot I}{H \cdot n \cdot a}$, therefore the pole distances in Fig. 186 vary directly as I. Having found H'_2, the pole p'_2 is located at this distance from the load line and on the ray C_1–p'_1 extended. Following the fourth slice, a number of the divisions of the moment diagram are under a portion of the beam having 90,000 for its I. Using this value of I, H'_3 is computed, and the pole p'_3 located on the ray $E_1–p'_2$ extended. When the moment of inertia changes again, a new pole distance is computed for the area under that portion of the beam having the new I, and a new pole is located in the same way as described above. The funicular polygon in Fig. 187, which represents the elastic curve, is now drawn, its strings being parallel to the corresponding rays in Fig. 186. The deflections are given as usual by measuring the vertical intercepts between the line $m–m'$ and the elastic curve to the scale of 1 inch = 1 inch and dividing by n.

The shear diagram is shown in Fig. 188, and because of its simplicity, needs no explanation. It might be well to call attention to the fact that a large part of the notation is left off in these figures because they are shown to a small scale. It is a good plan for the student to use more notation and put the scale on every figure in order to avoid mistakes.

The solution of the wooden beam shown in Fig. 189 is an interesting problem. This beam has a uniform depth, but a variable width, there being a uniform decrease from 12 inches at s to 4 inches at the left end. The beam carries, in addition to a uniform load, a concentrated load of 5600 pounds. Fig. 190 and the moment diagram Fig. 191 are drawn in the usual way, and the moment diagram is divided into small vertical slices from the centroids of which the vertical lines are drawn. The moment of inertia of the beam at various sections is now computed. I is found to be constant from t over to s and then to decrease uniformly down to the value I_r at r, as shown in Fig. 192. This figure is drawn so that the vertical intercept at any section measured to scale will give the I of the corresponding section of the beam. The areas of the slices of the moment diagram are laid off to scale along the load line in Fig. 193, starting with the areas at the right end of the beam. For computing the pole distance in Fig. 193, the ordinary formula is used, the E being the modulus of elasticity for wood. The pole distance H'_1 is used for all of the areas over as far as s, where the moment of inertia of the beam begins to decrease. The pole distance H'_2, for the area of the next slice, is computed by using the average I for the corresponding length of beam. Also, in computing the other pole distances, the mean I for the corresponding length of beam is used. These average I's may be obtained from intercepts in Fig. 192. From

Fig. 193 the elastic curve Fig. 194 is drawn, from which the deflection may be obtained.

61. Beams with One End Fixed.—If one end of a beam is fixed, and the other end supported, there is a very marked change in the reactions, in the moments produced, and in the shape of the elastic curve. The determination of the reactions, moments, and deflec-

FIG. 189.

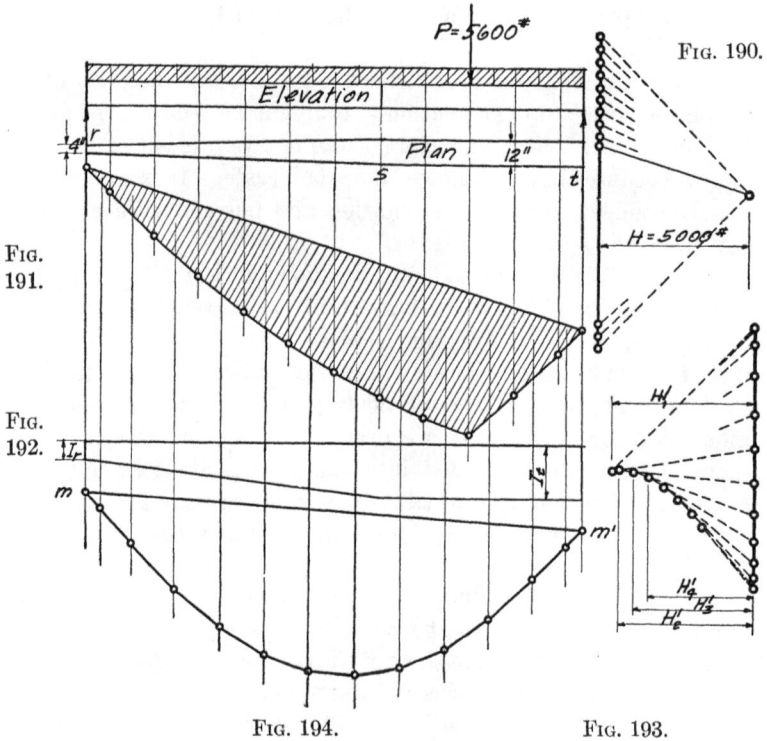

FIG. 190.

FIG. 191.

FIG. 192.

FIG. 194.

FIG. 193.

tions is a much more difficult problem, especially when the loading is irregular and consists of both concentrated and uniform loads.

Let the beam shown in Fig. 195, carrying a uniform load together with concentrated loads, be considered as having the right end fixed. The uniform load is split up into small divisions as usual, and all of the loads are laid off along the load line, Fig. 196. From this figure the funicular polygon in Fig. 197 is drawn, beginning at o and continuing around to u. If a line were drawn connecting o and

u, the closing string of a moment diagram for a simple beam would be obtained.   With the beam fixed at the right end, the tan to the neutral axis at this support must remain unchanged in direction during bending, and there must be a considerable fixing moment at this end.   The magnitude of this fixing moment at the right support is as yet unknown, but we do know that it is of a value such that, when the areas of the moment diagram are laid off along a load line and the elastic curve drawn, this curve will be tangent to the line $m–m'$ at the right support, as illustrated in Fig. 205.

Choose any point r, in Fig. 197, which seems to give a reasonable value for the fixing moment at the right support, and draw the temporary base line or closing string $r–o$.   The areas between the funicular polygon and the line $r–o$ are divided into slices and the vertical lines drawn as usual.   The areas of the slices are laid off in Fig. 198, and any convenient distance assumed for H'.   When the moment changes sign, the pole p' changes to p'_1 on the other side of the load line and at the same distance H', also $p'–p'_1$ must be a straight line.

From Fig. 198 the funicular polygon in Fig. 199 is drawn. This polygon represents the elastic curve of the beam for the case when the fixing moment is represented by $r–u$.   The line $m_1–m'_1$ shows the original position of the neutral axis, but the tangent to the elastic curve at m'_1 is $m'_1–Y_1$, which does not coincide with $m_1–m'_1$ extended, but makes an angle with it, so that at d_1 from R_2 the distance between them is $Y_1–W_1$.   Since Y_1 lies above W_1, the assumed fixing moment, represented by $u–r$, was not large enough.   The point s in Fig. 197 is therefore taken, making $s–u$ larger than $r–u$ by what seems a reasonable amount, and the line $s–o$ is drawn.   Fig. 200 is drawn in the usual way, using the same scale and pole distance that was used in Fig. 198, and the areas to the new base line $s–o$.   The elastic curve in Fig. 201 is next drawn and the intercept $Y_2–W_2$ obtained.   The point Y_2 is still above, which shows that the assumed moment is still too small.   The point t is now taken giving the new base line $o–t$, and Figs. 202 and 203 are drawn.   In Fig. 203, the point Y_3 is below W_3, which shows that the last assumed moment was too large.

In Fig. 197 draw $r–1$ perpendicular to R_2 and equal to $Y_1–W_1$; also draw $s–2$ equal to $Y_2–W_2$ and $t–3$ equal to $Y_3–W_3$.   Now draw the line 1–2–3 which gives the intersection X.   This point X locates the true base line, and the correct moment diagram is

FIG. 195.

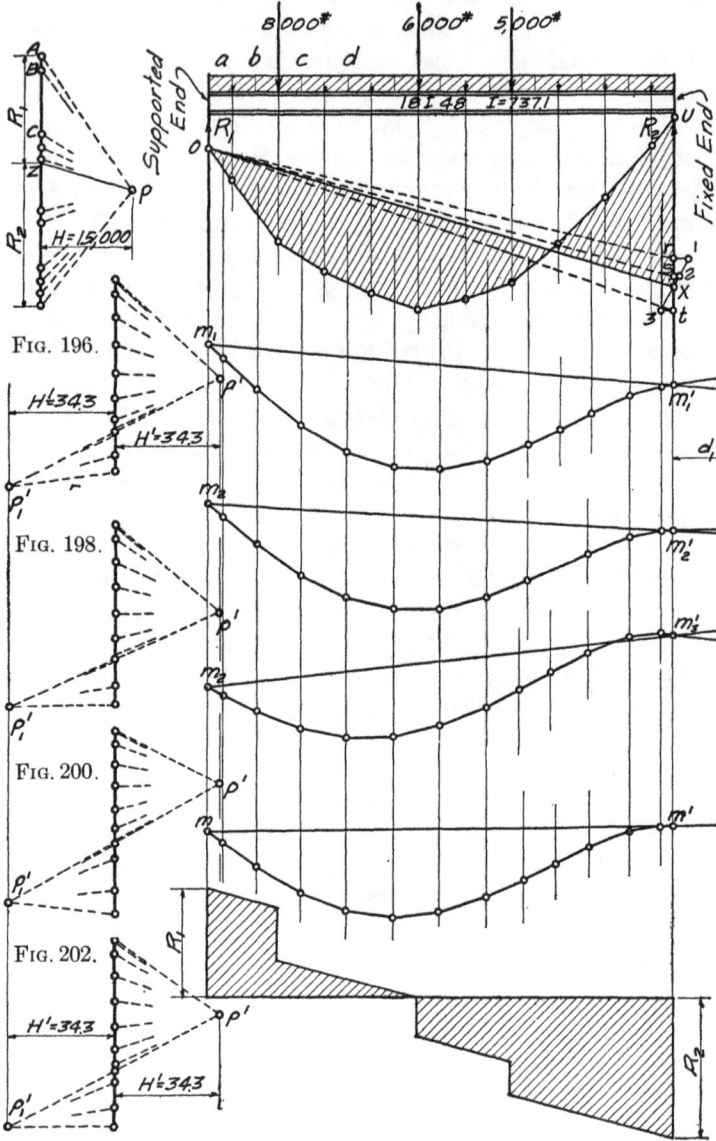

FIG. 196.

FIG. 198.

FIG. 200.

FIG. 202.

FIG. 204.

FIG.
197.

FIG.
199.

FIG.
201.

FIG.
203.

FIG.
205.

FIG. 206.

shown by the shaded area.   The maximum moment is at the support, and is numerically equal to $u–X$ times the pole distance H. The reactions may be obtained from Fig. 196 by drawing the ray $p–Z$ parallel to $X–o$.   It will be noted that the line 1–2–3 is practically a straight line and the intersection X could have been located approximately by a straight line 1–2 extended.   Or in other words, Figs. 202 and 203 are unnecessary.   In order to check the correctness of the results obtained, Figs. 204 and 205 are drawn. The areas of the slices are measured to the base line $X–o$, and are used in drawing Fig. 204.   From this figure the funicular polygon in Fig. 205, which represents the elastic curve, is drawn, and points Y and W are found to be very close together.   That is, the tangent to the elastic curve at the right support practically coincides with the neutral axis of the beam before bending.   Therefore, the problem has been solved with very small error, and $X–u$ times H is the fixing moment.   The beam may now be designed, or redesigned if required, and its I determined.   Then in the formula $H' = \dfrac{E \cdot I}{H \cdot n \cdot a}$ we know H', H, E, I, and a; the only unknown is n, and it can be found by solving the equation.   Then the deflection may be found for any section of the beam by measuring the proper vertical intercept in Fig. 205 and dividing it by the value of n.

The shear diagram is shown in Fig. 206.

62. Beams Fixed at Both Ends.—A beam which is fixed at both ends may be solved in a way somewhat similar to the method just explained.   The beam and loading shown in Fig. 207 will be considered.   After dividing the uniform load into small divisions, all the loads are laid off in order along the load line in Fig. 208, and the pole p chosen with the pole distance H.   It will be found convenient to make the pole distance H relatively small, so that the funicular polygon in Fig. 209 will be of convenient depth.

The problem now resolves itself into locating the base line $X–Y$ so that the moment given by $o–Y$ times H will be just enough to fix the left end of the beam, and that given by $(u–X) \cdot H$ will just fix the right end.   When these fixing moments are of correct value and, the elastic curve is drawn, the line corresponding to $m_1–m'_1$ of Fig. 211 should be tangent to the elastic curve at the supports, or in other words, the distances corresponding to W_1 and W'_1 should be approximately zero.   By inspection, the trial base $r–r'$ is assumed as one which looks reasonable; the areas of the

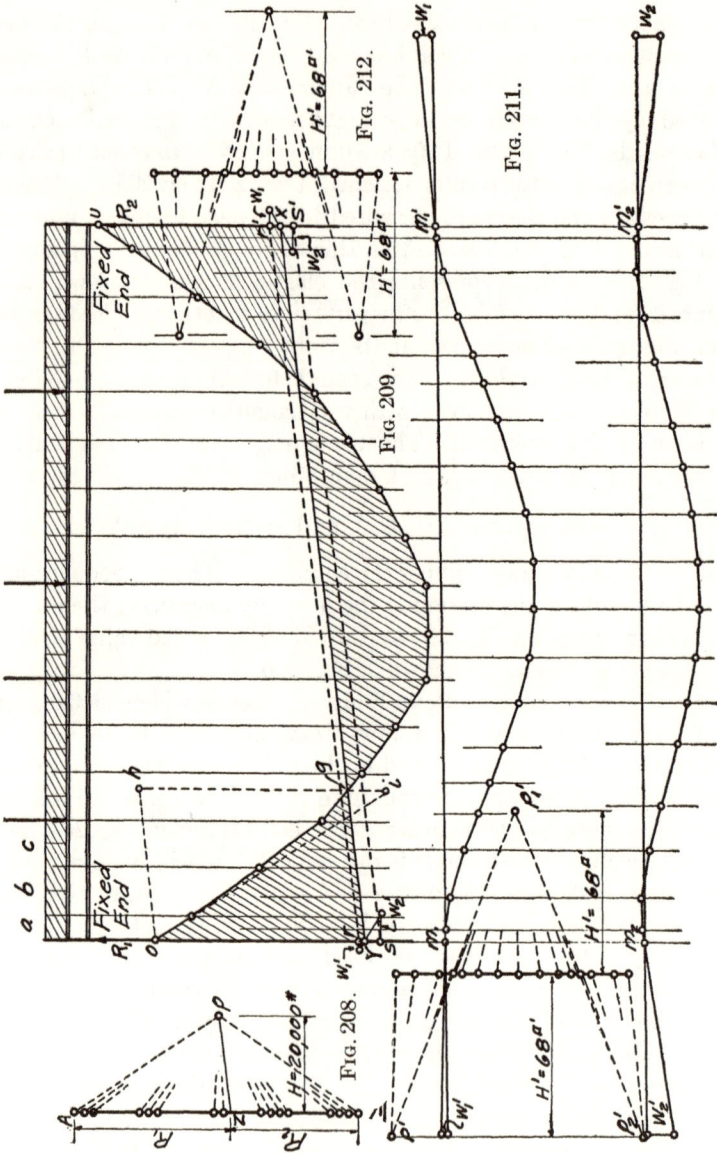

Fig. 207.

Fig. 208.

Fig. 209.

Fig. 210.

Fig. 211.

Fig. 212.

Fig. 213.

moment diagram are divided into slices and the vertical lines
drawn as usual.   The areas of these slices are laid off along the
load line of Fig. 210, and the pole distance H' assumed.   The pole,
which first is taken at p' on the left side of the load line, changes
over to p'_1 on the right side when the sign of the moment changes,
and moves back again to p'_2 on the left side, when the moment
changes sign again.   The funicular polygon of Fig. 211, which
represents the elastic curve for the case when r–r' is the base line
of the moment diagram, is now drawn.   Fig. 211 shows that the
base line r–r' is not quite correct because W_1 and W'_1 are not equal
to zero.   In this case the fixing moments are not large enough,
because W_1 and W'_1 are above the base line m_1–m'_1 extended.
The trial base line is, therefore, moved down to s–s', and the areas
of the slices, measuring to this new base line, are laid off in Fig.
212, from which the elastic curve in Fig. 213 is drawn.   It is found
that the tangent to the elastic curve at each support makes an
angle with the line m_2-m'_2.   The intercepts W_2 and W'_2 are below
m_2–m'_2.   Therefore, the base line s–s is too low.   Now from r'
measure off W_1 on one side of R_2, and from s' measure off W_2
to the same scale but on the other side.   If W_1 and W_2 were both
on the same side of the lines m_1–m'_1 and m_2–m'_2, they would be
measured off on the same side of R_2.   By connecting the extremes
of these two lengths, as shown, the intersection X is obtained,
which locates approximately the right end of the true base line.
In a similar way the left end, Y, is located.   A check on the correctness of this base line could be obtained by constructing an
elastic curve similar to the ones shown in Figs. 211 and 213, and
seeing if the intercepts W and W' are approximately equal to zero.
If the error found is considered too large, it may be corrected by
moving the base line Y–X a small amount in the direction indicated
by the error.

Now to find the reactions: Let p–Z in Fig. 208 be drawn parallel
to the base line Y–X; then the point Z marks off the two reactions,
R_1 and R_2.   It can be shown that Z–A is the correct reaction at the
left end of the beam in the following way: Draw the line o–h parallel to p–Z, which is itself parallel to Y-X, and draw the vertical
line g–h from g.   At the point of contra flexure g, the bending
moment is zero; that is, the moment of the loads and reaction to
the left of g about g is just balanced by the fixing moment o–Y
times H.   When R_1 has the value Z–A, its moment about g is

h–i times H, while the moment of the loads to the left of g about g is given by g–i times H. Then the moment of R_1 less the moment of the load is g–h times H. But this value is just equal to the fixing moment because h–g equals o–Y, since o–h was drawn parallel to X-Y. Therefore Z–A is the correct value for R_1.

63. Continuous Beams of Two Spans.—An interesting and somewhat difficult problem arises when a continuous beam of two spans is considered. When the spans are of equal length and the load is uniform, the problem can be conveniently handled by analytical methods; but when the spans are of unequal length, and there is a uniform load together with a number of unequal, irregularly spaced, concentrated loads, an exact analytical solution is more difficult. It is in the solution of such problems that graphical methods are of special value.

Fig. 214 shows a continuous beam which has two unequal spans and carries a complicated loading. As usual, the uniform load is divided into small slices for each of which an equivalent concentrated load is substituted. In this problem it will be found especially desirable to take rather small divisions. All of the loads, including the large concentrated loads, are laid off in order along the load line of Fig. 215, and using a rather short pole distance H, the funicular polygon in Fig. 216 is drawn. The object of the short pole distance is to make the vertical distance in Fig. 216 of more convenient length. Now if the moment at R_2 was known, the closing strings u–X and o–X could be drawn, and the true moment diagram would be obtained. But the position of X is as yet unknown, so the principal part of the problem is to find its location. First a point w is chosen, which is nothing more than an intelligent guess at the location of X, and the temporary closing lines w–o and w–u are drawn. Now the areas between these lines and the funicular polygon would be the moment diagram for the beam, if the elevation of R_2 were such as to give a moment at the center support equal to w–t times H. For it is evident that the moment at R_2 will increase as the elevation of the center support increases above the line connecting the other two supports, and will decrease as the elevation of the center support decreases below this line. Also, for every value of moment at the center support, there is a corresponding elevation of the center support. The point X is a point such that the moment X–t times H is the moment produced when all of the supports have the same elevation, or when R_2 is on the straight line connecting the other two supports.

The temporary moment diagram, already drawn, is divided into slices and the vertical lines drawn in the usual way. The areas of these slices are computed, and Fig. 217 is drawn. The pole distance H' is computed as already explained, or, if the beam has not been designed, any convenient length may be assumed. The important thing is to use the same pole distance in all three Figs. 217, 219, and 221. It will be noted that the pole is changed from one side of the load line to the other, as the sign of the moment changes. Using Fig. 217, the funicular polygon of Fig. 218 is drawn, which represents the elastic curve for the beam when the moment at the center support is w–t times H. The straight line m_1–m'_1, connecting the two end supports, is drawn and at the center support the elastic curve is found to be a distance Y_1 above it. This means that too large a moment was assumed at the center support. So w_1 is chosen making w_1–t smaller than w–t by what seems a reasonable amount. The lines w_1–o and w_1–u are drawn, and the areas of the slices in the moment diagram measured to them, are laid off in Fig. 219. From Fig. 219 the funicular polygon in Fig. 220 is drawn, which represents the elastic curve of the beam when the moment at the center support is w_1–t times H. By drawing the line m_2–m'_2, it is found that in order to have the assumed moment at the center support, this support must be below the other two supports. Therefore, the distance w_1–t is too small.

In Fig. 216, w–r is drawn perpendicular to R_2 and made equal to Y_1, and w_1–s is made equal to Y_2, s and r being on opposite sides of R_2. The points r and s would be taken on the same side of R_2 if the intercepts Y_1 and Y_2 were on the same side of the lines connecting the two supports. The line r–s is drawn, and its intersection with R_2 locates X. The lines X–o and X–u are now drawn, giving the true moment diagram shown by the shaded area. The line p–Z_1 in Fig. 215 is drawn parallel to o–X and p–Z_2 is made parallel to X–u, thus the reactions R_1, R_2, and R_3 are obtained.

As a check on the correctness of the location of point X, and also for the purpose of finding the deflections, the elastic curve of Fig. 222 is drawn in the same way that Fig. 220 was drawn, except that the areas in the moment diagram are measured to the lines X–o and X–u. The line m–m' is drawn and the distance Y, which

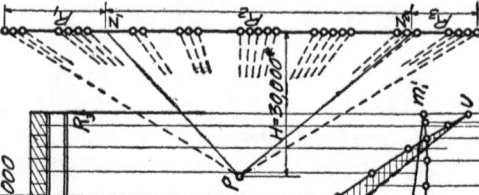

Fig. 215.

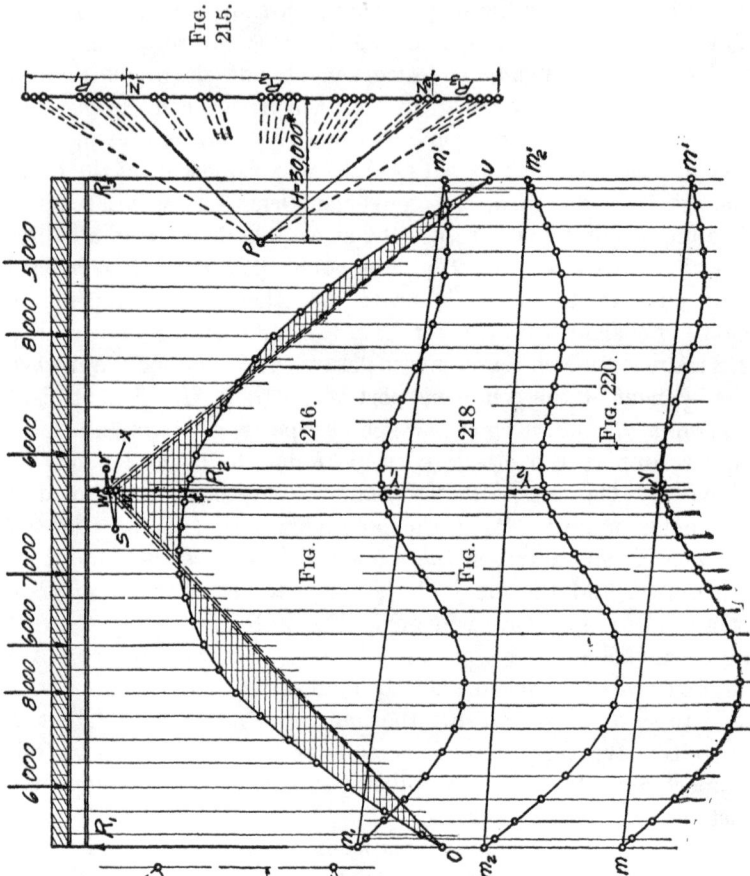

Fig. 214.

Fig. 216.

Fig. 218.

Fig. 220.

Fig. 222.

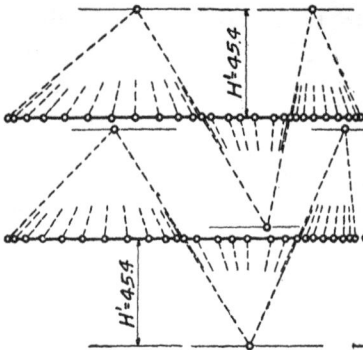

Fig. 219.

Fig. 217.

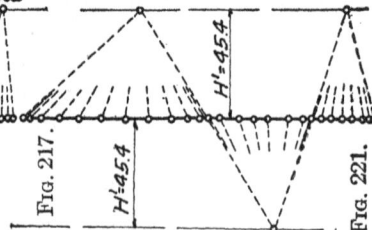

Fig. 221.

should be zero in this case, is found to be very small. The location of X, and the value X–t times H for the moment at the center support will, therefore, be considered satisfactory.

The shear diagram has not been drawn, but since the reactions are known, its construction would be comparatively easy.

Figs. 223 to 231 show the solution of a continuous beam of two spans by a construction slightly different than that which has just been employed. The only real difference, however, is in Fig. 224 and in the moment diagram. It will be observed that the vertical intercepts in the moment diagram, Fig. 216, are relatively small, and yet the figure as a whole has rather large vertical dimensions. In Fig. 224, the loads on the beam are laid off in order as usual, but in place of taking only one pole, as in Fig. 215, two poles p and p_1 are taken, one being for the loads over one span of the beam and the other being for the loads over the other span. Using the pole p, which is for the loads on the left span, the funicular polygon is drawn in Fig. 225 from point o around to point t. Then using the pole p_1 the polygon is continued until the point u is reached. It is, of course, necessary that the pole distance of p_1 be the same as that of p.

The point w is now chosen, as in Fig. 216 and Fig. 226 drawn. Then Fig. 227 is drawn from which Y_1 is obtained and by the use of Y_1 the point r is located in Fig. 225. The point w_1 is next chosen, Figs. 228 and 229 are drawn, and the point s located in Fig. 225. The line s–r locates X, and X–t times H is the correct moment at the center support for the case when the three supports are all on the same straight line. Fig. 230 is next drawn and from it Fig. 231, which represents the elastic curve for the case when the moment at the center support is H times t–X. In this case the distance Y, which should be zero, is found to be very small, thus checking the work. In order to obtain the reactions, the ray p–Z_1, in Fig. 224 is drawn parallel to X–o, and the ray p_1–Z_2 is made parallel to X–u, after which R_1, R_2, and R_3 may be scaled off.

It is interesting to see how this method of analyzing a continuous beam can be used to determine how the distribution of bending moment will be affected by an unequal settlement of the supports. Let it be assumed that the center support of the beam shown in Fig. 223 settles $\frac{1}{4}$ inch more than the other two, so that it is $\frac{1}{4}$ inch below the straight line connecting them. When the center support settles, the moment there decreases. Therefore,

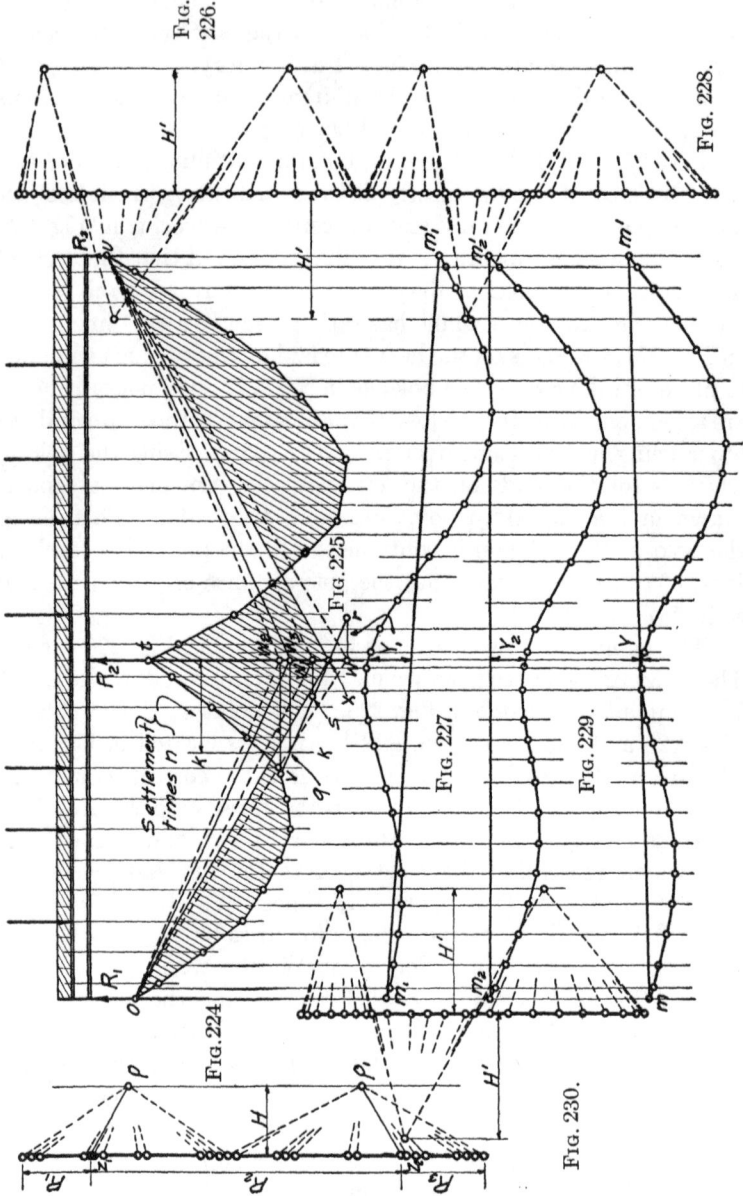

FIG. 223.
FIG. 224.
FIG. 225.
FIG. 226.
FIG. 227.
FIG. 228.
FIG. 229.
FIG. 230.
FIG. 231.

choose W_2 and draw figures corresponding to Figs. 226 and 227, using the areas to the base lines $o-w_2$ and $u-w_2$. From these figures which are not shown, Y_3 is obtained, and the line $v-w_2$ is made equal to it, thus giving the curve $v-s-r$ which should be practically a straight line. Now the excess settlement at R_2, which was just assumed to be $\frac{1}{4}$ inch, is multiplied by n and the distance thus obtained is measured to the left of R_2 and the vertical line $k-k$ located. The value used for n is the same value that was used in the formula $H' = \dfrac{E \cdot I}{H \cdot n \cdot a}$. From the intersection of the line $k-k$ with the line $v-s-r$, draw a horizontal line giving the intersection w_3 with the action line of R_2. Now w_3-t times H is the moment at the center support, also $o-w_3$ and $u-w_3$ are the base lines of the moment diagram when the center support has settled the assumed amount. The reactions may be obtained by drawing lines from p and p_1 parallel to w_3-o and w_3-u.

64. Continuous Beams of Three Spans.—A continuous beam of three spans can be satisfactorily solved by graphical constructions similar to those used for two span beams. In order to demonstrate the fact that the more complicated cases can be handled without any great amount of work, the illustrative problem chosen deals with a beam having three unequal spans and carrying, in addition to a uniform load, a number of irregularly spaced unequal concentrated loads. A diagram of this beam, which we will assume has a constant I, is shown in Fig. 232.

The uniform load is divided into small divisions as usual and all of the loads are laid off along the load line of Fig. 233. The poles p_1, p_2, and p_3 are all chosen with the same comparatively small pole distance H, and are used for the loads that are applied at the left, center, and right spans respectively. From Fig. 233, the funicular polygons of Fig. 234 are drawn. The pole p_1 is used for the polygon from O to S, the pole p_2 for that portion from S to T and p_3 for the portion from T to U.

It will be considered that all three supports have the same elevation or are on the same straight line before loading, and that they all settle the same amount under load. Under these conditions a certain definite moment, $S-X$ times H, exists at R_2, and another definite moment, $T-X'$ times H, at R_3. The problem resolves itself into locating the points X and X'.

As a preliminary step the point v is chosen, which in the judgment of the student is close to where he would expect to find X, and the point w is chosen close to where X' is expected to be. Preliminary base lines $O–v$, $v–w$, and $w–U$ are drawn which give a preliminary moment diagram. This diagram is divided into slices, the areas of which are laid off along the load line of Fig. 235, and the poles p', p'_1, p'_2, etc., are chosen, all with the pole distance H'. It should be remembered that the pole in Fig. 235 changes sides only when the moment in Fig. 234 changes sign, and further-more that the lines $p'–p'_1$, $p'_1–p'_2$, etc., are all straight lines. From Fig. 235, Fig. 236 is drawn and the intercepts Y_1 and Y'_1 obtained, both of which are above the line $m_1–m'_1$. This means that either the moment assumed at R_2, or the moment assumed at R_3, or both, were too large. For the next step the moment at R_2 is not changed, but that at R_3 is reduced to $T–w_1$ times H, w_1 being chosen at what seems a reasonable distance above w. Considering the base lines $O–v$, $v–w_1$ and $w_1–U$, the areas of the slices in Fig. 234 are laid off along the load line in Fig. 237, from which Fig. 238 is drawn, giving the intercepts Y_2 and Y'_2, both of which are below the base line $m_2–m'_2$.

Next the point v_1 is chosen a short distance above v and using the base lines $O–v_1$, $v_1–w$, and $w–U$, Fig. 239 is drawn. Fig. 240 is then drawn and the intercepts Y_3 and Y'_3 obtained, both of which happen to be above the line $m_3–m'_3$.

In Fig. 241 two axes are drawn at right angles to each other, the vertical one being called the $R_3–R_3$ axis and the horizontal the $R_2–R_2$ axis. The coördinates of the point w are Y_1 and Y'_1 from Fig. 236. When the moment at R_2 is kept the same and the moment at R_3 is decreased by changing from w to w_1, the inter-cepts Y_2 and Y'_2 are obtained in Fig. 238. These intercepts are the coördinates of the point w_1 in Fig. 241. As the intersection of the base lines with R_3 in Fig. 234 varies from w to w_1, the inter-cepts along R_2 and R_3 will vary approximately as the coördinates of the corresponding points of the line $w–w_1$, in Fig. 241.

When the points v_1 and w are used, in Fig. 234, the intercepts Y_3 and Y'_3 are found in Fig. 240. These intercepts are the coör-dinates of the point v_1 in Fig. 241. The line $v–v_1$ is moved down until it passes through the origin o. During the movement, it is kept parallel to its original position, and the point v follows along the line $w–w_1$.

Now as the moment at R_3 is decreased, w in Fig. 234 moving

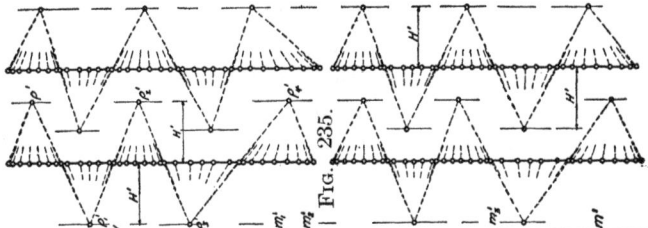

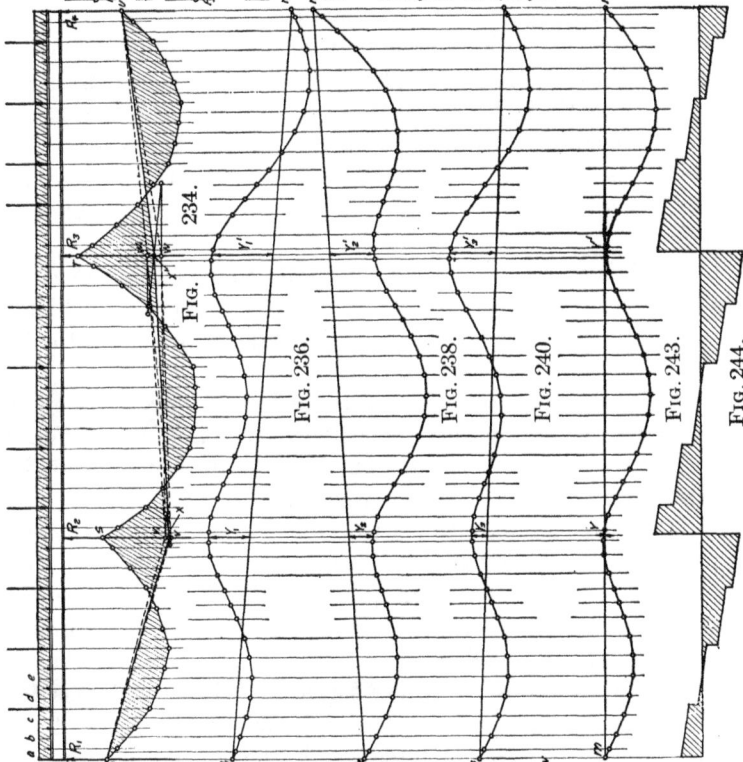

FIG. 237.

FIG. 242.

FIG. 239.

FIG. 235.

FIG. 232.

FIG. 234.

FIG. 236.

FIG. 238.

FIG. 240.

FIG. 243.

FIG. 244.

FIG. 233.

FIG. 241.

towards w_1, the intercepts Y_1 and Y'_1 will decrease as the coördinates of a point moving from w, in Fig. 241, towards w_1. But by varying the moment at R_3, the intercepts Y_1 and Y'_1 could never both be made equal to zero at the same time, since the line $w–w_1$ in Fig. 241 does not pass through the origin. The moment at R_3 is, therefore, decreased until the intercepts Y_1 and Y'_1 have values equal to the coördinates of v' in Fig. 241. Then the moment at R_2 is varied until these intercepts have values equal to the coördinates of o, or in other words, they are zero.

In Fig. 234 the point X is located so that it divides the line $v–v_1$ in the same proportion that o divides the line $v'–v'_1$ in Fig. 241; also the point X' is located so that it divides the line $w–w_1$ in the same proportion that v' divides the line $w–w_1$ in Fig. 241. The base lines $O–X$, $X–X'$, and $X'–U$ are then drawn giving the true moment diagram. Using these base lines, Fig. 242 is drawn, and from it the elastic curve shown in Fig. 243. The intercepts Y and Y' are found to be small showing that the results are satisfactory.

The deflection at any section may be obtained from the vertical intercepts in Fig. 243, and the reactions can be found in Fig. 233 by drawing the rays $p_1–Z_1$, $p_2–Z_2$ and $p_3–Z_3$ parallel to the lines $O–X$, $X–X'$ and $X'–U$, respectively.

In case the intercepts Y and Y', in Fig. 243, are so large that the accuracy of the results is questioned, the base lines in Fig. 234 should be shifted a small amount in the direction indicated by Fig. 243, and another elastic curve drawn.

Attention should be called to the fact that a small movement in the position of the base lines in Fig. 234 produces a large variation in the intercepts in Figs. 236, 238, 240 and 243. Therefore, when the intercepts Y and Y' are small, the error in the moment must be decidedly small.

When continuous beams are solved by this method, it is desirable to use a comparatively short pole distance H in Fig. 233, in order to make the vertical intercepts in the moment diagram relatively larger. Also a rather large scale should be used for the beam in Fig. 232. If this is done, the areas of the slices in the moment diagram can be measured with greater ease and with a smaller percentage of error.

It should be noted that a variation in the moment of inertia along the length of the beam would cause little trouble and would add very little to the amount of work.   It would be taken care of by varying the pole distance in Figs. 235, 237, 239 and 242 as in Fig. 186.

The methods here given for solving continuous beams may be used to advantage in investigating continuous reinforced concrete beams, when a relatively exact solution is desired and is hard to obtain analytically, as is the case when the spans are unequal and the loading is complicated.

CHAPTER V

TRUSSES

It is assumed that the reader is acquainted with the various forms of roof construction and that he knows, in a general way, what a truss is. This chapter will therefore deal with the determination of loads and stresses for various types of trusses, rather than enter into descriptions and definitions.

It is desirable to keep in mind that the loads are assumed to be applied at the panel points, and that these loads produce only direct stresses, as would be the case in a pin-jointed truss. A bending stress produced in any way must be combined with the direct stress in order to obtain the resultant stress which must be resisted.

In this chapter attention will be given chiefly to the solution of different types of trusses by stress diagrams. From these stress diagrams the stresses produced in the various members may be scaled, and recorded in a table or marked on the truss, as illustrated in Chapter IX in connection with design.

Fig. 245 shows the outline roof plan of a small, simple, rectangular building, while Fig. 246 represents a section, and Fig. 247 the outline of one of the trusses. Now the truss must carry its own weight, the weight of the rafters, purlins, sheathing, and roof covering, as well as the weight of the ceiling, should one be attached to the truss; all of these taken together make up what is sometimes called the dead load. In addition to these, there are the snow load, wind load, and any special loads which may be applied to the truss.

Trusses must be held rigidly in position and the compression chord especially must be well supported laterally. The purlins and roof covering usually form adequate lateral support for the upper chord. When there are no ceiling beams the lower chords are often spaced by a few light struts. For convenience in erection and in order to secure lateral stiffness it is usually desirable to use a certain amount of rod or light angle bracing. When the horizontal component of the wind cannot be taken by the supports at the ends of each truss, but must be transmitted to the ends of the building or some other distant part by the horizontal bracing acting as a horizontal truss, this bracing may become rather heavy.

65. Weight of Trusses.—After the truss is designed it is a very simple matter to obtain its weight by computing the weight of the various members, and summing them up. But since the weight of the truss is a part of the load which the truss must carry, it is important that at least an approximate estimate of its weight

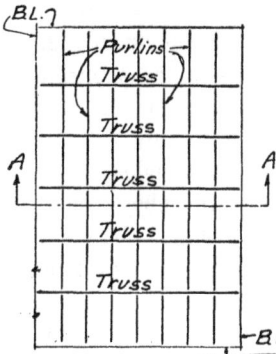

FIG. 245.

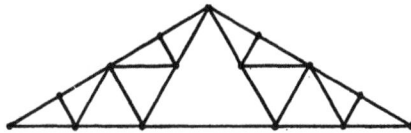

FIG. 247.

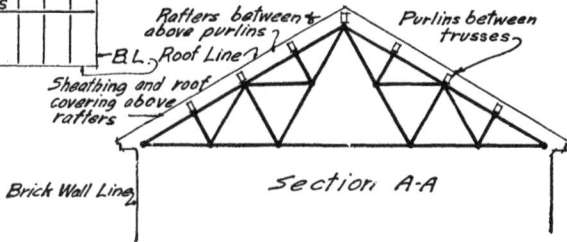

FIG. 246.

be made before solving for the stresses that are to be used in determining the size of the members.

There are a number of standard formulas which may be used for estimating the weight of trusses, two of which are given below:

$$(1) \quad w = \frac{1}{2}\left(1 + \frac{L}{10}\right) \text{ for wooden trusses (Ricker).}$$

$$(2) \quad w = \frac{P}{45}\left(1 + \frac{L}{5\sqrt{A}}\right) \text{ for steel trusses (Ketchum).}$$

In these w is the weight of the truss in pounds per square foot of horizontal projection of roof, L is the span of the truss in feet, A is the distance in feet from center to center of trusses, and P is the capacity of the truss in pounds per square foot of horizontal projection. It follows that the total weight of one truss is $w \cdot A \cdot L$.

Such formulas are only partially satisfactory, and of course give only approximate results. However, the weight of roof trusses of ordinary proportions is usually small compared with the loads they carry, therefore a large percentage error in the estimated weight of the truss makes only a small error in the loading for which the truss is designed.

66. Other Weights.—Fig. 245 shows a small scale roof plan of a simple rectangular building. Fig. 246 shows a section of this building and an outline of one of the trusses is shown in Fig. 247. The truss carries purlins at the panel points. These purlins are sometimes of wood, but more often of steel. When the roof is of wood there are usually light rafters spanning from purlin to purlin, which carry sheathing to which the shingles are nailed. Sometimes the rafters are omitted and concrete slabs, cement tile, or some other type of construction used instead. There are a great many different forms of trusses, and types of roof construction, but they will not be considered in detail here.

The weight of the purlins, rafters, sheathing and roof covering can of course be computed after the roof has been designed, but the following values will be found convenient in many cases, especially for preliminary work.

Steel purlins weigh from $1\frac{1}{4}$ to 4 pounds per square foot of roof surface and wooden purlins from $1\frac{1}{2}$ to 3 pounds per square foot. Rafters will weigh from $1\frac{1}{2}$ to 3 pounds per square foot.

WEIGHT OF ROOF COVERINGS

	Pounds per sq. ft. of roof.
Corrugated steel, without sheathing	1 to 3
Felt and asphalt, without sheathing	2
Tar and gravel, without sheathing	8 to 10
Slate $\frac{3}{16}$ inch to $\frac{1}{4}$ inch, without sheathing	7 to 9
Tin without sheathing	1 to $1\frac{1}{2}$
Skylight glass $\frac{3}{16}$ inch to $\frac{1}{2}$ inch including frames	4 to 10
White pine sheathing 1 inch thick	3
Yellow pine sheathing 1 inch thick	4
Tiles, flat	15 to 20
Tiles, corrugated	8 to 10
Tiles on concrete slab	30 to 35
Plastered ceiling	10

See Ketchum's Structural Engineers' Handbook.

67. Snow Load.—The maximum snow load which it is reasonable to expect a roof will have to support, varies with the latitude and the pitch of the roof. Fig. 248 shows the variations according to Ketchum. An ice and sleet load of 10 pounds per square foot is assumed for all pitches. It should be noted that the diagram gives loads in pounds per square foot of horizontal projection, and not per square foot of roof surface.

FIG. 249.   FIG. 248.

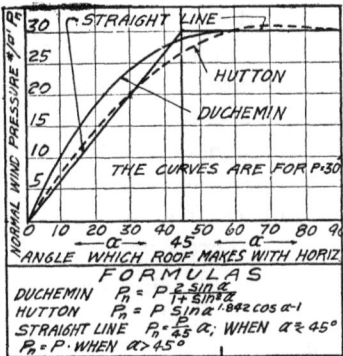
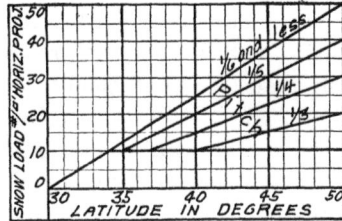

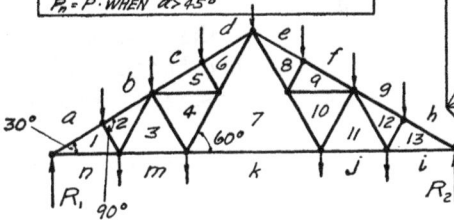
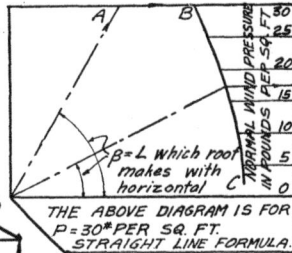

FIG. 251.   FIG. 250.

68. Wind Loads.—In some cases the stresses produced by wind are quite important. This is especially true of high, steep, trusses or trusses supported on steel columns. The wind pressure is usually figured in pounds per square foot of roof surface, and is assumed to act normally. Experiments in connection with wind pressure on roofs have not been very complete. Fig. 249 shows two curved lines which represent the results of two different experimenters, and the straight line represents what is sometimes called the straight-line formula. The latter is simple to use and does not differ greatly from the curves derived from experimental results. Fig. 250 gives a chart which may be used for finding the pressure on roof surfaces of various slopes. It is based on the straight-line formula, and is for a wind that will produce a pressure of 30 pounds per square foot on a vertical surface. The effect of lighter winds may be found by proportion. Ketchum expresses the relation between wind velocity and the pressure it will produce on a flat normal surface by the following formula:

$$P = 0.004V^2$$

in which P is the pressure in pounds per square foot and V is the velocity in miles per hour.

69. Notation.—For convenience it is desirable to adopt some form of notation so that reference to various members can be easily made. The method to be used in this chapter is illustrated in Fig. 251. Letters are used on the outside of the truss and figures on the inside. Each external load is designated by two letters, each web member by two numbers, and each section of the upper or lower chord by a letter and a number. The left reaction is called R_1 and the right R_2. Lower case letters will be used on the truss diagram but when the loads are laid off along the load line capitals will be used. Also capitals will usually be used in the text when referring to loads or members.

70. Stresses Obtained Analytically.—First the reactions may be found by using moments. Then the truss may be taken, a joint at a time, and the two unknowns computed, using $\Sigma F_x = o$ and $\Sigma F_y = o$.

Let the upper chord loads in Fig. 251 each be 6000 pounds and the lower chord loads 3000 pounds each. Since the truss and the loading are both symmetrical, each reaction will equal one-half the total load or 27,000 pounds. Consider the left end of the truss. In order to have summation F_y equal to zero, the vertical component of A–1 must be numerically equal to the reaction. The stress in A–1 is, therefore, 27,000 ÷ .50 = 54,000 pounds compression. In order to have summation F_x equal to zero, the stress in 1–N must be equal to the horizontal component of A–1. It follows that the stress in 1–N is 54,000 times .866 = 46,764 pounds tension. For the joint where the force A–B is applied it will be convenient to rotate the axes so that the x–x axis coincides with the upper chord. Then the stress in member 1–2 equals 6000 times .866 = 5200 pounds compression; and the stress in B–2 = 54,000 − (6,000 · .5) = 51,000 pounds compression. The stresses in other members may be found in a similar way. When the loads are all vertical and the truss simple the analytical solution is easy, but it becomes very complicated when the loads are inclined and the truss irregular.

71. Stresses by Analytical Moments.—The reactions may be found as above. Sections are then taken and the unknowns computed, keeping in mind the fact that if any portion of the truss is cut off by a section, the moment, about any point, of all the external forces plus the moment of the stresses in the members cut by the section, is zero. There are, of course, various other analytical methods and short cuts which are of more or less value. The student will do well to remember that the graphical method is of greatest value in solving long complicated problems, especially when wind as well as vertical loads is considered.

72. Stress by Graphical Moments.—Fig. 252 shows a small truss with an upper chord loading. These loads are laid off in Fig. 253, from which the funicular polygon of Fig. 254 is drawn. When the closing string is drawn this figure forms a moment diagram, that is, a vertical intercept at any section between the funicular polygon and its closing string, times the pole distance H gives the moment at that section. Now take any vertical section x–x cutting the truss members A–1 and 1–Y. The moment of the external forces about this section is r–s times H, and this moment is of course balanced by the internal forces. Taking o as the center of moments, we have the stress in member A–1 times the distance o–$o_2 = s$–$r \cdot H$ or the expression $(r$–$s \cdot H) \div o$–o_2 gives the stress in A–1, o–o_2 being the perpendicular distance from o to A–1. In like manner taking moments of the internal forces about o_1 the stress in 1–Y is found to be $H \cdot s$–r divided by o_1–o. The stress in 3–Y may be easily found by use of the section x_1–x_1. The moment of the external forces about this section is given by r_1–$s_1 \cdot H$. If the center of moments for the internal forces is taken at o_3, o_3–o_4 is the arm for the stress in 3–Y. The stress is therefore obtained by dividing r_1–$s_1 \cdot H$ by o_3–o_4. The section x_2–x_2 cuts three members, but the stress in one of them is known. The moments of the external forces about this section is r_2–$s_2 \cdot H$. First let o_5 be taken as the center of moments for the internal forces. There are two internal forces or stresses producing moment about this center, namely, the stress in the member 1–2 and that in Y–1. If the left half of the truss is considered, the stress in

1–Y, already found to be tension, tends to produce rotation counter-clockwise about o_5. It happens that the moment of 1–Y about o_5 is greater than r_2–$s_2 \cdot H$, the moment of the external forces; so the stress in 1–2 must tend to produce rotation about o_5 in a clock-wise direction, and its magnitude is the stress in 1–Y times o_5–o_7 less s_2–$v_2 \cdot H$ and the remainder divided by o_5–o_8. It will act up, as shown by the arrow, producing compression in 1–2, because only by acting thus would it tend to produce clockwise rotation about o_5.

The method of graphical moments is not of any special value.

FIG. 252.

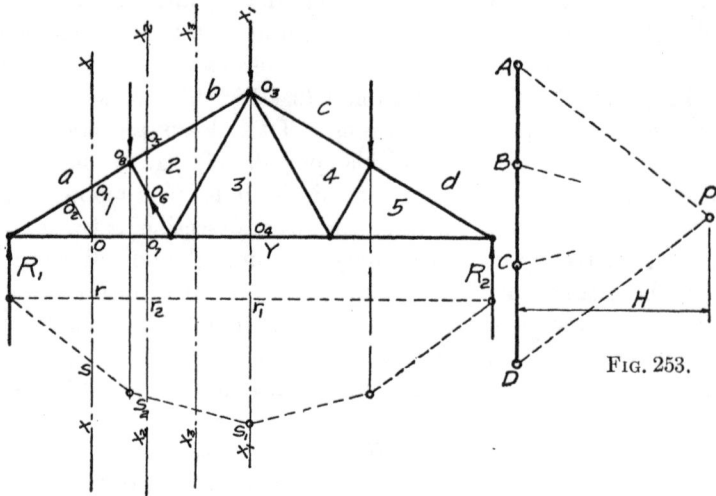

FIG. 253.

FIG. 254.

In most cases the stress diagram, which will soon be explained, is much to be preferred, so this discussion will not be extended. The student should, however, keep the method in mind, so that he may use it whenever desirable.

73. Stresses Obtained Graphically by Joints.—Consider the simple truss shown in Fig. 255, with the vertical loads at the upper chord panel points. The loads B–C, C–D, and D–E are all equal, and A–B is equal to E–F. The reactions can easily be found. In fact, since the truss is symmetrical, it can be seen by inspection that they will each be equal to half the total loading. Now at R_1 there are four forces acting, the reaction, which is known, the load $A-B$, which is known, and the two stresses $B-1$, and $1-Y$, each of which is known in direction but not in magnitude or sense. In Fig. 256 the reaction $Y-A$ is laid off to some convenient scale. Its sense is up, and going around joint I in a clockwise direction, we read the force as $Y-A$. Therefore the letter Y is placed at the bottom and A at the top of the vector just drawn in Fig. 256. Taking them in order, the next force which acts on the joint is $A-B$, the vector for which is shown starting at A in Fig. 256 and extending down to B. Now through B in Fig. 256 draw a line parallel to $B-1$, and through Y a line parallel to $Y-1$. These lines intersect at 1, thus completing the force polygon. The vector $B-1$, measured to the proper scale, gives the magnitude of the stress in the member $B-1$, and the vector $1-Y$ measured to the same scale, gives the stress in the member $1-Y$. With regard to the sense it will be recalled that the forces being in equilibrium, the sense going around the polygon must be continuous. The sense of R_1 is up. Therefore starting at Y in Fig. 256 move up to A then down to B, over to 1, and back to Y, the movement is in the direction of the sense, as shown by the arrows. The stress in $B-1$ is found to act downward towards joint I, thus producing compression, while that in $1-Y$ is found to act away, producing tension. The student should remember that we are dealing with joints, or if you wish, the joint and a small portion of the ends of the members. If the effect of the stress in a member, on the joint, can be represented by a force acting towards the joint, the stress is compression, when by a force acting away from the joint, the stress is tension.

It should be noted that the force $A-B$ produces no stress in the members of the truss, but simply goes into the reaction. If $A-B$ were left off the reaction would be just that much less or $Y-B$, but the stress in $B-1$ and in $1-Y$ would be the same. It is important to remember that the forces at joint I, or at any other joint for that matter, could be laid off in the force polygon in any order, and the correct result obtained. The reason for here taking them in order, going around the joint in a clockwise direction, is to have the polygons for the various joints of such shape that they will fit together and form a stress diagram.

Now take joint II, at which there are four forces acting, $1-B$ and $B-C$, both known, also $C-2$ and $2-1$, known in direction but not in magnitude or sense. In Fig. 257 lay off a vector for the

stress in 1–B, and then one for the load B–C as shown. Through
C draw a line parallel to C–2, and through 1 a line parallel to 2–1.
The intersection 2 closes the polygon. The sense of 2–1 and C–2,
shown by the arrows in Fig. 257, was determined so that the sense
going around the polygon is continuous. Both 2–1 and C–2 are
found to have compression.

Joint III may now be considered, there being only two

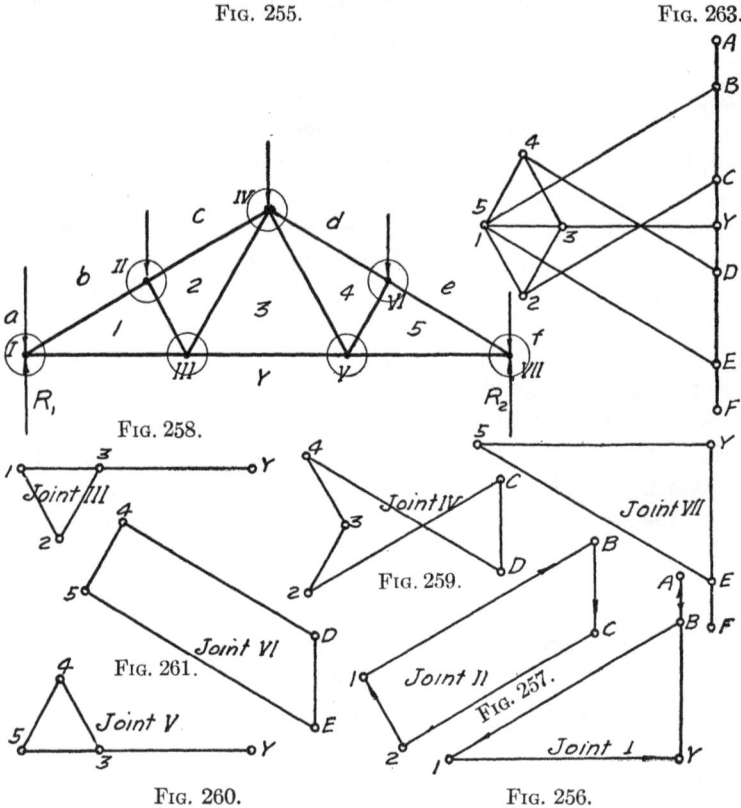

FIG. 255.   FIG. 263.

FIG. 258.

FIG. 262.

FIG. 259.

FIG. 261.

FIG. 260.   FIG. 256.

unknowns. The force polygon is drawn in Fig. 258 and the
unknowns obtained as for joints I and II. In a similar way, the
force polygons for the other joints may be drawn and all the
stresses obtained. These polygons are shown in Figs. 259 to 262,
inclusive.

74. Stresses Obtained by Stress Diagram.—Let all of the
force polygons Figs. 256 to 262, inclusive, be piled up on the

polygon of Fig. 256 in the following way. Let the polygon of Fig. 257 be placed upon that of Fig. 256 so that the vector B–1 of the second polygon coincides with B–1 of the first. See Fig. 263. Make Y–1 of the polygon shown in Fig. 258 coincide with Y–1 of the polygon for joint I, etc., Fig. 263 is the result, and it is often called a stress diagram.

The stress diagram may be drawn direct in the following way: First lay off the vectors for all of the loads, going around the truss in a clockwise direction and taking them in order. These vectors extend from A to F, R_2 brings us back to Y and R_1 back to the starting point A. Now from B draw a line parallel to the member B–1, and from Y a line parallel to Y–1. These lines intersect at 1, thus determining their length and the stress in the two members B–1 and Y–1. Now from 1 draw a line parallel to the member 2–1, and from C a line parallel to the member C–2. These two lines intersect at 2, which intersection fixes their length and therefore determines the stress in the members C–2 and 1–2. From 2 draw a line parallel to the member 2–3 and from Y a line parallel to the member Y–3. These intersect at point 3. This process is continued until the stress diagram is complete. The magnitude of the stress in any member may be found by scaling the corresponding line in the stress diagram. The kind of stress in any member may be found in the following manner: Consider any member, say 2–3. Going around the joint III in a clockwise direction the member is read 2–3. Now in the stress diagram, go from 2 to 3 along the vector 2–3. If the vector were placed just above the member 2–3, the movement from 2 to 3 would be away from joint III, therefore the member has tension. Now consider the member 1–2, still reading around joint III in a clockwise direction. In the stress diagram the movement from 1 to 2 is towards joint III, therefore the member 1–2 has compression. Just to show that this method works in any case, take the member just considered and read it going around joint II in a clockwise direction. It is read 2–1. Now if we go from 2 to 1 in the stress diagram the movement is towards joint II, and we get compression as before.

This method of finding the kind of stress is very important, and the student should endeavor to get as good a grasp of it as possible. In order to find the kind of stress in any member, consider a joint at one end of the member, and read the member going around the joint in a clockwise direction. Now in the stress diagram, go from the first letter or member to the second. If the movement is towards the joint considered, the member has compression, if away from it, the member has tension. In order to determine whether the movement is away from or towards a joint, consider the vector from the stress diagram placed over the member in question, then move along the vector as indicated by the way the member is read.

75. Stress Diagram, Upper and Lower Chord Loads.—Consider the truss shown in Fig. 264 which supports both upper and lower chord loads. The stresses produced by these loads may be found conveniently in two different ways. A diagram may be drawn for the upper chord loads, another for the lower chord loads, and the stresses obtained added together; or one stress diagram may be drawn for all the loads and the total stresses obtained direct by scaling the vectors of the stress diagram.

Fig. 265 shows a diagram for the upper chord loads. It was drawn in a way similar to that used for Fig. 263. The upper chord loads were laid off in order, starting with $A-B$ and going around the truss in a clockwise direction.

After $G-H$, the last upper chord load, R_2 was laid off. Since no lower chord loads are to be considered in this diagram M, N, O, P, Q and S are all the same point, and R_1 closes the force polygon bringing us back to the starting point. The reactions may be found by analytical computation, or by the use of a funicular polygon. When the truss and loading are symmetrical, no computation or construction is necessary for it is evident that the reactions are equal, each being equal to half the loading. With the force polygon completed for the loads and reactions, considering only the upper chord loads, the stress diagram may be drawn as already explained in connection with Fig. 263. The stresses produced by the upper chord loads may now be found by scaling the vectors in Fig. 265.

Fig. 266 shows the stress diagram for the lower chord loads. It is drawn in the following way. The loads are laid off in order, starting with $M-N$ and going around the truss in a clockwise direction. After the last lower chord load, the reactions R_1 and R_2 close the force polygon, bringing us back to the starting point M. These reactions are found considering only the lower chord loads. It should be noted that points A, B, C . . . H all fall on one point in Fig. 266, because no upper chord loads are considered. Now

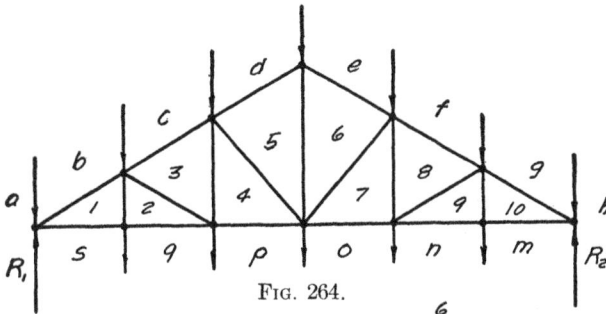

FIG. 264.

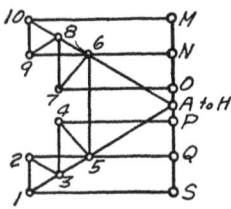

FIG. 266

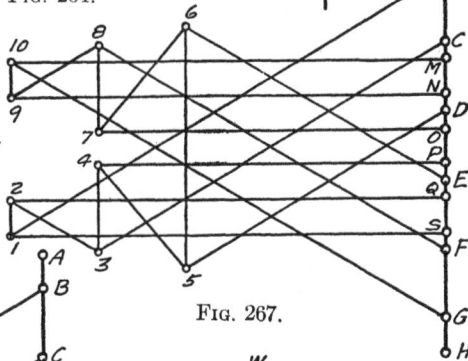

FIG. 267.

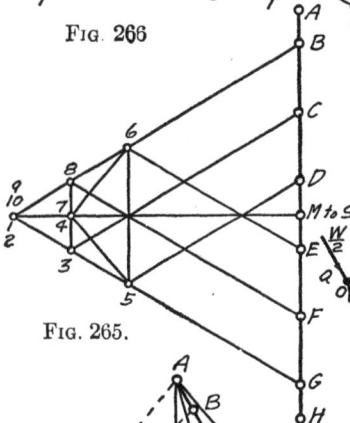

FIG. 265.

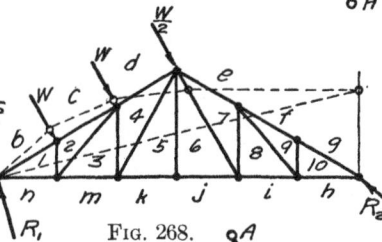

FIG. 268.

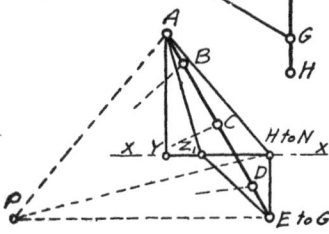

FIG. 269.

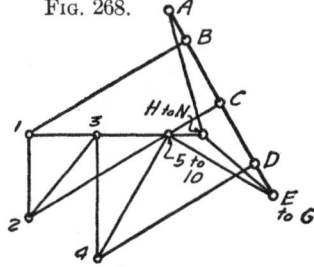

FIG. 270.

from B draw a line parallel to member B–1, and from S a line parallel to the member S–1. These lines, by their intersection, locate point 1 which determines their length. Now from point 1 draw a line parallel to the member 1–2, and extend it until it intersects the line from Q parallel to Q–2. This intersection locates point 2. From 2 draw a line parallel to 2–3, and extend it until the line from C parallel to C–3 is intersected, locating point 3. This process is continued until the stress diagram is completed. The lengths of the vectors in Fig. 266 give the stresses produced by the lower chord loads, and if they are added to the stresses obtained from Fig. 265, the stresses produced by the total loading will be obtained.

The total stresses may be obtained by the use of one large stress diagram, see Fig. 267. The reactions R_1 and R_2 are found for the total loading, and the force polygon drawn, starting with A–B and taking the loads and reactions in order, going around the truss in a clockwise direction. After the load G–H, the reaction R_2 is the next in order, then the load M–N, etc., and finally R_1 closes the force polygon. The reactions may be read H–M for R_2 and S–A for R_1. With the force polygon, often called the load line, furnished, the construction of the stress diagram is easy. From B draw a line parallel to the member B–1, and extend this line until the one from S parallel to S–1 is intersected, locating point 1. From 1 draw a line parallel to 1–2, and from Q one parallel to Q–2. Their intersection locates 2. This process is continued until the stress diagram is completed.

There are two things to which special attention should be called. First the loads and reactions are laid off in order, going around the truss in a clockwise direction. When a reaction is reached it is laid off, not skipped. Second the stress diagram is built up by drawing lines parallel to unknown members from letters or figures whose location in the stress diagram is known, and which are in each case one of the two figures or letters which denote the unknown member.

76. Wind Loads: Reactions and Stress Diagrams.—In Fig. 268 a small truss is shown carrying a wind load. For a discussion regarding the computation of wind loads, see the section headed *Wind Loads*. Before a stress diagram can be drawn, the reactions must be determined, and before this can be done some assumption must be made regarding the distribution of the horizontal component of the loading between the reactions. The way this horizontal component is actually divided depends to a large extent upon the nature of the supports. The support which is the more rigid laterally will take the larger proportion, while if they are of equal rigidity the division may be almost equal. If one reaction is very rigid, and the other rather weak laterally, the rigid support may reasonably be assumed to take most of the horizontal component. See the section on truss reactions in Chapter I.

In Fig. 269 lay off the wind loads to some convenient scale and choose any pole p. Now to start with, let it be assumed that the reaction R_2 is vertical. The funicular polygon will therefore be started at o, the left end of the truss, because that is the only known point on the action line of R_1. Starting at o then, the first string should be drawn parallel to p–A and extended to the action line of the force A–B. But since the action line of this force passes through o, this string will be only a point. The next string is parallel to the ray p–B, and is extended until the action line of B–C is intersected. The construction of the polygon is thus continued until the vertical line from R_2 is intersected. Then the closing string may be drawn, and parallel to it a line from p which gives the intersection with the vertical from G. This intersection may be called H to N, because there are no loads from H to N, and therefore H, I, . . . N all fall on the same point. Draw the line connecting N and A, thus completing the force polygon. Now if the supports of the truss are so constructed that R_2 will be vertical, then the reactions are determined in magnitude, direction and sense; R_2 by the vector G–H, and R_1 by the vector N–A. Through N draw the line x–x parallel to a line connecting the two supports of the truss, and from A draw the vertical line A–Y which gives the intersection Y. Now the horizontal distance from Y to H is the horizontal component of the loading, and when this component is to be divided into any given proportion between the reactions, the reactions may be obtained by locating a point Z on Y–H which divides Y–H in the same proportion, G–Z then gives R_2, and Z–A, R_1. Suppose the supports are constructed so that R_2 will be about twice as rigid laterally as R_1, and we therefore wish to assume that R_2 takes two-thirds of the horizontal component. Locate Z_1 on the line Y–H so that H–Z_1 is twice as large as Y–Z_1. Then G–Z_1 gives R_2, and Z_1–A, R_1 for the above condition.

Now that the reactions have been determined, the stress diagram may be started. The load line of Fig. 269 could be used, but the new one shown in Fig. 270 is clearer. The reactions used are for the last condition considered above. From B draw a line parallel to B–1, and extend it until the line from N parallel to N–1 is intersected, and point 1 located. From 1 produce a line parallel to 1–2 until the line from C parallel to C–2 is intersected and point 2 located. In a similar way points 3 and 4 are located. Now suppose we continue the construction of the stress diagram by starting in at the right end of the truss. From G draw a line parallel to G–10 and extend it until its intersection with the line from H, parallel to 10–H, locates point 10. Now from F draw a line parallel to F–9 and extend it until the line from 10 parallel to 9–10, is intersected, locating point 9. But F and G fall on the same point, and F–9 has the same direction as G–10, therefore F–9 falls on G–10 and intersects the line from 10, parallel to 9–10, at point 10. In the same way it can be shown that points 8, 7, 6 and 5 all fall on 9. Now from 5 draw a line parallel to 4–5 and extend it until the line from D parallel to D–4 is intersected, locating point 4. This intersection, locating point 4, should be at the same point as the intersection of the line 3–4 with the line from D parallel to D–4. Any variation indicates an error, but if these two intersections fall on approximately the same point the diagram is said to close. This check on the accuracy of the construction, which is obtained without extra effort whenever a stress diagram is completely drawn, is one of the important properties of the stress diagram. The last line drawn in any stress diagram should intersect each of two other lines at a point which has already been located, and in addition, it should be parallel to the corresponding member of the truss, the direction of which is fixed. Any variation indicates an error, which, if small, may be neglected, but if large, indicates a serious error, and the construction should be carefully gone over.

Attention should be called to the fact that the web members on the leeward side of the truss have no stress due to the wind.

77. Stress Diagram Combined Loads.—Let it be required to find the stresses produced in the truss of Fig. 271 by the use of one stress diagram. Note that the truss supports vertical upper and lower chord loads as well as wind loads. The first step in the solution is to find the reactions. These may be found either analyt-

FIG. 271.

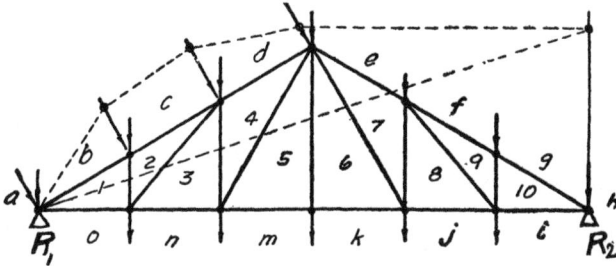

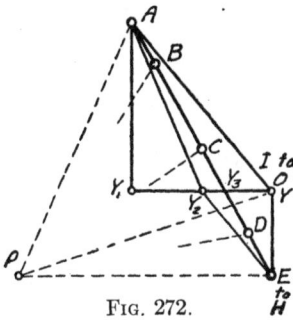

FIG. 272.

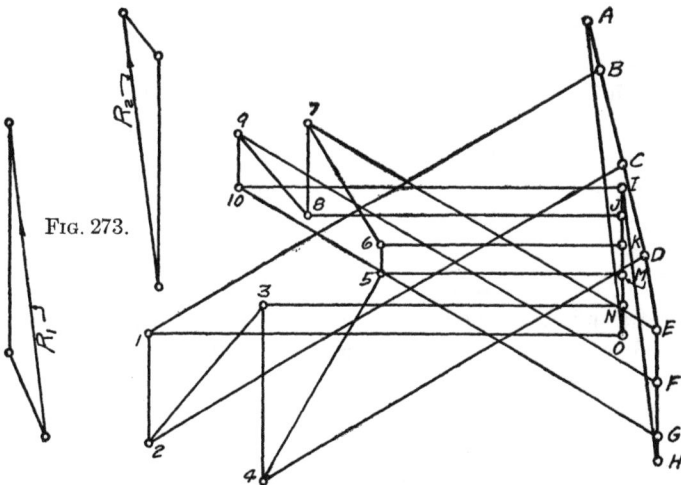

FIG. 273.

FIG. 274.

ically or graphically, but the graphical solution is often more convenient for the wind loads, especially if the truss is of unusual shape. In this problem, the vertical upper and lower chord loading is symmetrically placed, also the truss is symmetrical, therefore it is noted by inspection that the reaction from these loads will be equal, or in other words each one is equal to half the sum of the loads. Now if the reactions from the wind are found, and added to these reactions from the vertical loads, the total reactions will be obtained. By the use of Fig. 272 and the funicular polygon of Fig. 271 the reactions from wind are found, assuming that the horizontal component is divided equally between the two reactions. Fig. 273 shows the addition of wind and vertical load reactions in order to obtain the total reactions, the scale being different than that used in Fig. 272, in order to keep Fig. 273 of convenient size.

In Fig. 274 the loads carried by the truss are laid off, going around the truss in a clockwise direction and using the reactions obtained from Fig. 273 in their proper order. Note that the ceiling loads come in between the two reactions as shown. The stress diagram is now drawn in the usual way. From B a line is drawn parallel to the member $B-1$, and extended until it is intersected by a line from O parallel to the member $O-1$. Their intersection locates point 1, from which a line parallel to member $1-2$ is drawn and extended until the line from C, parallel to member $C-2$, is intersected, and point 2 located. From point 2 a line is drawn parallel to member $2-3$, and extended until the line from N, parallel to member $3-N$, is intersected, locating point 3. Note that the lines from O and N do not coincide, but are separated a distance equal to the load $N-O$. This process may be continued until the stress diagram is complete, closing at the right end of the truss; or the stress diagram may be started at each end and closed near the middle. The later method is often desirable because it tends to reduce the accumulation of small errors in directions and intersections.

78. Maximum and Minimum Stresses, Reversals.—The loads supported by most trusses vary more or less, and it is important when designing to know the maximum stresses which may be produced in each member, as well as the stresses produced by dead loads. It is also desirable in some cases to know the minimum stresses. The difference between the maximum and minimum stresses gives the range of stress. When a maximum stress is reduced under certain conditions, until it passes through zero and becomes a stress of the opposite kind, we have what is called a reversal. It is important to know which members, if any, have reversal of stress, because, while a good many compression members are able to take tension, many tension members are able to take but little compression.

Consider the truss shown in Fig. 275. The truss carries dead loads on the upper chord and also on the lower chord panel points. These loads are constantly acting, and in addition there may be a wind of 30 pounds per square foot on the vertical, acting from either side. A snow load exists at times, but it is hardly reasonable to consider that the maximum snow load will be acting when a high wind is blowing. However, a certain amount of snow or ice and sleet may stick to the roof when there is a high wind, therefore we will consider that an ice and sleet load of 10 pounds per square foot of horizontal projection may exist at the same time as the maximum wind load.

Let R_1 be vertical then R_2 takes all of the horizontal component from the wind loads. The dead loads will be considered first. The truss is symmetrical, also the upper chord dead loads, so the reactions due to them will each be equal to one-half of their sum. The lower chord dead loads are not symmetrical, so the reactions which they produce are found by the use of the force polygon of Fig. 276 and the lower funicular polygon of Fig. 275. These reactions are added to those produced by the upper chord dead loads, and the total dead load reactions obtained, and shown as R_1 and R_2 in Fig. 277.

The load line and stress diagram for the dead loads are drawn in the usual way, in Fig. 277, and the stresses produced by this loading may be obtained by scaling the proper lines. The stress diagram for the maximum snow load is shown in Fig. 278, and the one for the ice and sleet load, which we are assuming is the maximum snow load that can exist at the same time as the maximum wind load, is given in Fig. 279.

By the use of Fig. 280 and the upper polygon in Fig. 275, the reactions produced by a wind load from the left are determined. The wind loads were laid off extending from A to F, any pole p was chosen, and the upper funicular polygon in Fig. 275 drawn. The intersection of its first and last strings locates a point x on the action line of R, the resultant of the wind loads. The direction of this resultant is given by the line A–F in Fig. 280. Now this

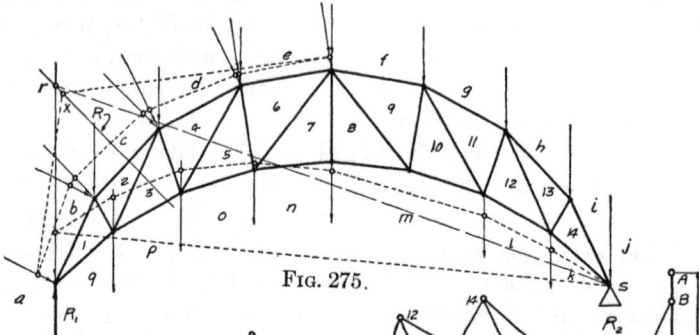

Fig. 275.

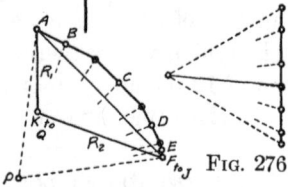

Fig. 276.

Fig. 280.

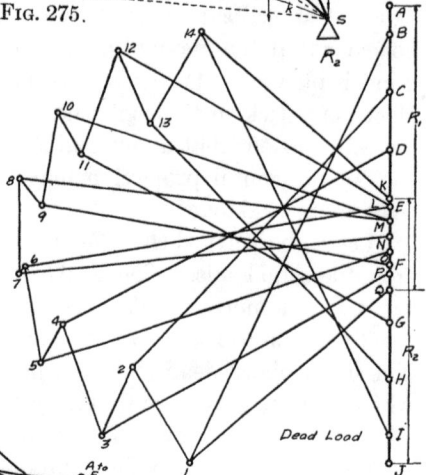

Dead Load

Fig. 277.

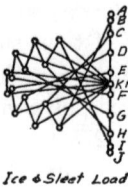

Ice & Sleet Load

Fig. 279.

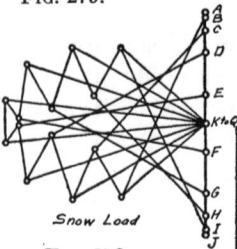

Wind Load Right

Fig. 282.

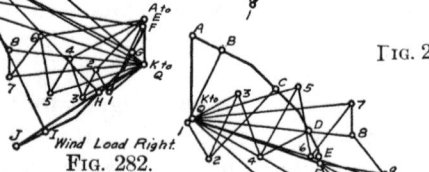

Snow Load

Fig. 278.

Wind Load Left

Fig. 281.

Member	B-1		C-2		6-7	
1 D.L.	47,100	C	37,700	C	1,000	T
2 S.L.	19,600	C	16,200	C	2,200	C
3 I&S.L.	11,900	C	9,700	C	700	C
4 W.L.R	8,000	C	7,000	C	5,100	C
5 W.L.L	8,100	C	9,700	C	6,600	T
1+2	66,700	C	53,900	C	3,200	C
1+3+4	67,000	C	54,400	C	6,800	C
1+3+5	67,100	C	57,100	C	4,900	T
Max.	67,100	C	57,100	C	6,800	C
Rev.	- - - -		- - - -		5,600	T

C. Denotes Comp. T. Denotes Tension

Fig. 283.

resultant and the two reactions which balance it make three forces which are in equilibrium, and therefore they must be concurrent. Since R_1 was assumed vertical, extend its action up until the intersection r is obtained. It follows that s–r must be the action line of R_2. Now in Fig. 280 draw from A a line parallel to the action line of R_1, and from J one parallel to that of R_2. The intersection of these lines locates the points K to Q, and determines the reactions. The reactions just found might have been obtained by using a large funicular polygon and its closing string, but the method just illustrated is sometimes more convenient as in this particular case.

The loads and reactions are now laid off in order in Fig. 281, going around the truss in a clockwise direction, and the stress diagram drawn in the usual way. Fig. 282 shows the stress diagram for the wind loads when the wind is blowing from the other direction. The construction for determining the reactions is not shown, but one similar to that used in determining the reactions for Fig. 281 would answer the purpose. The stresses produced by the various loads are now scaled from the five different stress diagrams just drawn, and recorded in a table a part of which is shown in Fig. 283. The five upper lines in this table show the results scaled from the stress diagrams, the next three show three different combinations of loadings. The maximum stresses are shown in the next line and reversals, when there are any, are shown in the last line.

The first line of combinations gives $1 + 2$, that is the maximum snow load stresses added to the dead load stresses. The dead loads produce stresses of the same kind, therefore the stresses produced by them acting together will be greater than produced by the dead load only. The second line of combinations gives $1 + 3 + 4$, which is the sum of the dead, wind from the right, and ice and sleet load stresses. The last line of combinations gives $1 + 3 + 5$. One of these three combinations will usually give the maximum stress, and this value is recorded in the line marked maximum. It will be noted that the wind load from the left produces tension in the member 6–7, while all the other loads produce compression; also that this tension stress is large compared with the other stresses. Since this tension stress is larger than the dead load stress and is of different sign, it will produce a reversal. Combining the dead load stress of 1000 pounds C and the wind load left stress of 6600 T, the maximum reversed stress that can be produced in 6–7 is found to be 5600 pounds T. This value is recorded in the

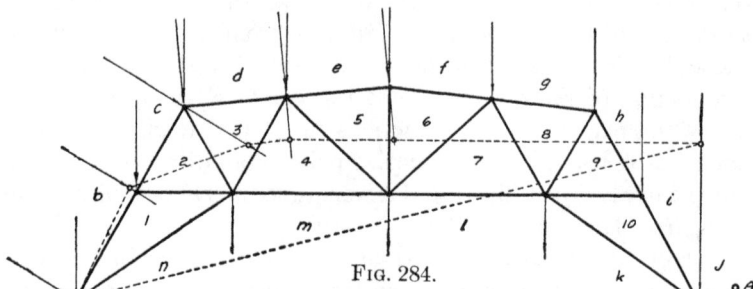

Fig. 284.

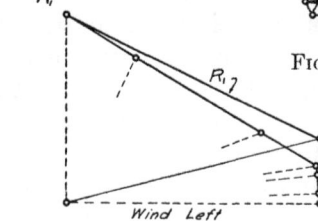

Fig. 286.

Fig. 289.

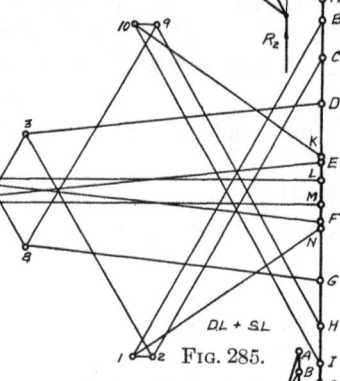

Fig. 285.

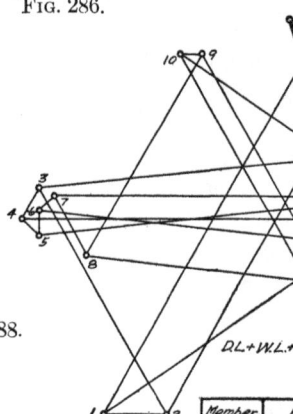

Fig. 288.

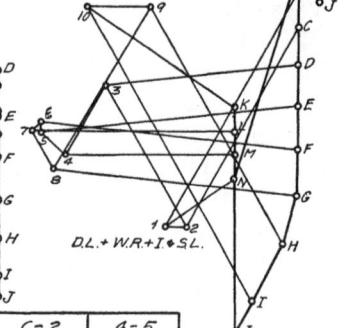

Fig. 287.

Member	B-1		C-2		4-5	
D.L.+S.L.	48,600	C	43200	C	1100	T
D.L.+W.R.+I.+S.L.	34,400	C	29100	C	4200	T
D.L.+W.L.+S.L.	52,300	C	44300	C	3000	C
Max	52,300	C	44300	C	4200	T
I.+S.L.					50	T
Rev					3050	C

Fig. 290.

line for reversals. The maximum stress of the same kind as the dead load stress, that can be produced is given by the maximum value of this kind shown in the three lines of combination, and is recorded in the line for maximums.

It has now been found that the stress in 6–7 varies from a maximum compression of 6800 pounds to a maximum tension of 5600 pounds, and this should be kept in mind when the member is designed.

The attention of the reader is called to the fact that five diagrams were drawn in Figs. 275 to 283, and the combination made in the table. Now it is possible to make the combinations in the stress diagrams so that the results given in the 6th, 7th, and 8th lines of the table, Fig. 283, may be obtained direct from the diagrams by scaling. The first five lines of the table would then be omitted. Figs. 284 to 290 illustrate how this may be done. The truss is shown in Fig. 284. Dead, snow, wind, and ice and sleet loads will be considered. In Fig. 285 the stress diagram for the dead and snow loads is drawn, and the stresses scaled from it are recorded in the upper line of Fig. 290, which corresponds to line 6 in Fig. 283. Fig. 287 gives the stress diagram for the three following loads: Dead, wind right, and ice and sleet. The stresses scaled from it are recorded in the second line of Fig. 290, which corresponds to line 7 in Fig. 283. The diagram for the third combination, dead, wind left, and ice and sleet, is shown in Fig. 288, The stresses from it are recorded in the third line of Fig. 290, and correspond to those given in line 8 of Fig. 283. The reactions for Fig. 288 were found by adding to the dead and ice and sleet load reactions, the wind load reactions found by the use of Fig. 286 and the funicular polygon in Fig. 284. No construction was necessary for finding the reactions due to dead and ice and sleet loads, because the truss is symmetrical and these loads are symmetrically placed. It follows that the reactions due to these loads are each equal to one-half the sum of the loads.

The reactions for Fig. 287 may be obtained in a similar way, but the figure corresponding to 286 is not given.

When the combinations are made in the diagrams, there are fewer diagrams to draw, but the ones that must be drawn are more complicated. There is less scaling and the lines corresponding to the first five of Fig. 283 are omitted. Also the adding or subtracting necessary to make the combinations from the results given in these five lines is eliminated.   In Fig. 290 the values for the maximum stresses are obtained in each case by taking the largest of the three values directly above.

One objection to the method of making the combinations in the diagrams is that it is not quite so easy to detect reversals. When reversals do occur it is necessary to draw a fourth stress diagram in order to find out the maximum value for the reverse stress.   Reversals of course occur only when a wind load, or some special live load, produces a stress of different kind than the dead load stress, and of greater magnitude.   Now we find that in member 4–5, the dead load, wind load left, and the ice and sleet load all taken together produce a stress of different kind than that produced by the dead and snow loads.   The maximum value for 4–5 is 4200 T, but the maximum compression that can be produced in this member may be greater than 3000, depending upon whether or not the ice and sleet load produces the same kind of stress as the dead load.   In Fig. 289 the stress diagram for the ice and sleet load is shown.   The stress in 4–5 is very small, almost too small to measure; let it be called 50 T.   This stress is of the same kind as the dead load stress, therefore, if the ice and sleet load had been omitted from the stress diagram of Fig. 288, we would expect the stress in member 4–5 to be increased 50 pounds, making 3050–C. The latter value is recorded in the line for reversals in Fig. 290.

Unusual trusses with special live or moving loads should be investigated with great care.   When the problem is complicated and there is danger of a number of reversals, a method making all, or at least a part of the combinations in the table is to be preferred.   When there are special live loads other combinations than those shown in Figs. 283 and 290 may be desirable.

79. Cantilever Trusses.—A small cantilever truss is shown in Fig. 291, carrying upper and lower chord dead loads, and also wind loads.   An interesting point in connection with this truss is the fact that the stress diagram can be drawn without determining the reactions.   However, no check or closing point is obtained until these reactions are found.   The loads are laid off along the load line in Fig. 292, starting with A–B and taking the loads in order, going around the truss in a clockwise direction.   At the left end of the truss there are only two unknowns.   Therefore in Fig. 292 a line is drawn from E parallel to the member E–1 and extended until the line from D, parallel to the member D–1, is intersected, and point 1

located. Then from 1, a line is drawn parallel to member 1-2
and extended until it intersects the line from F parallel to the
member F-2, locating point 2. This process is continued until
point 7 is located and from 7 a line drawn parallel to 7-K. Let it
be assumed that the supports are of such a nature that R_2 takes all
of the vertical component, and R_1 is horizontal. From I a hori-

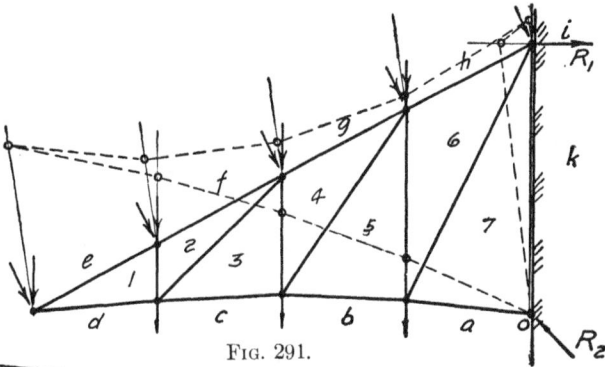

FIG. 291.

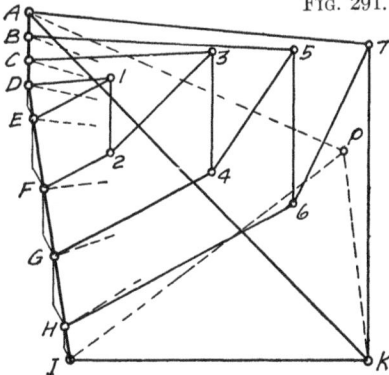

FIG. 292.

zontal line may now be drawn, and extended until the vertical
from 7 is intersected. This intersection may be called K, and the
points K and A connected. The student may now think he has
found the reactions, and it is true that he has found them in a certain indirect way. But any error which may have been made in
the construction of the stress diagram will produce an error in
these reactions, since they have simply been put in of such direction and magnitude as to make the polygon close. In other words,

no check has been obtained on either the diagram or reactions. However, a satisfactory check may be obtained by choosing a pole *p* in Fig. 292, and determining the reaction by the use of the funicular polygon of Fig. 291. This funicular polygon is started at *o*, because *o* is the only known point on the action line of R_2.

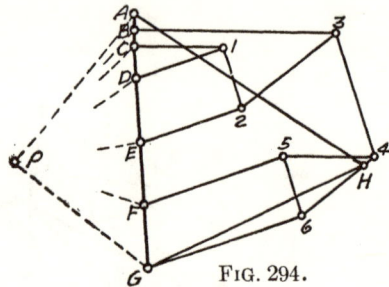

FIG. 294.

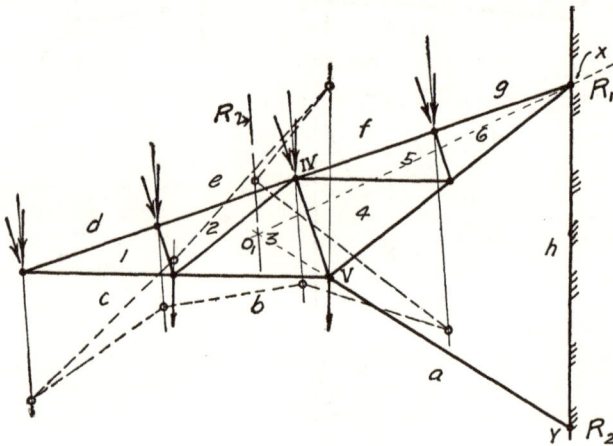

FIG. 293.

From *p* a line is drawn parallel to the closing string, and Fig. 292 checks, if this line intersects the horizontal from *I* at *K*.

Another cantilever truss is shown in Fig. 293, and its stress diagram in Fig. 294. The stress diagram may be started at the left end of the truss, and continued until point 3 is located, but since there are 3 unknowns at each of joints IV and V, it cannot be continued until the reactions are found. Any pole *p* is chosen in Fig. 294, the funicular polygon drawn in Fig. 293, and the action line

of the resultant of the loads located. Now this resultant and the
two reactions make three forces which are in equilibrium and they
must therefore be concurrent. The action line of R is known, and
the action line of R_2 must be along the member $A–H$, since it is the
only member at Y. These action lines intersect at o_1 determining
a second point on the action line of R_1, which may now be drawn.
Parallel to it a line is drawn from G, and parallel to the action line
of R_2 a line is drawn from A. The intersection of these lines
locates H, which determines the reactions. With H located, the
stress diagram may be completed. The reactions might have been
determined by starting the funicular polygon at point x, and using
a closing string in place of the resultant.

80. Mill Bent.—A fink truss supported on steel columns and
knee braced to them, as illustrated in Fig. 295, is sometimes called a
mill bent. Under dead loads only, the determining of the reactions
and the drawing of the stress diagram is relatively simple, but
when wind loads are added the problem is complicated. It makes
a decided difference whether the columns are fixed at the base
and have a point of contra-flexure, or are assumed hinged at the
base. In Fig. 295 they are assumed hinged at the base.

In order to facilitate the construction of the stress diagram,
the three members $A–1$, $B–2$, and $1–2$ are added temporarily at
the left end, and similar ones at the right end. The addition of
these members takes the bending stress out of the columns, but
does not affect the stress in any of the other members. Therefore
these temporary members may be put in, the stress diagram drawn,
and the correct stresses found in all the members except the columns. Then the column stresses may be determined as a separate
problem. In order to show that the addition of these extra members does not affect the stresses in the members of the truss, consider section $x–x$. The stress in the member $D–5$ may be found by
dividing the resultant moment of R_1 and the loads applied at the
left of section $x–x$ about point Y, by the perpendicular distance from
Y to the member $D–5$. Neither these loads nor the reaction were
changed in magnitude, action line, or sense by the addition of the
temporary members, therefore the stress in $D–5$ was not affected.
Now consider the joint at the lower end of $D–5$. Neither the
stress in $D–5$ nor the loads applied at this joint, were affected by
the additional members, therefore the remaining two stresses
at this joint, namely the stress in $C–4$ and in $4–5$, could not be

affected.   In a similar way by considering various sections and joints it can be proved that the stress in each truss member is not affected by the additional members.   The temporary members, A–1 and K–19, should, of course, be drawn to the points of contra-flexure when the columns are fixed at the base.

The reactions will now be determined.   Since the dead loads are symmetrically placed, their resultant R''' will have its action line at the center of the truss.   The wind loads applied to the sloping surface have their resultant R'' as shown, with action line normal to the slope.   The horizontal wind loads A–B and B–C were laid off in Fig. 296, p_1 chosen, and the action line of R' located by the use of the small funicular polygon in Fig. 295.   The reactions could be found by using the individual loads in place of the resultants R', R'', and R'''; but the use of these resultants is often convenient and more accurate.   The resultant loads just determined are now laid off in Fig. 296, and any pole p chosen.   For the purpose of finding the reactions it will be found convenient to assume temporarily that R_2 is vertical, then r is the only known point on the action line of R_1.   The funicular polygon is, therefore, started at this point and continued in the usual way until completed.   From p a line is drawn parallel to the closing string, and the point Z_1 located.   Now for the case when the right reaction is vertical, K–Z_1 is the vector for R_2, and Z_1–A for R_1.   But, since the columns of the bent are just alike, it would seem more reasonable to assume that the horizontal component is divided equally between the two reactions than to assume that it is all taken by R_1.   Draw through Z_1 a horizontal line, and from A project vertically down locating Z_2.   The line Z_1–Z_2 gives the magnitude of the horizontal component of the loading.   Locate Z the mid-point of Z_1–Z_2, then K–Z gives R_2, and Z–A gives R_1 for the case when the horizontal component is divided equally between the two reactions.

The loads at the various panel points are now laid off in order along the load line of Fig. 297, going around the truss in a clockwise direction and taking the reactions obtained from Fig. 296 in order.   Then the stress diagram is drawn in the usual way. From Q a line parallel to member Q–1 is drawn and extended until the line from A parallel to A–1 is intersected, etc.   This process is continued until point 6 is reached, when difficulty is encountered, as joints VII and VIII each have three unknowns.   Now the truss would be stable if the two members 7–8 and 8–9 were taken out and a new member u–v substituted in their place. Such a temporary substitution would not affect the stress in the member 10–N, or the

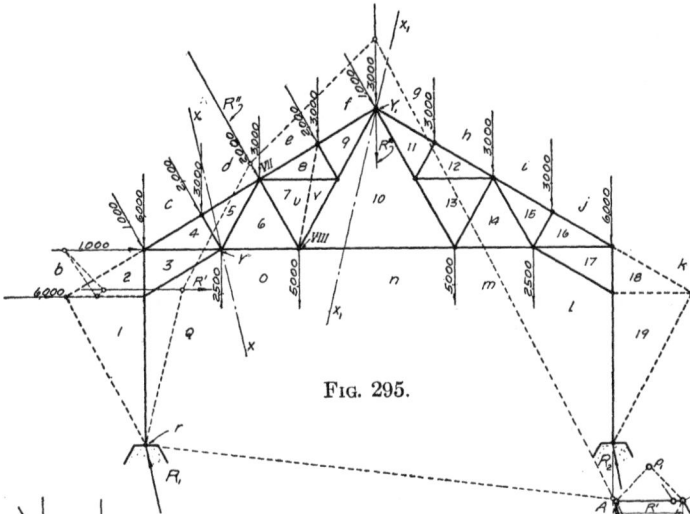

FIG. 295.

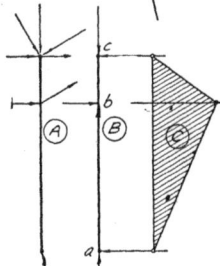

FIG. 298.

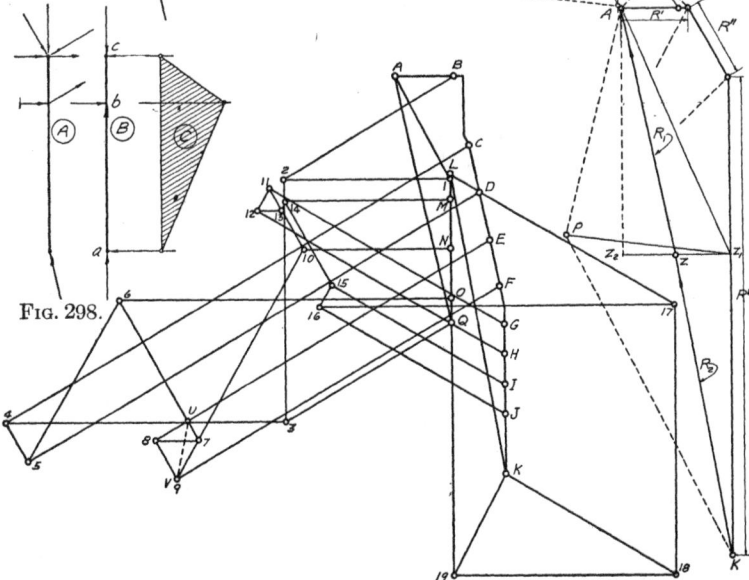

FIG. 297.

FIG. 296.

location of point 10 in the stress diagram, because it could have no effect on the moment of R_1 and the loads to the left of section

x_1–x_1 about point Y_1. This substitution is therefore made and the stress diagram continued until point 10 is located. With point 10 determined and the stress in 10–N known, the original members 7–8 and 8–9 are put in again. Now there are only two unknowns left at joint VIII and points 7, 8, and 9 can be located. The diagram may be completed and the stresses for all of the members of the bent, except the columns, obtained by scaling.

Let the left column be considered first. There are eight forces acting on this column as shown in Fig. 298 (A). There is the reaction at the bottom, while higher up we have the knee brace stress, and the wind load at the same level. At the top there are the three loads and the two stresses, one from C–4 and the other from 3–4. It will be noted that the three temporary members having served their purpose have been discarded. The eight loads applied to the column, as indicated in Fig. 298 (A) are now broken into their horizontal and vertical components as shown by the diagram in Fig. 298 (B). At the base there are the horizontal and vertical components from the reaction. Where the knee brace comes in, there is a vertical component from the knee brace while the horizontal force is produced in part by the load and in part by the horizontal component from the knee brace. At the top the vertical force is the resultant of the dead load, the vertical component of the wind load and the vertical component of the stress in C–4, while the horizontal force is the resultant of the stress in 3–4, the horizontal wind load and horizontal component of the stress in C–4 and of the inclined wind load. The direct stress in the column may be found from the vertical forces. The direct stress in the length from a to b is given by the vertical component of the reaction, while the stress from b to c is given by the resultant vertical component at the top. The difference between these two stresses should be just equal to the vertical component thrown in by the knee brace at b. The horizontal components produce bending, the moment diagram being shown in Fig. 298 (C). The maximum moment occurs at the level of the knee brace, and is equal to the horizontal component of the reaction times the length a–b. The resultant horizontal component at the top times c–b should give the same result. As far as the moments produced are concerned, the column may be considered as a beam, supported at the ends and carrying a concentrated load at b. The stresses in the right column may be found in like manner.

81. Three-hinge Arch.—A small three-hinge steel arch is shown in Fig. 299. It may be regarded as two separate trusses pinned together at the top. The stress diagram will be drawn considering both wind and dead loads, part of the dead loads being on the inside.

All the loads applied to the right half of the truss are laid off in Fig. 300, the pole p chosen, and the funicular polygon drawn. This, by the intersection of its first and last strings, locates the action line of R_r, the resultant of all the loads applied to the right half of the truss. On the left half of the truss there are, in addition to the loads applied to the right half, the wind loads; also S–T is twice as a large as P–Q. There are two horizontal wind loads and their resultant R' is located by placing its action line so that it divides the distance between these loads inversely as their magnitudes. The inclined wind loads are uniformly applied along the sloping surface, therefore the action line of their resultant R'' will pass through the mid-point of the sloping surface, and normal to it. In Fig. 301, R' and R'' are laid off and their resultant R''_L found. The action line of this resultant passes through the intersection of the action lines of R' and R''. R'_L is the resultant of R''_L and $\frac{1}{2}$ S–T, and its action line passes through the intersection of their action lines as shown. Now R'_L is the resultant of all the loads that are applied to the left half of the arch, in addition to those corresponding to the loads applied to the right half. The vertical line R is drawn with d_2 made equal to d_1, and in Fig. 301 the vector R is equal in magnitude to R_r. The resultant of R'_L and R gives R_L, the resultant of all the loads on the left half, and its action line passes through the intersection of R'_L and R. The resultant loading on each half has now been determined in magnitude, direction and action line.

These resultants R_L and R_r are laid off to scale in Fig. 302. Now assume that the right half of the arch is loaded and that the left is not. This makes three forces acting on the right half, the resultant loading R_r and the reaction at either end. Since these forces are in equilibrium, they must be concurrent. On the left half there are only two forces acting, one at Z and the other at W. Since these two forces are in equilibrium, they must be of equal magnitude and have the same action line, which action line must pass through Z and W. This line Z–W must also be the action line of one of the three forces acting on the right half, because the force applied to the left half at W is nothing more than the reaction at W from the right half. In other words, the upper end of the right half is supported by the left half at W, and the left half is in turn supported by the pin at Z. It follows that the force which the right half delivers to the left half at W, and the reaction which the left half delivers to the right half at W, also the force which the left half delivers to the pin at Z, and the reaction which the pin delivers to the truss at Z, must all have the same magnitude and action line. As said above all this is for the case when the right half is loaded and the left half is not. The line $Z–W$ is extended until it intersects the action line of R_r. Now if this intersection is connected with Y, the action line of $R_2{}^r$, the reaction at Y due to R_r, is obtained. From X, in Fig. 302, a line is drawn parallel to $W–Z$, and from M another line parallel to $O_1–Y$, thus obtaining the intersection X'. Now $M–X'$ gives the magnitude of the reaction at Y from the loading on the right half, and $X'–X$ gives the magnitude of the reaction at W and also at Z due to the load on the right half.

Now let it be assumed that the left half is loaded and that the right half is not. The magnitude and direction of the reactions at Z, at W, and at Y, due to the loading on the left half, are desired. From what was said in connection with the right half, it is evident that the action line of the reaction at W and also of that at Y, produced by the loading on the left half, will be along $W–Y$. Now there are three forces acting on the left half, the resultant load R_L, which is known in direction and magnitude, the reaction at W, which is known in direction, and the reaction at Z which is unknown in direction and magnitude, but has its action line passing through Z. Since these forces are in equilibrium they must be concurrent, but they intersect off the paper. Therefore for convenience, the pole p will be chosen in Fig. 302 and a funicular polygon drawn for the three forces in Fig. 299, starting at point Z. From X a line is drawn in Fig. 302 parallel to $Y–W$, and extended until it is intersected by the line from p parallel to the closing string of the funicular polygon. Then line $X–X''$ gives the magnitude of $R_2{}^L$, the reactions at Y and also at W, due to the load on the left half; and $X''–A$ gives the magnitude of $R_1{}^L$, the reaction at Z due to the load on the left half.

The reaction at Z due to the load on the left half is $R_1{}^L$ and that from the load on the right half is $R_1{}'$. The resultant R_1 of these two reactions is the total reaction from the pin at Z. The

FIG. 299.

FIG. 300.

FIG. 301.

FIG. 302.

FIG. 303.

FIG. 304.

FIG. 305.

magnitude and direction of the R_1 is found in Fig. 302. The vector for $R_1{}^L$ is shown as $X''-A$ and the one for $R_1{}^r$ as $X'-X$. From X'' draw the line $X''-X_1$ parallel to $X-X'$ and of the same length. Then $X''-X_1$ may be considered a vector for $R_1{}^r$, and X_1-A, which closes the force triangle, gives the magnitude and direction for R_1. The total reaction R_2, from the pin at Y, is obtained in magnitude and direction from Fig. 302 in a similar way. From X' a line is drawn parallel to $X-X''$. This line passes through X_1 and $X-X''-X_1-X'$ is a parallelogram. Therefore $X'-X_1$ has the same length as $X-X''$, and may be considered a vector for $R_2{}^L$.

Since the reactions have been found, the stress diagram may be drawn. The loads and reactions are laid off in order, going around the truss in a clockwise direction, and the stress diagram drawn in the usual way. There will be two closing points, one for each half of the arch. In fact, two separate stress diagrams might have been drawn, one for the left and the other for the right half. For the left half the reactions would be R_1 and R_3, while for the right half they would be R_2 and R_3. The force R_3, which is the thrust or reaction on the crown pin, is given in Fig. 302 by $X-X_1$, and shown in Fig. 303 as $X-R$. This reaction would, of course have one sense when considering one half, and the other sense when considering the other half, but when considering the arch as a whole, it may be thought of as a certain compression in the pin.

The reactions R_1 and R_2, which have been considered thus far, are the reactions at Z and Y between the trusses and the supporting pins. If there is no tie connecting the two lower pins, R_1 and R_2 will be transmitted to the footings, and must be provided for in the footing design. In many cases, however, it is desirable to relieve these footings of some of the horizontal component of the reactions. If there were no wind loads, the tie would simply take out the horizontal component of each reaction, leaving the resultant load on each footing vertical. But when there is a wind load, its horizontal component is carried down, and must be divided in some way between the two footings. Suppose the horizontal component from the wind is divided equally between the two footings. Drop a vertical from A to X_2 and locate X_3 midway between X_2 and X_4. Then X_3-X_4 represents one-half the horizontal component from the wind. It follows that R'_f is the reaction at the right footing and R_f the reaction at the left footing. In order to produce equilibrium, it is evident that the tension in the tie must be equal to X_3–X_1.

82. A Large Three-hinge Arch.—A large three-hinged steel arch is shown in Fig. 306 with loading as indicated. The pin X_1 is a short distance below the top of the roof, and the two members R–41 and S–41 carry the top purlin. This top purlin produces stresses in the members R–41 and S–41, as determined by the force triangle R–S–41 shown in Fig. 306 (A). These stresses are now substituted, for the members, as loads on the two halves of the arch. The resultant of all the loads applied to the right half is found in magnitude and direction by the use of the force polygon of Fig. 307, and the funicular polygon in the upper right-hand part of Fig. 306. A number of the loads have been grouped together, and the lower chord loads mixed in with the upper chord loads for convenience, but all of them have been considered. If this method of grouping the loads together when finding resultants is confusing to the student, he should take them one at a time.

On the left half of the arch there is a vertical load corresponding to every load on the right half, and of the same magnitude; also there are the wind loads. R' is therefore drawn as the action line of the resultant of all the loads on the left half of the arch, except those from wind. It has the same location with respect to the left half as R_r has with respect to the right half and its magnitude is the same as that of R_r. R_w the resultant of all the wind loads is found by the use of Fig. 308 and the funicular polygon shown in the upper left-hand part of Fig. 306. Some of the different loads were grouped together for convenience before drawing the polygon. The action line of R_w is located by the intersection of the first and last strings of this polygon, and is drawn in, as shown in Fig. 306.

In Fig. 309, R_w and R' are laid off and R_L, the resultant of all the loads on the left half of the truss, found in magnitude and direction. The action line of R_L is, of course located by the intersection of the action lines of R_w and R' in Fig. 306.

The resultant load on each half of the arch has now been determined in magnitude, direction, and action line. Fig. 309 is therefore drawn, corresponding to Fig. 302, and the reactions at the pins Z_1 and Y_1 determined. The loads and reactions are laid off in Fig. 310, going around the truss in a clockwise direction,

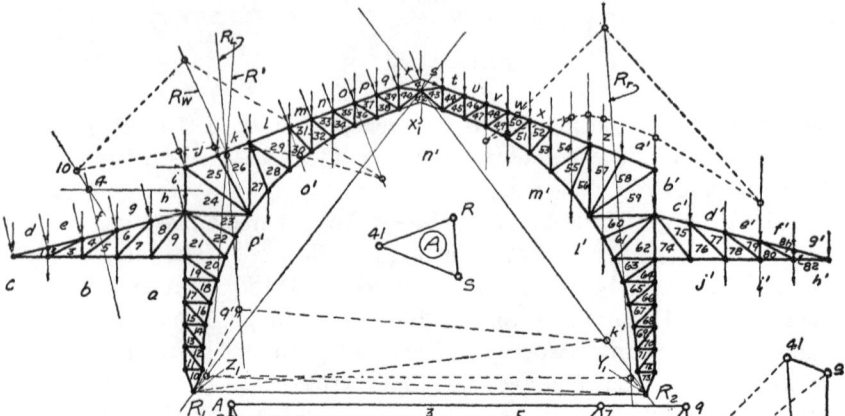

FIG. 306.

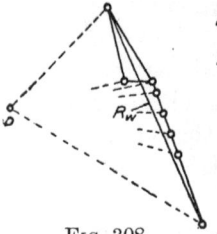

FIG. 308.

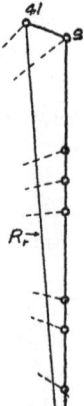

FIG. 307.

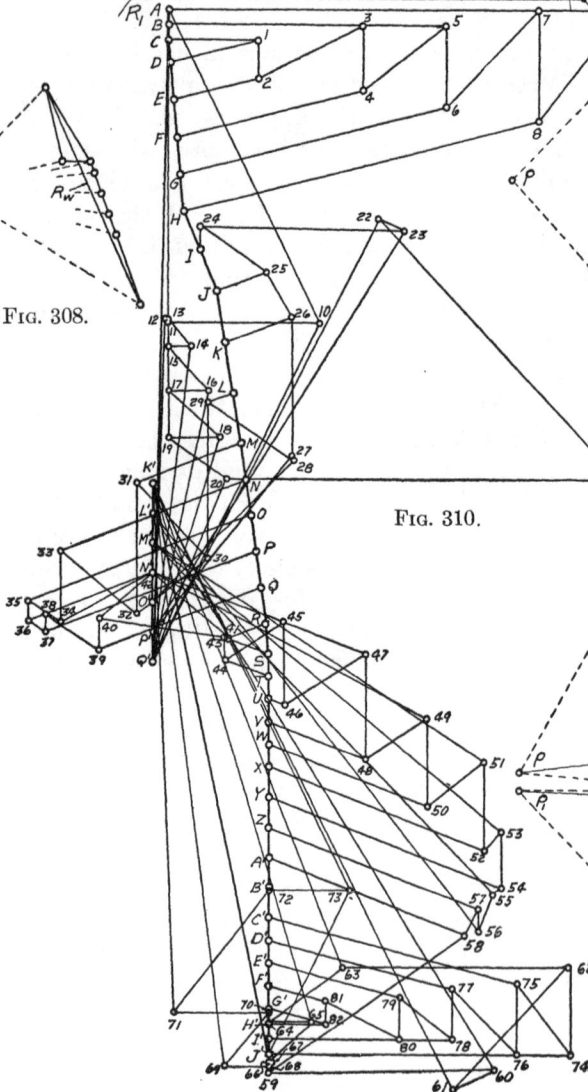

FIG. 310.

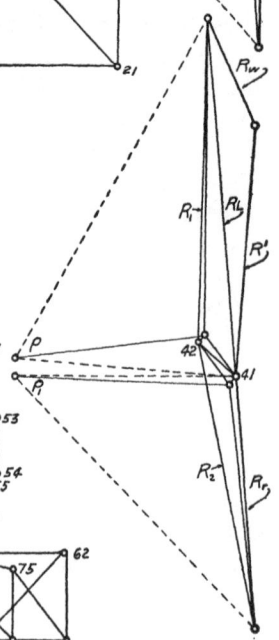

FIG. 309.

and taking each force in order. The stress diagram may now be drawn in the usual way. There will be two closing points, as in Fig. 303. Also the stress diagram may be started at six different points, three for each half; at the lower pin, at the upper pin and at the cantilever end.

The reaction on the pin X_1 is 41–42, and the stress in the tie can be found after making some assumption regarding the division of the horizontal component from the wind between the two supports. A method similar to that explained in connection with Figs. 302, 304 and 305 might be used.

83. A Large Mill Bent.—A large mill bent with a complicated loading and with columns fixed at the base is shown in Fig. 311. Note that in addition to the ordinary dead loads there are wind loads, and loads such as piping or shafting loads hung from the interior panel points of the truss. All of the vertical loads are symmetrically placed and the truss is symmetrical, therefore they will be divided equally between the two reactions. The reactions from the wind loads will now be found. The wind loads are laid off in Fig. 312, and the funicular polygon drawn in Fig. 311, assuming that the right reaction is vertical. This polygon is started at Z_1 the point of contra-flexure of the left column, because it is the only known point on the action line of the left reaction. From p, in Fig. 312, a line is drawn parallel to the closing string and point X_1 located. Then if the right reaction is vertical, it is represented by $V–X_1$, and $X_1–A$ gives the left reaction from the wind loads. But it would be more reasonable to consider the horizontal from the wind divided equally between the two supports, rather than consider that the left support takes it all. Therefore locate X_3 midway between X_1 and X_2, then $V–X_3$ gives the right reaction from the wind, and $X_3–A$ the left reaction. Now if one-half the vertical loading is added to each one of these reactions, as was illustrated in Fig. 273, the reactions R_1 and R_2 are obtained. The loads and reactions are now laid off along the load line of Fig. 314, taking them in order going around the truss in a clockwise direction. The interior loads are taken care of by extending them down, as shown by the dotted lines, and laying them off as lower chord loads. The lines 11–12, 20–21, etc., will be considered as members of the truss for the present, but they will be thought of simply as hangers having no connection to the lower chord. The three temporary members are put in at each end and the stress diagram started, as in

Fig. 297. It should be noted that the temporary members B–1 and V–42 intersect the columns at the point of contra-flexure, which, when the columns are fixed, is often assumed midway between the base and the foot of the knee brace. The portion of the column below the point of contra-flexure may be considered as a vertical cantilever carrying the reaction from the structure above, at its top. After point 7 has been located in the stress diagram, a condition similar to that encountered in Fig. 297 develops, and it is desirable to use the temporary member a'–b' until point 12 is located. The problem here is a little more complicated than the one in connection with Fig. 297, because of the load or stress 11–12 shown in the load line of the stress diagram as E_1–F_1. Only the member 8–9 will be omitted, 9–10 being left in, and called b'–10 while the temporary member a'–b' is being used. At present there are only four members at joint IX, namely; 10–12, 12–11, 11–b', and b'–10. The members 10–12 and 11–b' lie on the same straight line, so the component of the stress in 12–11, normal to this line, must be just balanced by the component of b'–10, normal to this line. The stress in 11–12 is, of course, just equal to the load E_1–F_1. The magnitude of load 11–12 is laid off extending down from joint IX, the normal component determined and the stress in b'–10 found to be tension of a certain magnitude as shown. From F a line is drawn parallel to member F–a' and extended until the line from 7, parallel to the member 7–a', is intersected. Now from a' a line is drawn parallel to the member a'–b' and we know that b' is a point somewhere along this line. From G a line is drawn parallel to the member G–10 and point 10 must lie some place along its length. It is also known that the distance between b' and 10 represents the stress in member b'–10 and must therefore be equal to the length shown in Fig. 311 for this stress. In other words, since the member b'–10 is at right angles to member G–10, the point b' must be distant from G–10, in the stress diagram, an amount representing the stress in b'–10. Two lines are now drawn in Fig. 314 parallel to G–10, and distant from it on either side an amount equal to the stress in b'–10. Both of these lines intersect the line from a' parallel to a'–b', but the lower intersection is b', because the upper one would give compression in b'–10, which is impossible since that stress has already been found to be tension. From b' the line parallel to member b'–10 is drawn and point 10 located. From

10 a line parallel to member 10–12 is drawn and extended until the line from E_1 parallel to member 12–E_1 is intersected, locating point 12. Point 11 is located in a similar way by drawing lines from b' and F_1, with points 11 and 12 located, the member 8–9 is put back in, and points 8 and 9 located as illustrated in Fig. 297.

The stress diagram has now been drawn from the left side up to point 12, and can be drawn from the right side as far as point 31 in a similar way. In order to continue the stress diagram it will be necessary either to compute the stress in the central part of the lower chord so that point 20, 21, 22, and 30 may be located, or some substitution of members must be made which will not affect the stress in the lower chord, but which will permit the completion of the stress diagram. The latter method is chosen and the members 13–14, 13–16, 15–. . ., to 19 are omitted, for the present, and members $d'-e'$, $e'-f'$, $f'-g'$, and $g'-19$ are substituted for them. It can be proved by a method similar to that already explained that this substitution does not affect the stress in the lower chord. Using these substitute members the stress diagram can be continued until point f' is located, when two difficulties are encountered, the stress in $g'-19$ and the stress in the monitor. Let it be assumed that the diagonal web members of the monitor, 43–44 and 45–46, are light ties unable to take compression. From K in Fig. 314 a line is drawn parallel to member K–43, and from L a line is drawn parallel to L–44. Now with the wind from the left as shown, we may assume that 43–44, since it cannot take compression, has zero stress. Therefore points 43 and 44 lie together, or at the intersection of the lines just drawn. With points 43 and 44 found, points 45 and 46 can easily be located and points 43 and 46 used when the stress diagram for the truss is continued. If the student wishes, he can think of the monitor as having been removed, and loads K–43, 43–46, and 46–N substituted at the joints affected in its stead. An enlarged portion of the load line from K to N, with points 43 to 46 located, is shown in Fig. 313.

The stress in $g'-19$ is found in the same way as the stress for $b'-10$, but in this case the stress is larger than the load, because of the small angle that $g'-19$ makes with 19–21. In Fig. 314 a line is now drawn from f' parallel to member $f'-g'$, and we know that the point g' must be somewhere along its length. Also the point 19 must lie somewhere along the line from 43 parallel to

FIG. 313.

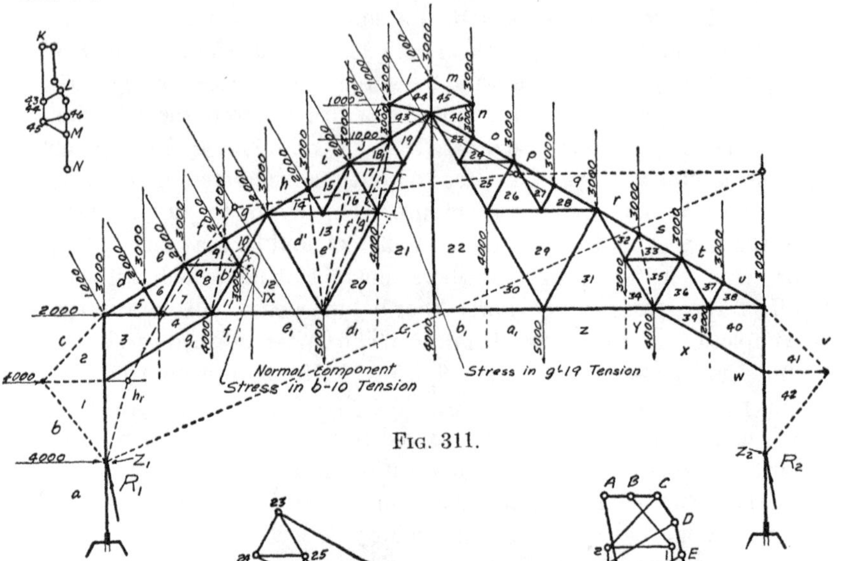

FIG. 311.

FIG. 312.

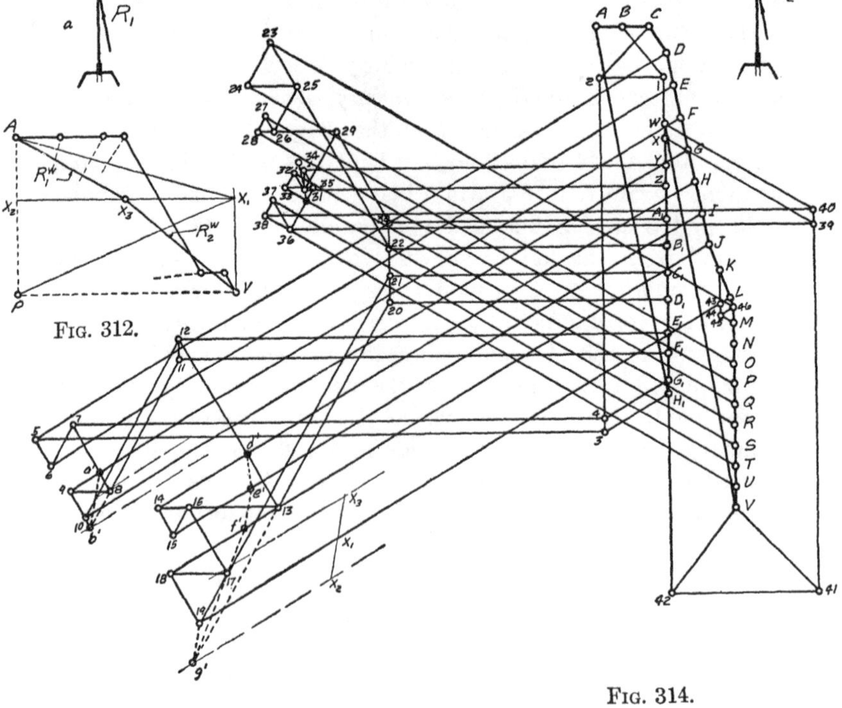

FIG. 314.

member 43–19. In addition points 19 and g' must be on some line parallel to member 19–g', and be separated by a length of line representing the stress in 19–g', to the scale of the stress diagram. The stress in member g'–19 has been found to be tension of magnitude as shown in Fig. 311, therefore from x_1, any point along the line from 43, the line x_2–x_3 is drawn parallel to member g'–19, and the two lengths x_1–x_2 and x_1–x_3 are each made equal to its stress. From x_2 and x_3 lines are drawn parallel to the line from 43, and extended until the line from f' is intersected. The lower intersection locates point g', because the upper one would require compression in g'–19. The construction may now be continued and points 20 and 21 located, after which the original members are put back in, and the points 13 to 19 located. With points 21 and 31 determined, the construction may be continued, and the stress diagram completed without any more substitution or difficulty.

The stress in the columns above the point of contra-flexure may be found as explained in connection with Fig. 298. The compression in the column below the point of contra-flexure is equal to the vertical component of the reaction, and the moment is that which would be produced by a horizontal force equal to the horizontal component of the reaction, acting at the point of contra-flexure, considering the column below as a vertical cantilever.

84. Cantilever Truss with Four Supports.—An interesting truss with four supports is shown in Fig. 315. In reality it consists of three trusses, the two outside trusses have cantilever ends projecting towards the center and carry the third truss as shown. Let the two center supports R_2 and R_3 be long slender columns unable to take any horizontal component, then all of the horizontal component due to the wind is divided between the two outside supports, suppose we say equally between them.

In Fig. 316 the resultant horizontal and inclined wind loads are laid off, and the magnitude of the resultant horizontal component of all the wind loads applied to the truss is found to be x–x_1. Half of this length then gives the horizontal component of R_1 and also of R_4.

After grouping them together as much as possible for convenience, the loads applied to the center truss are laid off in the force polygon of Fig. 317. Any pole p is then chosen, and the

center funicular polygon in Fig. 315 drawn, assuming R'' vertical
for the present.   Parallel to the closing string a line is drawn from
p, and the intersection Z_1 obtained.   The reactions are now
determined for the case when R'' is vertical.   But R_4 has been
assumed to have a horizontal component equal to x_2-x_1 of Fig.
316.   There are only two forces, acting on the right truss which
can have horizontal components, R_4 and the reaction from the
center truss.   Therefore these two forces must have equal horizontal components, and it follows that R'' must have a horizontal
component with magnitude equal to x_2-x_1.   In Fig. 317, Z is
located on the horizontal line from Z_1, and $Z-Z_1$ is made equal to
x_2-x_1 of Fig. 316.   If different scales are used for the two figures,
due consideration must be given to the change in scales;   the
point is that Z_1-Z should represent a magnitude equal to half
the resultant horizontal component of the wind loading.   Then
$Q-Z$ gives the magnitude and direction of the reaction at Y_2,
and $Z-H$ the magnitude and direction of that at Y_1.

    Now consider the left truss.   The loads including the reaction
from the center truss are laid off in Fig. 318. R_6 represents the resultant of all of the vertical and inclined upper chord loads, including the vertical dead and inclined wind loads at Y_1. R_7 represents
the resultant of the two lower chord loads, and R_8 the resultant
of the horizontal wind load at Y_1 and the load or reaction from
the center truss.   The pole p is chosen and the funicular polygon
drawn in Fig. 315, starting at Y and keeping in mind the fact
that R_2 is vertical.   A line is now drawn from p, parallel to the
closing string and extended until the vertical from H is intersected, locating the intersection Z.   Then $H-Z$ gives the magnitude and direction of R_2, and $Z-A$ the magnitude and direction
of R_1.   Note that, in this particular case, R_2 is greater than the
total loading, while at Y the reaction is a pull, that is, this end
of the truss must be held down.

    The reactions R_3 and R_4 for the right truss are found in a
similar way, using Fig. 319 and the right funicular polygon in
Fig. 315.   All the loads applied to the truss including the reaction
from the center truss are considered.   This load from the center
truss is given in magnitude and direction by R''.   The reactions
R_1, R_2, R_3, and R_4 now being known, the loads and reactions are
laid off along the load line of Fig. 320, going around the entire
truss in a clockwise direction.   The stress diagram is then drawn

FIG. 315.

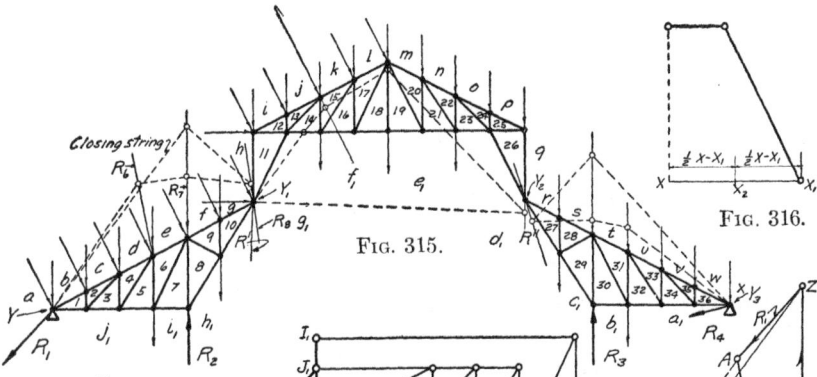

FIG. 315.

FIG. 316.

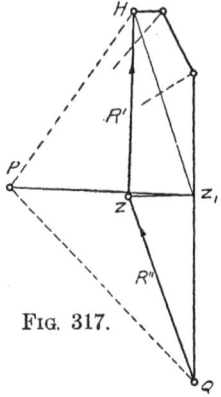

FIG. 317.

FIG. 318.

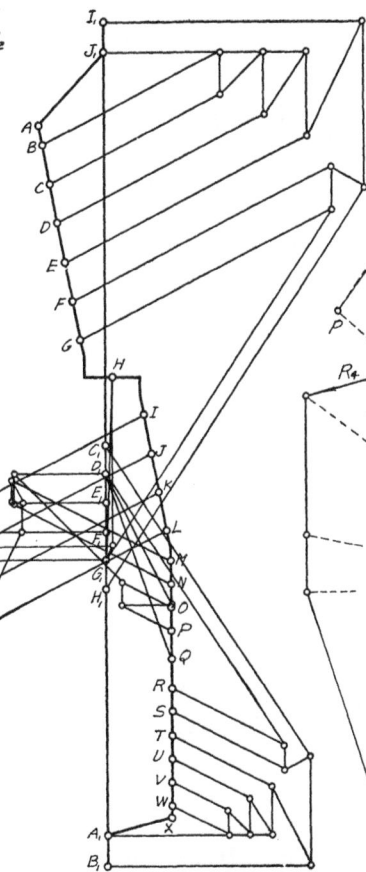

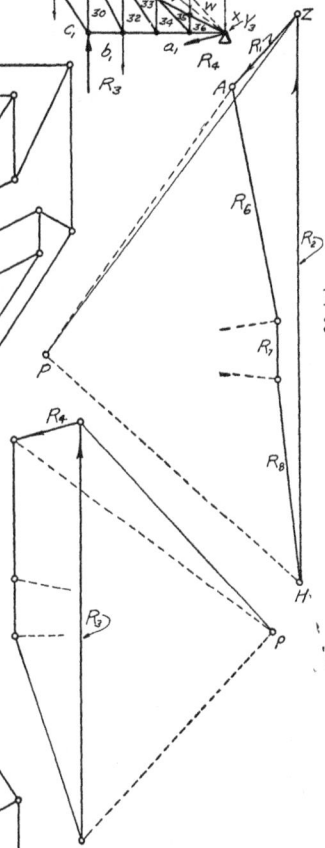

FIG. 320.

FIG. 319.

in the usual way. It may be started at 6 different points, at each end of each of the three small trusses, and three closing points will be obtained, one for each of the three trusses. In fact, the problem could be solved by drawing three separate stress diagrams one for each of the small trusses. The big stress diagram shown in Fig. 320 is nothing more than all three of these smaller diagrams put together.

85. Combination Truss, Three-hinge Arch and Mill Bent.— A sort of combination three-hinged arch and mill bent is shown in Fig. 321. A small three-hinge arch is supported on a truss which, in turn, is supported by and knee braced to steel columns that are fixed at their base. Vertical upper chord loads and a wind load from the right will be considered. The reactions for the three-hinge arch are found by a method, already explained, using Fig. 322. The stress diagram is then drawn in Fig. 323, and the stress obtained.

The truss supporting the three-hinge arch is next considered. The reactions from the arch may be used as loads on the larger truss, when finding R_1 and R_2. However, it seemed more desirable in this particular case to use the original loads applied to the entire truss which, of course, give the same results. In Fig. 324 the resultants, obtained after grouping the loads together as much as convenient, are laid off as shown, and the large funicular polygon in Fig. 321 drawn, assuming for the present that R_1 is vertical, and therefore starting the polygon at Y_2. From p a line is drawn parallel to the closing string, and the point Z_1 located. Then if Z is taken as the mid-point of Z_1–Z_2, X_1–Z will give the magnitude and direction of R_1, and Z–X the magnitude and direction of R_2, for the case when the horizontal from the wind is divided equally between the two reactions. The loads and reactions are now laid off along the load line of Fig. 325, going around the truss in a clockwise direction. After load D–E, the load at Y'_1 plus the reaction from the arch, may be laid off, and point 25 located. Then the reaction from the arch at Y'_2 plus the loading at Y'_2 is laid off and L located; after which L–M, M–N, etc. are also laid off. Or just the loading at Y'_1 may be laid off, and F located, then G, H, etc., in which case L and M will of course fall on the same points as before. In Fig. 325, F, C, H, etc., are located and also point 25, after which the stress diagram is drawn in the usual way. The stress diagram

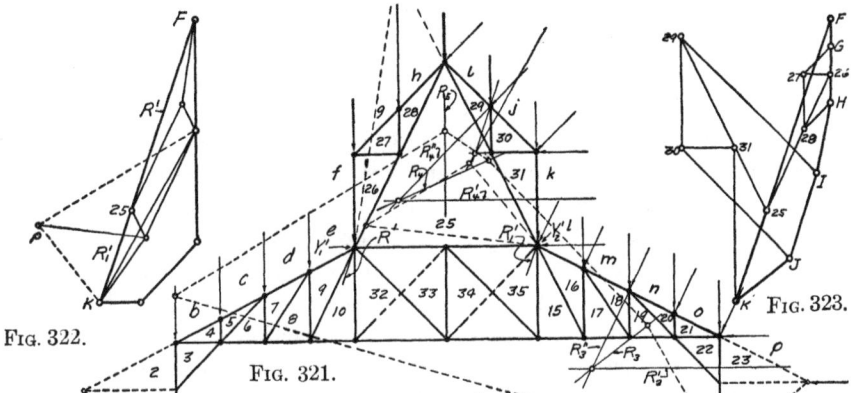

FIG. 322.

FIG. 321.

FIG. 323.

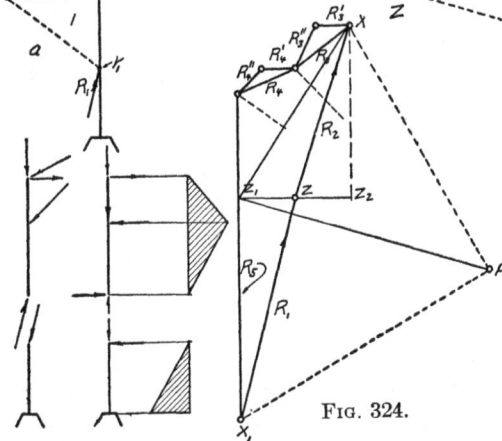

FIG. 324.

FIG. 326.

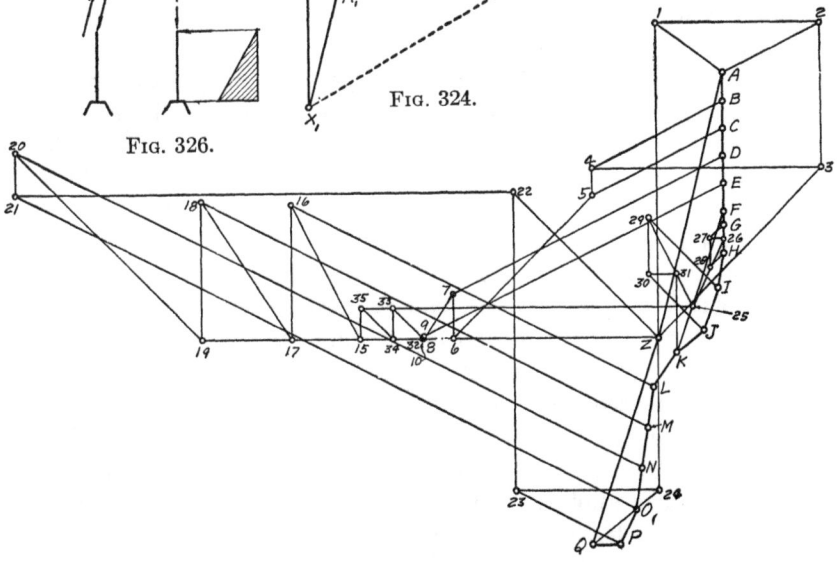

FIG. 325.

for the three-hinge arch is shown in Fig. 325, but it was unnecessary, as far as the construction of the larger diagram was concerned, as only the location of point 25 was required.

Fig. 326 illustrates how the stresses in the left column may be investigated. At the top there are three forces acting, for which the resultant horizontal and vertical components are found. Where the knee brace comes in there is just one force acting, and it is broken into horizontal and vertical components. At the point of contra-flexure there is the reaction for which the horizontal and vertical components are easily found. The portion below the point of contra-flexure receives the reaction and acts somewhat as a vertical cantilever, the horizontal component producing moment and the vertical component producing compression. The direct stress in the column below the knee brace is therefore just equal to the vertical component of the reaction, while from the knee brace to the top it is just equal to the resultant vertical component at the top. The moment diagrams are shown at the right of the figure. As far as moment is concerned the part of the column above the point of contra-flexure acts very much as a vertical beam supported laterally at the top and bottom and loaded with the horizontal component from the knee brace.

The stresses in the right column can be found in a similar way, but the additional loads from the wind would make the solution slightly more complicated.

86. Bridge Trusses of the K Type.—In many cases stress diagrams can be used conveniently to find the stresses in bridges, especially those stresses produced by dead loads. In Fig. 327 a truss is shown, supported at the center and at the left end, which is similar to one of the trusses in the new Quebec bridge. The reaction from the suspended span is applied at the right end, and in this case is found to be sufficient to produce a tendency for uplift at the left end. This end must therefore be anchored down. The reactions R_1 and R_2 are found analytically by taking moments about R_2. Since a large number of the loads are symmetrically placed about R_2, they balance each other and thus simplify the computation. The dead load has been divided between the upper and lower chord panel points, while a live load has been applied to the verticals just below the floor line. These interior loads are carried down below the lower chord and laid off with the lower chord loads. Moving these loads down

FIG. 327.

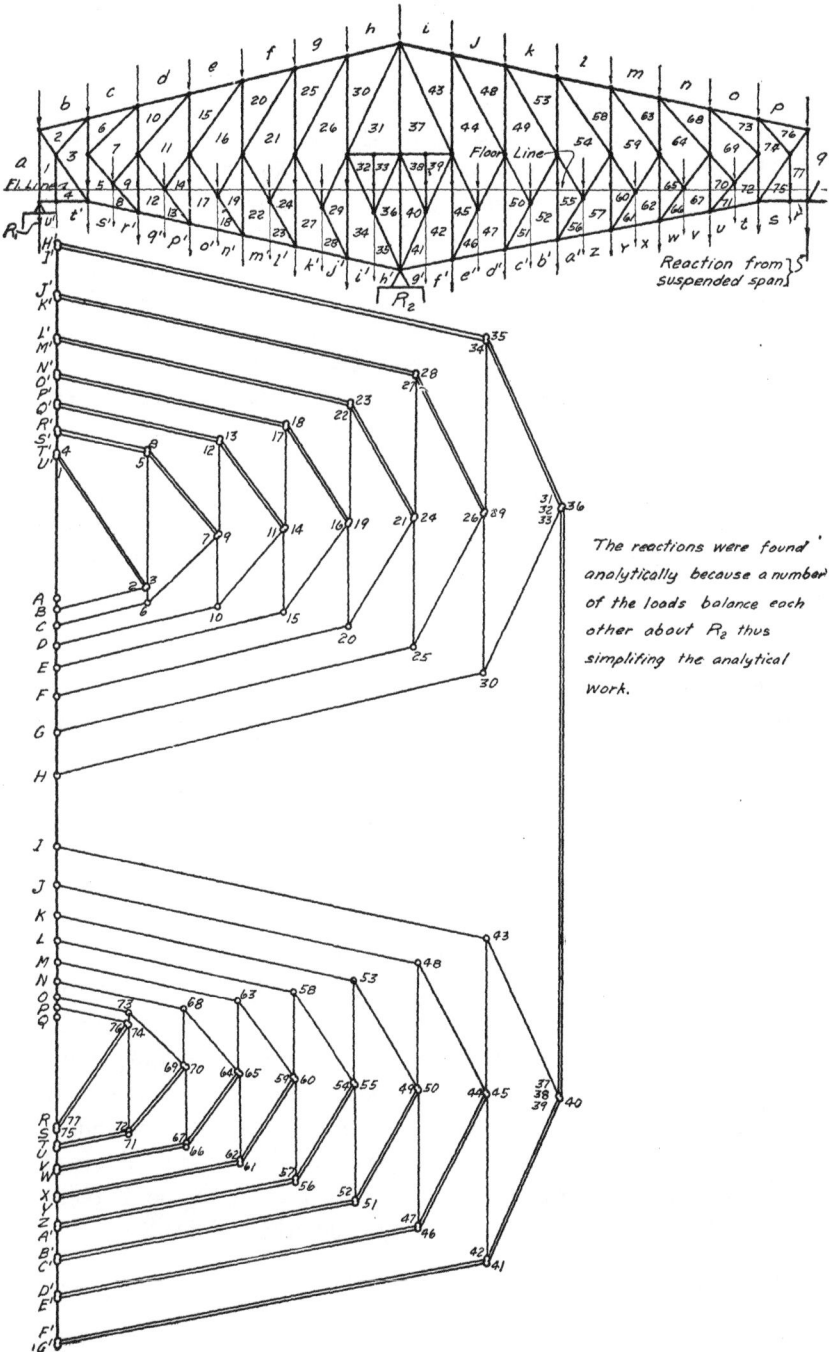

The reactions were found analytically because a number of the loads balance each other about R_2 thus simplifing the analytical work.

FIG. 328.

to the lower chord has very little effect upon the stresses in the members of the truss. When the interior loads, applied to the long verticals such as 19–22, 24–27, etc., are moved down to the lower chord, the stress in the entire member becomes the same as it was in the portion above the floor line. Therefore the stress diagram gives the correct stress for the upper part of the member. The direct stress in the lower part differs from the stress in the upper part an amount equal to the load thrown in by the floor construction. The new members 22–23, 27–28, etc., which were put in in order to carry down the remaining interior loads may be considered as hangers with no connection to the lower chord. Their use will therefore in no way affect the stresses in the various members of the truss. The stress diagram then gives the correct stresses for the various members, also the stress in the hangers, which will be eliminated when the loads are moved back to their proper place.

The loads and reactions were laid off in order along the load line of Fig. 328, going around the truss in a clockwise direction. The stress diagram was then drawn in the usual way, starting at each end of the truss.

87. Trussed Dome.—The plan of a large dome supported by eight crescent trusses is shown in Fig. 329, while Fig. 330, directly above, shows one of the trusses. A plan of that part of the roof carried by the truss of Fig. 330 is shown by the shaded area in Fig. 329. The area of roof surface carried at each panel point may be determined approximately by projecting down from the panel point, measuring the corresponding width of this shaded area, and multiplying by half the sum of the two adjoining panels. Taking a special case; the area carried by panel point IX is given approximately by d_1 times d_2. Knowing the area of roof surface carried by each panel point, the dead and wind loads can be easily computed. The loads are now laid off in order in Fig. 331, the pole p chosen, and the funicular polygon drawn in Fig. 330, assuming temporarily that R_1 is vertical. Parallel to the closing string a line is drawn from p, and the intersection Y_1 obtained. It seems reasonable to suppose that the horizontal component is divided equally between R_1 and R_2 and the reactions are obtained by locating Y midway between Y_1 and Y_2. With the Y point located, the stress diagram can be drawn without difficulty.

88. Ring Dome. Dead Loads.—An interesting and often economical roof construction called the ring dome is shown in Figs. 332 and 333. It consists of a system of circular rings and equally spaced radial ribs. The areas carried by the various panel points may be determined by the same method that was

FIG. 330.

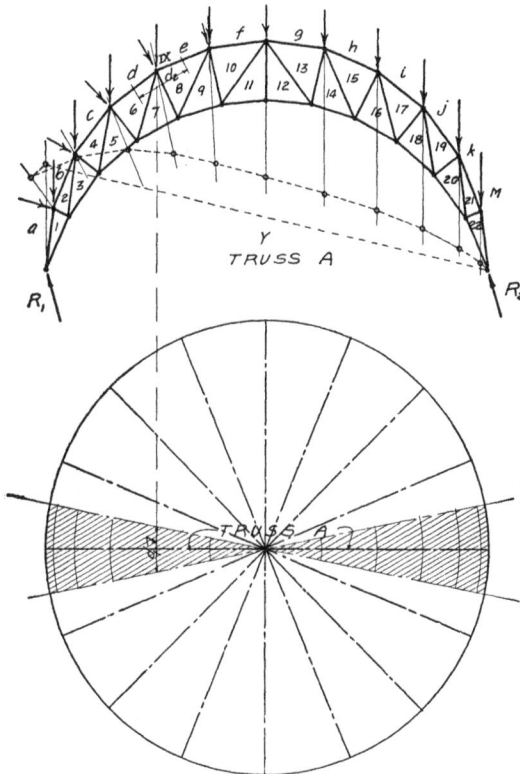

FIG. 329.

used in connection with the crescent truss just considered. A plan of the dome showing rings and ribs is shown in Fig. 332 and the elevations of one of the ribs in Fig. 333. It should be noted that the rings intersect the ribs at the places where the ribs, and also the rings, change direction. These intersections will be referred to as panel points, and the loading is applied only at these points.

A vertical dead load uniformly distributed over the surface

of the dome will be first considered. This means that every panel point of any one ring will have the same loading, but the panel points for different rings will have different loads, because they carry different areas of roof surface. The loads for the different panel points of any rib Z, shown in elevation by Fig. 333, are computed and laid off to some convenient scale along the load

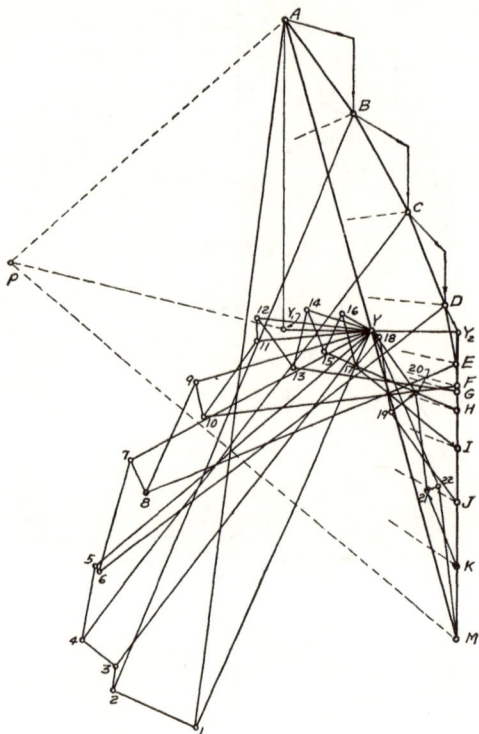

Fig. 331.

line of Fig. 334. For equilibrium the resultant of the two stresses F–6 and G–7 must just balance the load F–G, F–G being that portion of the load at the center which would go to one rib, or one-eighth of the total load. The stress diagram will therefore be started at the center by drawing a line from F in Fig. 334 parallel to member F–6 and extending it until the line from G parallel to member G–7 is intersected. This intersection locates point 6 which is also point 7, and the vectors F–6 and G–7 measured to scale give the stresses in the members *F*–6 and *G*–7. It is evident from inspection that these stresses are compression and this is verified by applying our ordinary rule for finding the kind of stress. To illustrate, considering the joint at the center and going around in a clockwise direction, the member at the right is read *G*–7. Now when we go from *G* to 7 in the stress diagram, the direction would be towards the joint if the vector were placed above the member.

The joint or panel point just to the left of the center, at which is applied the load *E*–*F*, will be considered next. The stress in member *F*–6 and the load *E*–*F* are known, and it follows that the stress in member *E*–5 must have just enough vertical component to balance the resultant vertical component of the other two, because the ring can resist only horizontal forces. In Fig. 334 a line is drawn from *E* parallel to member *E*–5 and extended until the horizontal line from 6 is intersected and point 5 located. The vector *E*–5 gives the stress in the member *E*–5, and 6–5 gives the horizontal component that must be thrown in by the ring in order to produce equilibrium. Going around the joint in a clockwise direction, this force is read 6–5, and going from 6 to 5 in Fig. 334 the movement is towards the joint, therefore the ring must have compression. The ring changes direction at the joint, and the ring member on each side of the joint must have a stress such that the resultant of the two stresses in the plane of rib *Z* will be just equal to 6–5. The direction of these two ring members, r_6 and r'_6, is shown in Fig. 332. Lines are now drawn parallel to these ring members from points 6 and 5 in Fig. 334 and extended until they intersect as shown, their length giving the stress in the ring members. The stress in r_6 is, of course, the same as the stress in r'_6, because these two ring members make the same angle with the plane of the rib *Z*. It makes no real difference whether the intersection is above or below the vector 6–5, but it will be found convenient to keep the intersections for the rings having compression above, and for those having tension below.

The next joint at the left, which carries the load *D*–*E*, will now be considered. From *D* in Fig. 334 a line is drawn parallel to member *D*–4, and its intersection 4 with the horizontal line from 5 obtained. The stress in member *D*–4 and the resultant of the stress in r_5 and r'_5 are determined by this intersection. Since 5–4 is the resultant of the stress in the two ring members

Fig. 335.

Fig. 336.

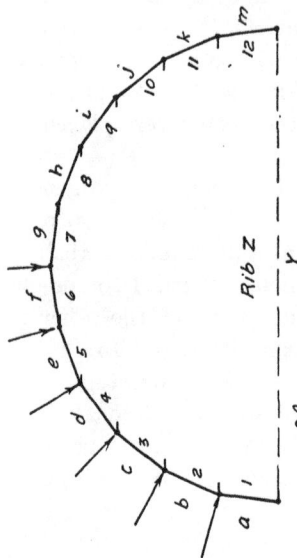

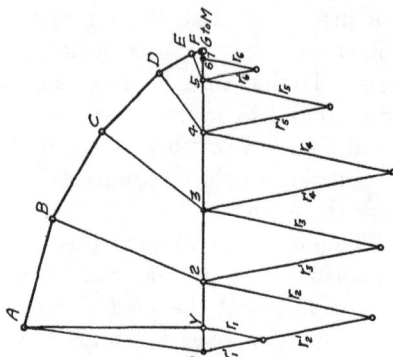

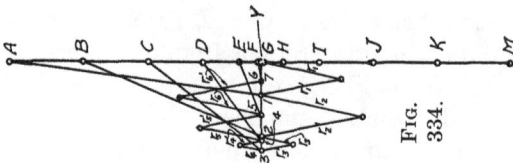

Fig. 334.

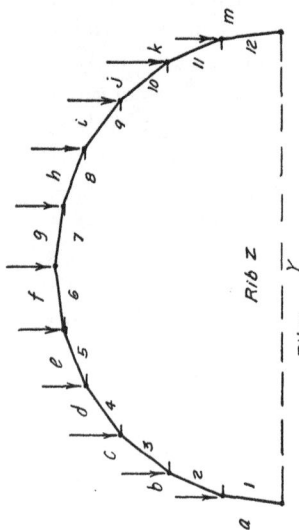

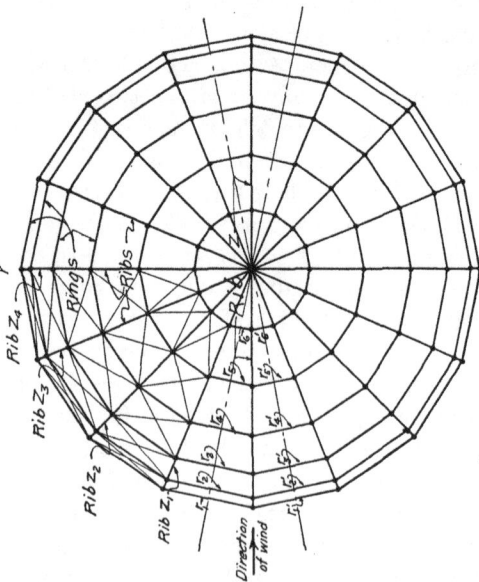

Fig. 333.

Fig. 332.

r_5 and r'_5, the stresses in these members can be obtained by drawing lines parallel to them from points 4 and 5 and extending these lines until they intersect as shown. In a similar way the stresses in the rest of the rib and ring members may be found. It should be noted that the ring members r_4 and r'_4 are the last which have compression, those of the lower rings having tension. The lines have not been drawn in Fig. 334 for the stresses in the right half of the rib Z, because, since the rib and loading are both symmetrical about the center, the stresses in the members of the right half will be the same as in the corresponding members of the left half. Also, since the loading on the panel points or joints of each rib is the same as on those of rib Z, the stress in every member of each ring will be the same as the stresses found by the construction of Fig. 334. The analysis of the stresses in the ring dome is therefore found to be rather simple under a uniformly distributed vertical load. But the problem becomes much more complicated when an attempt is made to follow through the stresses produced by a wind load or those produced by irregular loads.

89. Ring Dome. Wind Loads.—In Fig. 335 rib Z is shown with loads produced by a wind having direction as indicated by the arrow at the left of Fig. 332. In order to better study their effect in producing stress, only the wind loads will be considered for the present. Later the stresses which they produce may be combined with the dead load stresses in order to obtain the maximum and minimum stresses.

The wind loads indicated in Fig. 335 are laid off along the load line of Fig. 336, and extend from A to G. Since there is no load on the right half of the rib $H, J \ldots M$ all fall on point G. Now for equilibrium the resultant of the stresses in all the rib members intersecting at the center must be no greater than the load applied at the center. It therefore seems reasonable to assume that the resultant of the two stresses F–6 and G–7 just balance the load F–G. From G a line is drawn parallel to member G–7 and from F another parallel to F–6. The intersection of these lines locates point 6, which is also point 7, and thus determines the stress in member F–6 and also in G–7. The construction is now continued for the left half of the rib the same way as in Fig. 334, also the stresses in the ring members adjacent to the rib are found as shown. Note that these ring members all have compression except the lowest. Since there are no loads applied to the joints at the right half of the rib the stress in each of the right half rib members must be such that its vertical component is the same as the vertical component of G–7. Lines might therefore be drawn from G, H, . . . and M in Fig. 336 and extended until they intersect the horizontal line from 7, but they would be very short. The lines giving the stresses in the adjacent ring members might also be drawn, but it is evident that they too would be very small.

Now when the wind is blowing from the direction shown in Fig. 332, the wind load on rib Z_1 will be practically the same as on rib Z, the load on rib Z_2 will be slightly less, that on rib Z_3 will be very much less, and there will be almost no wind load on rib Z_4. Stress diagrams similar to Fig. 336 may be drawn for each of these ribs and stresses obtained for the various rib members and adjacent ring members. The only difficulty is that we find a stress of one magnitude thrown into certain ring members at one end and a stress of another magnitude thrown in at the other end. This unbalanced ring stress can be best taken up by diagonal tension members, as shown in a few of the pannels in Fig. 332. This diagonal X bracing should be put in all panels, because the wind may blow from any direction. The stresses produced in these diagonal members by the unbalanced ring stresses will, of course throw additional stresses into certain of the lower ring and rib members. These additional stresses will either increase or decrease the stresses already found.

In order to transmit these unbalanced ring stresses down to the base of the dome, one or more of the three following conditions will develop at certain panel points or joints: (1) An unbalanced ring stress which will produce tension in a diagonal and additional compression in a rib member: (2) An additional stress in the upper rib member which will produce additional stress in the rib member below and in the adjacent diagonal members: (3) A tension thrown in by a diagonal from above, which will produce an additional stress in the rib member below, and also in the adjacent ring members.

Fig. 337 Ⓐ shows one of the panels of the dome revolved into the plane of the paper. Let it be assumed that the compression thrown into the ring by joint a is greater than that thrown in by joint b by an amount F_1. The ring member a–b will have the higher stress, but at b the force F_1 will be balanced by tension in the diagonal b–d and additional compression in the rib member *b–c.* The magnitude of the tension produced by F_1 in the diagonal, and of the additional compression which F_1 produced in the rib member, may be found by drawing a force triangle similar to Fig. 337 ⓑ.

A slightly different condition exists when the ring has tension. Let it be assumed that the tension thrown into the ring at joint *a* is greater than the tension thrown in at joint *b* by an amount F_2. The ring member *a–b* would have the lower stress and, in

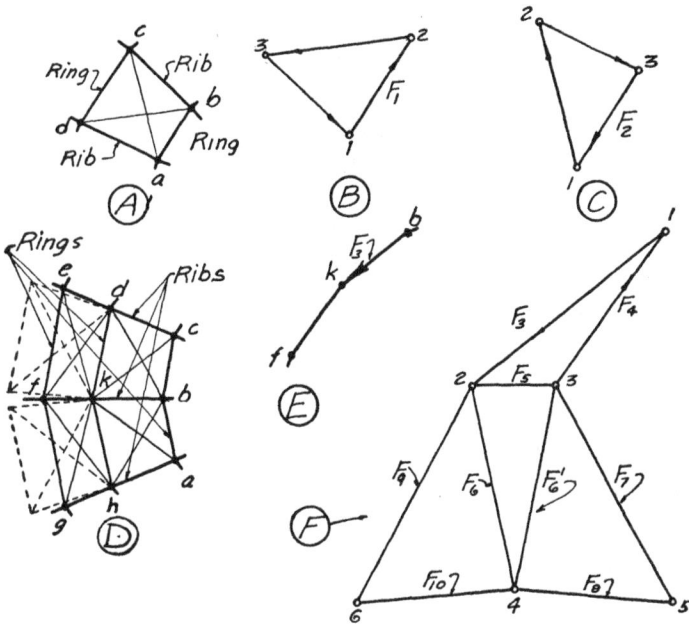

FIG. 337.

order to balance F_2, we would have tension in the diagonal *a–c* and additional compression in the rib member *a–d*. The magnitudes of these stresses which produce equilibrium may be found by drawing a force triangle similar to Fig. 337 ⓒ.

Fig. 337 ⓓ shows four panels of the dome in plan. At joint *k* a load F_3 is applied by an additional compression thrown into the rib member *b–k* from above. In Fig. 337 ⓔ an elevation of the two rib members *b–k–f* is shown and in Fig. 337 ⓕ the force triangle 1–2–3 is drawn to determine the effect of F_3. It is found that F_3, in addition to producing a compression of F_4 in *k–f*, re-

quires a horizontal force F_5, or the equivalent, applied at k in order
to produce equilibrium. A tension in the ring members k–h and
k–d of F_6 and F'_6 would produce such a horizontal component.
But if tension is thrown into the ring members k–h and k–d at
k and not at h or d, we have the condition considered in Figs.
337 Ⓐ and 337 Ⓒ. In this case it was found that the additional
tension did not go into the ring member, but went down the
diagonal, and by so doing increased the compression in the adjoining rib member.

The force triangles 3–4–5 and 2–4–6 in Fig. 337 Ⓕ, corresponding to Fig. 337 Ⓒ, are now drawn, after revolving panels f–e–d–k
and f–k–h–g into the plane of the paper, in order to get the proper
direction, as shown by the dotted lines. Then F_7 gives the stress
in diagonal k–e, and F_9 that in diagonal k–g. An additional compression F_8 is thrown into k–f in connection with F_7, and also of
F_{10} in connection with F_9. This makes the total additional
compression produced in k–f by F_3 equal to the numerical sum
of F_4, F_8 and F_{10}.

The condition would be somewhat different if F_3 were tension
in place of compression. Then F_4 would also be tension and the
ring members k–h and k–d would have compression equal in
magnitude to F_6 and F'_6. The compression in k–d would produce a tension in f–d and an additional compression in d–e, the
magnitude of which may be found by drawing a force triangle
similar to Fig. 337 Ⓑ. In the same way the effect of the compression in k–h upon h–f and h–g may be found.

The effect of an additional load applied to a panel point by
a stress in one of the diagonals above will now be considered. A
small portion of the dome is shown in plan by Fig. 338 Ⓐ, and
it is assumed that the diagonal a–k exerts a pull F_1 at joint k.
This pull is resolved into horizontal and vertical components in
Fig. 338 Ⓑ. The vertical component F_v may be thought of as
a vertical pull applied at k, while the horizontal component F_H
may be considered as a horizontal pull applied at k and acting
towards the right in the vertical plane containing a–k.

Fig. 338 Ⓒ shows an elevation of the two rib members b–k–f,
and in Fig. 338 Ⓓ F_v is broken into two components, F_2 parallel
to k–f and the other F'_H horizontal. In Fig. 338 Ⓔ the resultant
F_3 of F_H and F'_H is found, and then broken into two components,
one parallel to k–d and the other parallel to k–h. We have now

found that F_1 can be balanced by a tension F_2 in f–k, a compression F_4 in k–h, and a compression F'_4 in k–d. The additional compression thrown into the ring members can be balanced by stresses in diagonal and rib members, as already explained.

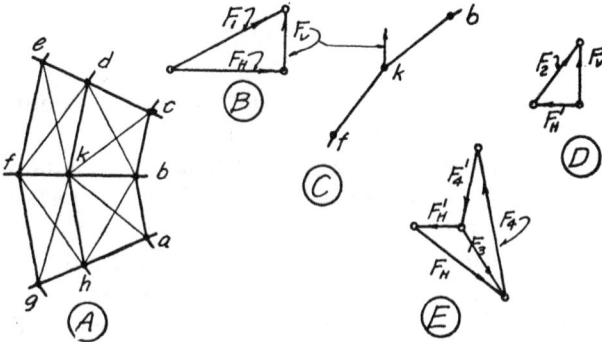

FIG. 338.

If the effect of the variation in the ring stresses, is followed down to the base of the dome, as indicated above, starting with the upper ring and taking a joint at a time, the results should be satisfactory. The amount of labor is not so great as one would at first suppose. After solving one joint for each ring, as indicated above, results at other joints can be found by proportion.

CHAPTER VI

MOVING LOADS

WHEN a beam is subjected to moving loads, it is often desirable to know the maximum moment that can be produced at any section. A diagram which gives this information is of value in designing the cover plates for plate girders, the reinforcement for reinforced concrete beams, etc. Not only are the maximum moments of value, but it is often desirable to know the maximum shears. Diagrams which give the maximum shears may be used in designing the web reinforcement for reinforced concrete beams, and for investigating the web stresses in other beams.

Runway girders for traveling cranes are a common example of beams carrying moving loads, and, of course, all bridges must carry moving loads.

90. Single Concentrated Moving Load.—The effect of a single concentrated load P, moving across the beam shown in Fig. 339, will be considered. In Fig. 340, lay off A–B equal to the load P to some convenient scale, and choose any pole p_1 with any desirable pole distance. It will be desirable to have the base line d–f in Fig. 341 horizontal, so in order to accomplished this, the line p_1–z_1 in Fig. 340 is made horizontal. In Fig. 341 the line d–f is now drawn parallel to p_1–z_1, d–e_1 parallel to p_1–A, and f–e_1 parallel to B–p_1. The triangle d–e_1–f thus obtained is the moment diagram for the case when the load P is at x_1, directly above the intersection e_1. By using the pole p_2 and the horizontal ray p_2–z_2, the triangle d–e_2–f is obtained in Fig. 341. This triangle is the moment diagram for the case when the load P is at the point x_2 directly above e_2. A number of other poles p_3, p_4, etc. are chosen and the corresponding moment diagrams drawn in Fig. 341, giving the points e_3, e_4, etc., Through these points the smooth curve d–e_1–e_2–e_3 . . . e_n–f is drawn. Now at any section of the beam the vertical intercept between this curve and the base line d–f, times the pole distance H, gives the moment which will be produced there when P is at the

168

section. Or if the number of feet which each inch in the space
diagram represents, is multiplied by H, and the intercepts in Fig.
341 measured to this new scale, the moment will be obtained direct
in foot-pounds. This moment is also the maximum moment
which can be produced at the chosen section by the load P,
because as P moves to either side, the moment at the section
decreases. The curve d–e_1–e_2–e_3 . . . e_n–f will therefore be called
the maximum moment curve.

Upon close inspection this maximum moment curve for a
single concentrated load proves to be a parabola with the maximum
ordinate at the center, and equal to $\dfrac{PL}{4}$, which is the moment
at the center of the beam when P is there. Thus the curve in
Fig. 341 might have been obtained without the use of Fig. 340,
by computing the maximum ordinate and constructing a parabola
as shown in Fig. 343. This maximum ordinate $\dfrac{PL}{4}$ may be laid
off to a scale of 1 inch equals any convenient number of foot-
pounds.

That the maximum moment curve for a single concentrated
load is a parabola may be verified by constructing a parabola over
the curve in Fig. 341, and finding whether or not the two curves
coincide, or by writing the equation for the maximum moment.

For any position of P, the shear between P and the right
reaction is constant and equal to R_2, also the shear between P
and the left reaction is constant and equal to R_1. But as P moves,
the shear at every section of the beam changes. The maximum
shear at any section will occur when P is either just to its right
or just to its left, because as P moves away from the section
towards either reaction, the reaction towards which it moves
increases and the other one, which gives the shear at the section,
must decrease.

When the load P is at the section x_1, by referring to Fig. 340,
R_1 is found to be equal to z_1–A and R_2 equal to B–z_1. From
these reactions and by using the base line m–n in Fig. 342, the
diagram m–r_1–u_1–v_1–s_1–n–m is drawn, which is the shear diagram
for the case when P is at x_1. With P in this position, the shear
just to the left of P is given by the line u_1–w_1, and the shear just
to its right is given by the line w_1–v_1. When P is at section x_2
the shear diagram is found to be m–r_2–u_2–v_2–s_2–n–m, and the shear

just to the left of P is u_2–w_2, that just to the right w_2–v_2.
Shear diagrams may be drawn for P when in other positions,

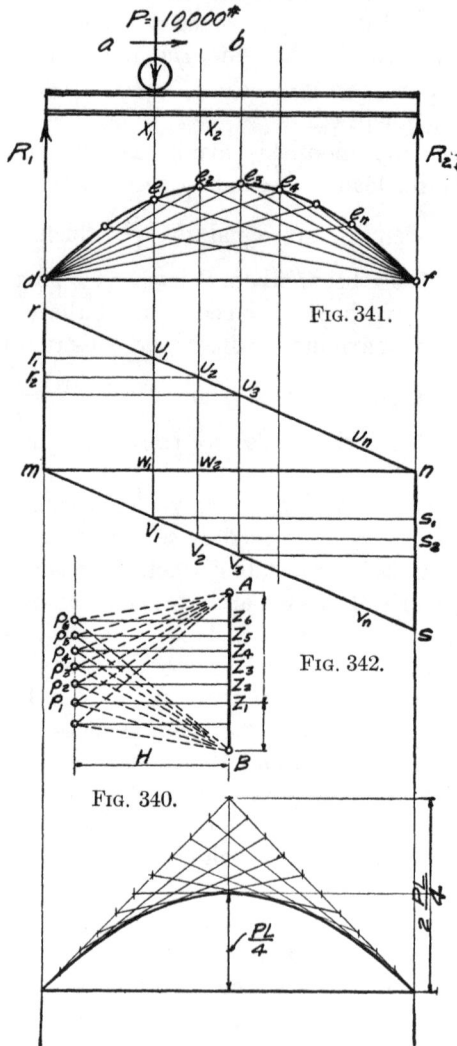

Fig. 339.

Fig. 341.

Fig. 342.

Fig. 340.

Fig. 343.

and other points $u_3 \ldots u_n$ and $v_3 \ldots v_n$ located. Then through
the points $u_1, u_2 \ldots u_n$ a line called r–n is drawn, which is found

to be a straight line.   In the same way the straight line m–s is drawn.   It should be noted that both r–m and n–s are equal to P, because when P is at either reaction, that reaction is equal to P. Now when P is at any section, the vertical intercept between the lines m–n and r–n at this section gives the shear just to the left of P, while the vertical intercept between the lines m–n and s–m gives the shear just the right of P.   The larger of these intercepts gives the maximum shear that can be produced at this section.

If the student wishes analytical proof that r–n is a straight line, he may write the equation for R_1 as P moves across the beam, R_1 being equal to the shear just to the left of P which in turn is equal to the corresponding ordinate of the line r–n.   This equation will be found to be that of a straight line.

91. Single Concentrated Moving Load and a Uniform Dead Load.—The beam shown in Fig. 344 is assumed to carry a uniform dead load of 1000 pounds per foot in addition to a concentrated moving load of 20,000 pounds.   The maximum shear and moment curves are desired.

Since the uniform load is a dead load, the line r'–s' of the shear diagram Fig. 345 is the maximum shear curve for the uniform load.   In Fig. 346 are shown the maximum shear curves for the concentrated load.   These lines may be drawn the same as the lines r–n and m–s in Fig. 342; but there is a shorter way.   Since r''–m_2 and s''–n_2 are each equal to P, and since the lines r''–n_2 and m_2–s'' are both straight lines, points r'' and s'' may be located and the lines drawn.   The same scale must be used in Figs. 345 and 346, because the intercepts in these two figures are now added in order to obtain the maximum shear curves for the total loading. When P is at the position x, the shear at this section due to the uniform load is given by the intercept Y_1, while the maximum positive shear that can be produced at x by the moving load P is given by the intercept Y_2.   The maximum negative shear that can be produced at x by the load P is given by the intercept Z_2.   Now the shears Y_1 and Y_2 are of the same sign and act at the same section at the same time, therefore they are added and give the intercept Y in Fig. 347, which is the maximum positive shear that can be produced at x.   The shear Y_1 is always acting and Z_2 is of different sign, therefore their difference Z gives the maximum negative shear that can be produced at x.   With

Fig. 344.

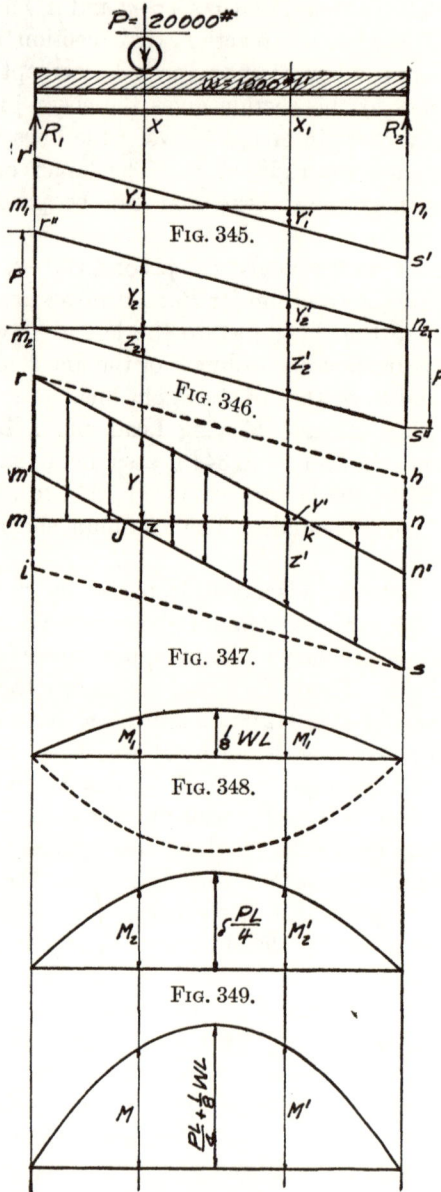

Fig. 345.

Fig. 346.

Fig. 347.

Fig. 348.

Fig. 349.

Fig. 350.

regard to the sign of the shear as indicated in the figures, it is important that the same kind of shear be shown above the base line in all of the figures. Usually it is convenient to show the postive shear above the base line and the negative shear below.

As we move from section x to the left, Y_1 increases and Z_2 decreases, which means that a point will be reached where Z is zero, and the line $s–m'$ crosses the base line. In other words no negative shear can be produced here, as the positive shear from the uniform load just balances the maximum negative shear that can be produced by P. As we move farther to the left, Z will increase but will be above the base line and therefore positive.

The intercept Y will increase with movement to the left, amounting to $r–m$ when the left reaction is reached. The line $r–m$ will be found to equal $r'–m_1$ plus $r''–m_2$, and $m–m'$ will equal $m_1–r'$.

As P moves to the right of section x, the intercepts under P vary, and when P reaches the section x_1 the intercepts are Y'_1, Y'_2, and Z'_2. The intercepts Y'_1 and Y'_2 are now of different sign, therefore their difference will be taken to give Y' which is positive since Y'_2 is here larger than Y'_1. Intercepts Y'_1 and Z'_2 are of the same sign and give the length Z'. With P at various other sections, lengths are found corresponding to Y and Z. Through the ends of these lengths lines are drawn which are found to be straight lines and are shown in Fig. 347 as $r–n'$ and $m'–s$. These lines are the maximum shear curves for the given loading; that is, for any section of the beam the intercept between the line $m–n$ and $n'–r$ at the section gives the maximum positive shear that can be produced there; or if no positive shear can be produced, it will give the smallest negative shear that can be produced there. Similarly, the intercept between the line $m–n$ and $s–m'$ gives the maximum negative shear or the minimum positive shear. It will be noted that no positive shear can be produced to the right of k and no negative shear can be produced to the left of j. Since $r–m$ is equal to $m_2–r''$ plus $m_1–r'$, and $n–n'$ is equal to $s'–n_1$, and since $r–n'$ is a straight line, the points n' and r could be located and the line $r–n'$ drawn at once. Also the line $s–m'$ could be drawn in a similar way.

When P is at any section the intercept between the lines $r''–n_2$ and $m_2–n_2$ at that section gives the value of R_1 due to P, and the intercept between m_2–n_2 and s''–m_2 gives the value of R_2 due to P. Now the uniform load produces an R_1 equal to r'–m_1. The line r–h is drawn, in Fig. 347, so that the intercepts between m–n and r–h are all just r'–m_1 greater than the corresponding intercept between r''–n_2 and m_2–n_2. Then when P is at any section, the intercept between r–h and m–n at the section gives R_1. Also the intercepts between m–n and s–i give R_2 for various positions of P.

The maximum moment curve will now be drawn. The maximum moment curve for the uniform dead load is, of course, given by its moment diagram. The curve of this moment diagram is a parabola with the maximum ordinate at the center and equal to $\frac{1}{8}WL$, in which W is the total uniform load. This parabola, the upper one in Fig. 348, is drawn as illustrated in Fig. 343. The maximum moment curve for the moving load P is a parabola with the maximum ordinate at the center and equal to $\dfrac{PL}{4}$. The value of this ordinate is laid off in Fig. 349, using the same scale that was used in Fig. 348, then the parabola is drawn, using the construction of Fig. 343.

Now when P is at any section x, the moment produced at this section by the uniform load is given by the intercept M_1, in Fig. 348, while the maximum moment that can be produced at the section by the moving load P, is given by the interept M_2. The sum of these moments gives the length M in Fig. 350, which is the maximum moment that can be produced at x. When P is at any other section x_1, the intercept M'_1 in Fig. 348 is added to the intercept M'_2 in Fig. 349, giving M' for Fig. 350. With P at other sections other lengths are obtained for Fig. 350 and through the upper ends of all of these lengths a smooth curve is drawn which is the maximum moment curve.

When P is at any section, the intercept between this curve and its base line, at the section, gives the maximum moment that can be produced there. It will be noted that the intercepts in Fig. 350 are equal to the sum of the corresponding intercepts in Figs. 348 and 349. Therefore the maximum moment curve in Fig. 350 is a parabola with the maximum ordinate at the center and equal to $\dfrac{PL}{4}$ plus $\frac{1}{8}WL$. Also attention should be called to the fact that if the parabola in Fig. 349 had been drawn upside down on the base line of Fig. 348, as shown by the dotted line, the intercepts in Fig. 348 between the two parabolas would have been the same as the intercepts in Fig. 350. It would therefore not be necessary to draw Fig. 350.

The student should remember that all vertical lengths and intercepts in Figs. 348, 349, and 350 are to be measured to the same scale, the scale being one inch equals some convenient number of foot-pounds or inch-pounds. A desirable scale may be chosen when the values $\frac{1}{8}WL$ and $\dfrac{PL}{4}$ have been computed and the construction of the parabolas started.

92. Uniform Moving Load Longer than the Span.—Consider the effect of a uniform moving load of 1000 pounds per foot on the beam shown in Fig. 351, the load being longer than the beam. Every increase of load on the beam increases the bending moment at every section of the beam, or in other words, as the load moves further on the beam, the moment at every part of the beam increases until the load covers the entire beam. Therefore the maximum moment curve for a uniform moving load longer than the beam, is the parabola given by the moment diagram for the case when the load covers the entire beam. The maximum ordinate of the parabola is at the center and is equal to $\frac{1}{8}WL$, in which W equals wL, w being the load per foot. This maximum moment curve is shown in Fig. 355.

The maximum shear curves, the constructions for which are shown in Figs. 352 to 354, are more difficult to obtain. When the right end of the load is at section x, the resultant of all of the load on the beam acts at Y, and the magnitude of this resultant, laid out to the scale in Fig. 353, is A–B. Any line d–f is now drawn in Fig. 352 and from d the line e–d, making any convenient angle with d–f. Points e and f are then connected. In Fig. 353, A–p is drawn parallel to e–d and B–p parallel to f–e, then from p, the line p–z is drawn parallel d–f, thus marking off the reactions. That is z–A is R_1 and B–z is R_2, when the right end of the moving load is at x. From these reactions the shear diagram m–r_1–v_1–s_1–n–m for the end of the load at x, is drawn.

When the right end of the load is at x_1, the resultant of all of the load on the beam acts at Y_1, and the magnitude of this resultant laid off in Fig. 353 is A–B_1. From A a line is drawn parallel to d–e_1, and from B_1 one parallel to f–e_1, their intersection

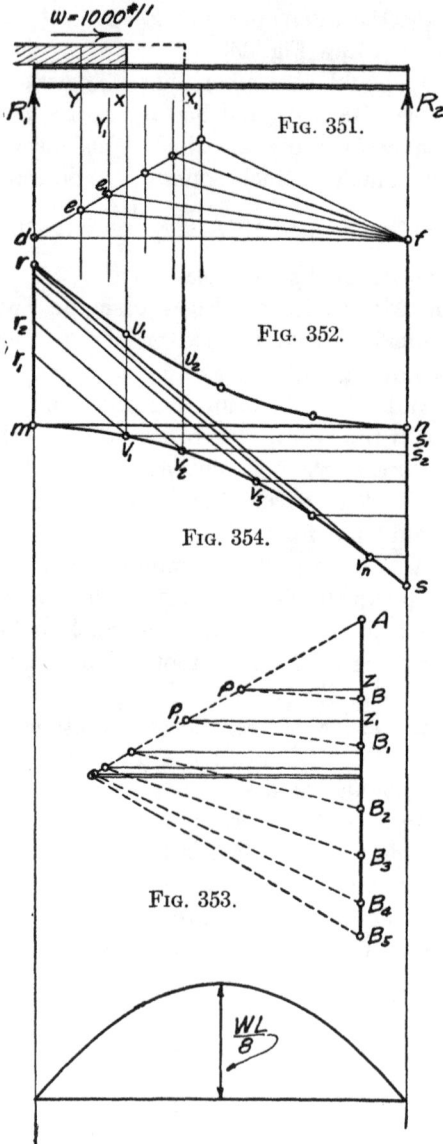

FIG. 351.

FIG. 352.

FIG. 354.

FIG. 353.

$\frac{WL}{8}$

FIG. 355.

giving p_1 from which the line p_1–z_1 is drawn, thus locating z_1 and marking off the reactions. From these reactions, for the case when the right end of the load is at x_1, the shear diagram m–r_2–v_2–s_2–n–m is drawn, which gives the point v_2 directly under x_1. With the right end of the load at various other sections the points $v_3 \ldots v_n$ are located and the curve m–v_1–$v_2 \ldots s$ is drawn. Now as the load moves across the beam, the shear just to the right of its right end is always negative and equal to R_2, but as soon as the right end of the load passes any section the shear there decreases. It is also true that the negative shear at any section will increase until the right end of the load reaches the sections, because R_2 increases as more load comes on the beam. Since the vertical intercept between the line m–n and the curve m–v_1–$v_2 \ldots s$ at any section gives R_2 for the case when the right end of the load is at that section, the curve m–v_1–$v_2 \ldots s$ is one of the maximum shear curves. The other maximum shear curve r–u_1–$u_2 \ldots n$, may be located in a way similar to that in which m–v_1–$v_2 \ldots s$ was located, shear diagrams being drawn in Fig. 354 for various positions of the left end of the load on the beam. Now the maximum positive shear that can be produced at any section, by the given loading, is given by the vertical intercept between the line m–n and the curve r–u_1–$u_2 \ldots n$ at the section, while the maximum negative shear is given by the intercept between the line m–n and the curve m–v_1–$v_2 \ldots s$, at the section.

93. Moving Uniform Load Shorter than the Span.—In Fig. 356, a uniform load of 1000 pounds per foot and 10 feet in length is shown moving across a 20-foot beam. It can be proved that the maximum moment curve for a moving load of this kind is a parabola with the maximum ordinate at the center and equal to $\frac{wl}{4}\left[L-\frac{l}{2}\right]$. This maximum ordinate is computed and laid off to scale in Fig. 360. The parabola is then drawn, using the method shown in Fig. 343. The analytical discussion proving that this maximum moment curve is a parabola is rather long and will not be given here. It may be found in " Elementary Applied Mechanics " by Alexander and Thompson.

The maximum shear curves, shown in Fig. 359, are obtained by the use of Figs. 357 and 358, the construction being practically the same as for a uniform moving load longer than the span. The load is placed at various positions on the beam, or partly on and partly off, the reactions are found for each position, and for each position a shear diagram is drawn in Fig. 359. When the

FIG. 356.

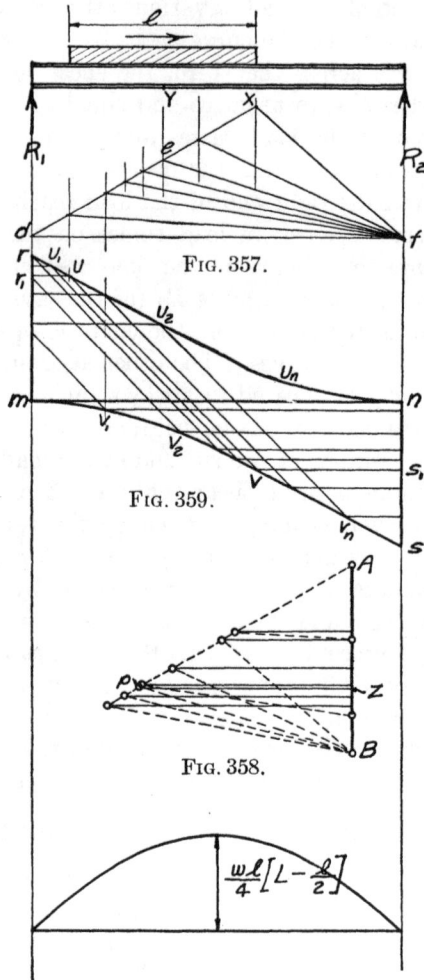

FIG. 357.

FIG. 359.

FIG. 358.

$$\frac{wl}{4}\left[L - \frac{l}{2}\right]$$

FIG. 360.

right end of the load is at section x, the resultant of the load on the beam acts at Y, and by the use of Figs. 357 and 358, R_1 is found to be equal to $Z–A$ and R_2 equal to $B–Z$. From these reactions the shear diagram m–r_1–u–v–s_1–n–m is drawn in Fig. 359, and the points u and v located. Other shear diagrams locate points u_1, u_2, ... u_n and v_1, v_2, ... v_n, and the maximum shear curves are drawn. These curves will be found to be straight lines for a part of their length, the curve portion being just as long as the load.

94. Uniform Dead Load and Uniform Moving Load Shorter than the Span.—Fig. 361 shows a 20-foot beam carrying a uniform dead load of 1000 pounds per foot in addition to a uniform moving load of 2000 pounds per foot and 8 feet in length. The maximum shear curve for the dead load is shown in Fig. 362, while the maximum shear curves for the moving load are shown in Fig. 363.

Fig. 363 was obtained in the same way as Fig. 359. Now the intercepts in Fig. 362 are added to the intercepts in Fig. 363, and the maximum shear curves for the entire loading are drawn in Fig. 364, in a way similar to that used for the shear curves in Fig. 347. The maximum moment curve for the dead load is shown in Fig. 365, and the one for the moving load is shown in Fig. 366. The intercepts in these two figures may be added and Fig. 367 drawn, giving the maximum moment curve for the entire loading; or the curve of Fig. 366 might be drawn upside down, using the base line of Fig. 365, as shown by the dotted curve. The intercepts between the curves in Fig. 365 will be the same as the intercepts in Fig. 367.

95. Two Concentrated Moving Loads.—A beam carrying two concentrated moving loads with a fixed distance a between them is shown in Fig. 368. In Fig. 369 the maximum shear lines for the smaller load P_2 are drawn, as already explained, and those for the larger load P_1 are drawn in Fig. 370. The maximum shear lines for the total loading will now be drawn in Fig. 371. The maximum shear at any section will occur when one of the loads is just to the right or just to the left of the section. Now when the loads move across the beam, the shear just to the left of P_2 is always equal to R_1, and the shear just to the right of P_1 is always equal to R_2. When P_2 is at section x, the reaction at the left end of the beam due to P_2 is given by the intercept Y_1, while the reaction at the left end produced by P_1 is given by the intercept Y_2. Therefore R_1, for the case when P_2 is at x, is equal to Y_1 plus Y_2, and it follows that the shear just to the left of

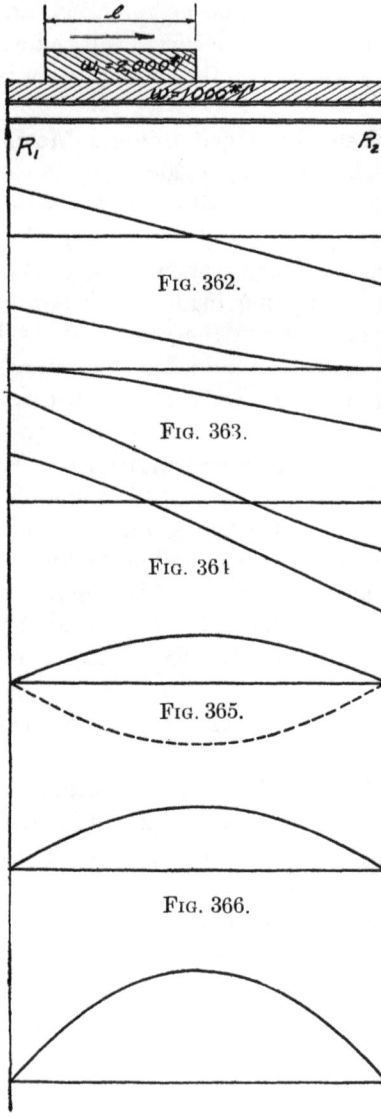

Fig. 361.

Fig. 362.

Fig. 363.

Fig. 364.

Fig. 365.

Fig. 366.

Fig. 367.

P_2 is also equal to Y_1 plus Y_2, which value is laid off in Fig. 371 as Y, and the point u_1 located.

When P_2 is at any other section of the beam, say at x_1, R_1 is equal to Y'_1 plus Y'_2, and the shear just to the left of P_2 is also equal to Y'_1 plus Y'_2. This value is laid off in Fig. 371 as Y', and the point u_2 located. When P_2 is at various other sections along the beam, other points $u_3, \ldots u_n$ are located in Fig. 371 and the line r–u_n–n is drawn. Notice that when P_1 passes off the right end of the beam, a break or change in direction occurs at u_n. Now the vertical intercepts at any section between the lines m–n and r–u_n–n give the shear which will exist at the left of P_2 when P_2 is at the section.

When P_1 is at x_2 the reaction at the right end of the beam due to P_2 is equal to the intercept w_1, while the reaction at the right end from P_1 is given by the intercept w_2. Therefore R_2, for the case when P_1 is at x_2, is equal to w_1 plus w_2 and, since the shear just to the right of P_1 is always equal to R_2, w is made equal to w_1 plus w_2, and the point v_1 is located in Fig. 371. When P_1 is at other sections of the beam, other points $v_2 \ldots v_n$ are located in Fig. 371, and the line m–v_n–s drawn. The intercept between this line and m–n at any section gives the shear just to the right of P_1 when P_1 is at the section.

Now the shear just to the left of P_1 is always equal to R_2 less P_1, or in other words, it differs from the shear just to the right of P_1 by P_1. In Fig. 371 the upper dotted line is drawn at a distance P_1 above the line m–v_n–s. The intercept between this and the base line m–n at any section gives the shear just to the left of P_1 when P_1 is at the section. In a similar way, the lower dotted line is drawn at a distance equal to P_2 below the line r–u_n–n, and the vertical intercept between this line and the base line m–n at any section gives the shear just to the right of P_2 when P_2 is at the section. These dotted lines are found to lie entirely between the lines r–u_n–n and m–v_n–s, therefore the lines r–u_n–n and m–v_n–s are the maximum shear curves for the given loading. When one of the loads is very small one of the dotted lines in Fig. 371 may cross one of the full lines. The part of the dotted line outside of the full line will be a part of one of the maximum shear curves.

The maximum moment curve for P_2 is shown in Fig. 372, and that for P_1 in Fig. 373. The maximum moment which can

Fig. 368.

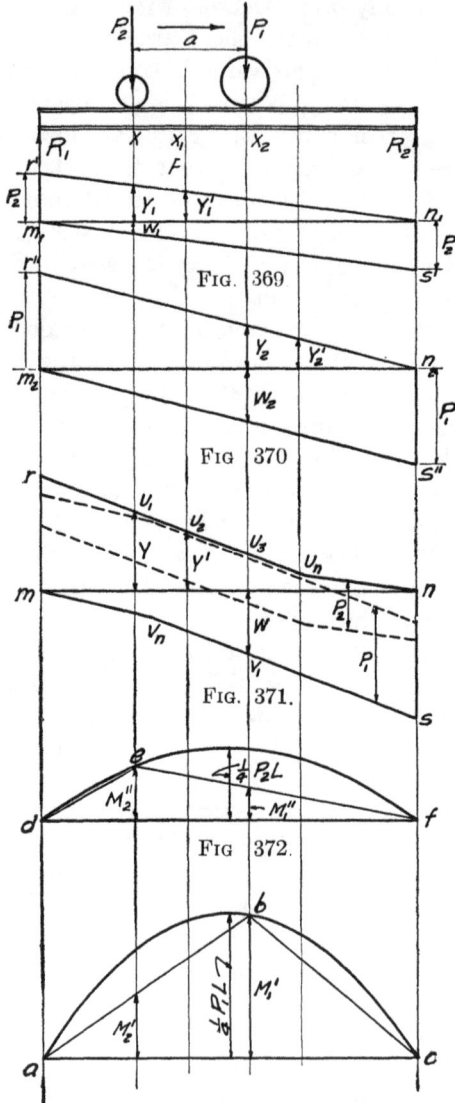

Fig. 369

Fig. 370

Fig. 371.

Fig. 372.

Fig. 373.

be produced at any section of the beam will occur when one of
the loads is at the section.   When P_2 is at section x, the moment
produced there by P_2 is equal to the intercept M''_2.   Also, when
P_2 is at x, the moment diagram for P_1 is the triangle a–b–c in
Fig. 373, and the moment produced at section x by P_1 is given

FIG. 374.

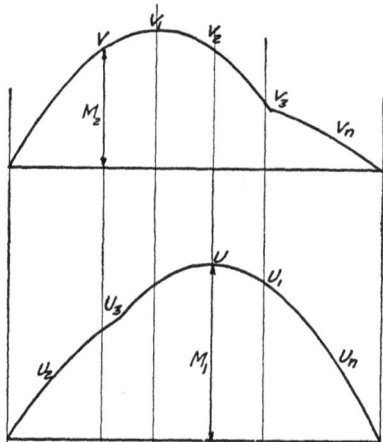

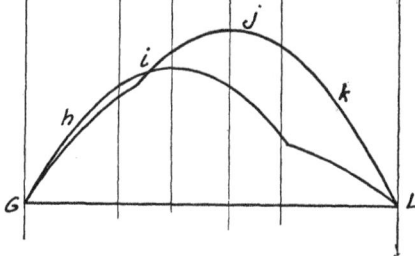

FIG. 375.

FIG. 376.

by the intercept M'_2.   It therefore follows that M''_2 plus M'_2
gives the total moment produced at section x when P_2 is there.
This value is laid off to scale as M_2 in Fig. 374, and the point v
located.   When P_2 is at various other sections the moment produced at each of these sections is found as just explained, and the
corresponding points v_1–v_2 . . . v_n are located in Fig. 374.   Through
these points a curve is drawn as shown.   Now the vertical inter-

cept between this curve and the base line at any section will give
the moment which exists at that section when P_2 is there.

When P_1 is at section x_2, the moment produced at this section
by P_1 is given by the intercept M'_1 in Fig. 373, while the moment
diagram for P_2 is the triangle d–e–f, and the moment produced
at section x_2 by P_2 is given by the intercept M''_1. It therefore
follows that the total moment produced at section x_2, when P_1
is there, is M'_1 plus M''_1. This value is measured off in Fig.
375 as M_1 and the point u located. When P_1 is at other sections
of the beam, the total moment under P_1 is found and the points
$u_1, u_2, \ldots u_n$ are located in Fig. 375. Through these points a
curve is drawn as in Fig. 374. The vertical intercept, in Fig.
375, at any section gives the moment which is produced at this
section when P_1 is there.

It should be noted that there is a break at v_3 in the curve of
Fig. 374, when P_1 passes off the right end of the beam, and also
one at u_3 in Fig. 375, when P_2 passes off the left end of the beam.
In Fig. 376, the curves of Figs. 374 and 375 are shown, drawn on
the same base line, and the curve G–h–i–j–k–L, given in part by
the curve of Fig. 374 and in part by that of Fig. 375, is the maximum moment curve for the given loading. That is, the maximum
moment that can be produced at any section is given by the
vertical intercept between the base line and the curve G–h–i–j–k–L
at the section. For sections over to i this maximum moment
occurs when P_2 is over the section, but for sections to the right
of i, the maximum occurs when P_1 is at the section. The student
should remember that the same scale must be used in Figs. 369,
370 and 371, also that the scale used in Fig. 372 should be used
in Figs. 373 to 376.

96. Three Concentrated Moving Loads.—When three concentrated moving loads with fixed distances between them are
considered, the method shown in Figs. 368 to 376 may be used,
but the construction would be rather complicated. A different
method is shown in Figs. 377 to 380.

Along the load line in Fig. 378 the three loads P_1, P_2, and P_3
are laid off to some convenient scale, and the pole p, with any
desirable pole distance H, is chosen. Using this pole p, the funicular polygon e–g–h–i–k is drawn in Fig. 379. When the loads
are at the position shown in Fig. 377, the moment diagram is
f–g–h–i–j–f, and the moment produced at any section is given by

the vertical intercept between the closing line f–j and the funicular polygon f–g–h–i–j, times the pole distance H. Now if the
loads move to another position, the moment along the beam will
vary, and a new moment diagram might be drawn. But to draw
this new moment diagram a new funicular polygon would have
to be drawn. In order to avoid this, it is found convenient in
this construction to move the beam in place of the loads. When
the beam is moved a distance x to the left, the same moments
are produced as would be obtained by moving the loads the same

FIG. 377.

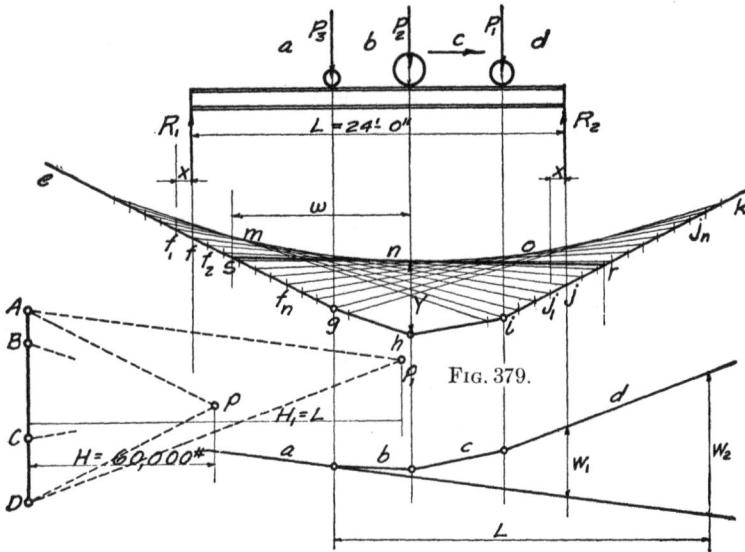

FIG. 379.

FIG. 378.　　　　　　　　　　FIG. 380.

distance to the right while the beam remained stationary. The
advantage of moving the beam is that the same funicular polygon
can be used, a new closing string giving the new moment diagram.
When the beam in Fig. 377 is moved to the left a distance x, R_1
is at f_1 and R_2 is at j_1, and the new closing string f_1–j_1 completes
the moment diagram f_1–g–h–i–j_1–f_1 for the new position of the
beam. The beam is moved to various other positions with respect
to the loads, and the corresponding closing strings drawn. Each
one of these closing strings completes a moment diagram, which
is the diagram for the beam when the loads P_3, P_2, and P_1 are

directly above g, h, and i, respectively, and the ends of the beam are directly above the ends of the closing string.

Let a smooth curve $m–n–o$ be drawn tangent to the various closing strings. Now the maximum vertical intercept between the curve $m–n–o$ and the funicular polygon $e–g–h–i–k$, times the pole distance H gives the maximum moment that can be produced by the given loads. At the top of this maximum vertical intercept Y, which is here found to be directly under P_2, let a tangent be drawn to the curve $m–n–o$ and extended until the funicular polygon $e–g–h–i–k$ is intersected at r and s. These intersections locate the ends of the beam with respect to the loads, for the case when the maximum moment occurs. In other words, the maximum moment that can be produced by the given loading is equal to the maximum vertical intercept Y, measured to the scale at which the beam was drawn, times the pole distance H, and it occurs under P_2 when the distance from P_2 to R_1 is w. In case the maximum moment that can be produced at any given section of the beam is desired, it may be found by a method which will be explained in connection with Fig. 383. In order to illustrate how R_1 may be found, Fig. 380 is drawn from Fig. 378, using the pole p_1 which has a pole distance H_1 equal to the length of the beam and measured to the scale at which the beam was drawn. Fig. 380 is simply a funicular polygon for the given loads with the a string extended back to the right. Now if we take the intercept w_1 and multiply it by the pole distance H_1, the first moment of the loads about R_2 will be obtained for the case when they are in the position on the beam shown in Fig. 377. If this moment is divided by the length of the beam, the reaction R_1 will be obtained. Usually a pole distance is measured in pounds to the scale at which the force polygon was drawn, and an intercept in a funicular polygon in feet to the scale of the space diagram. But the same numerical value would be obtained if the pole distance were measured in feet to the scale at which the beam was drawn, and the intercept in pounds to the scale at which the force polygon was drawn. Now as stated above, R_1 is equal to $\dfrac{H_1 \cdot w_1}{L}$, but H_1 was made equal to L to the scale at which the beam was drawn. Therefore when H_1 is measured in feet to the scale at which the beam was drawn, H_1 and L cancel, and w_1 measured in pounds to the scale at which the loads were laid off in Fig. 378, gives R_1 direct.

Suppose the reaction R_1 is desired for the case when P_3 is just to the right of the left support. When P_3 is just to the right of R_1, R_2 will be a distance L to the right of P_3. Then the vertical intercept w_2 is taken at a distance L to the right of P_3, and this intercept, measured to the scale at which the loads were laid out, in Fig. 378, gives R_1. The methods of obtaining the maximum reactions and the maximum shear for various sections will be explained in connection with Figs. 382 and 384.

97. Four Concentrated Moving Loads and a Uniform Dead Load.—Fig. 381 shows a 40-foot beam which supports a uniform dead load of 1000 pounds per foot in addition to four concentrated moving loads with fixed distances between them. The same method that was used in solving the problem of Fig. 377 will be used here. The uniform load is divided into small slices and an equivalent concentrated load substituted for each division. These loads together with the concentrated moving loads are laid off in order along the load line of Fig. 382, beginning at A and extending down to Y. Any pole p with any convenient pole distance H is chosen and the funicular polygon in Fig. 383 is drawn from e' on around to f'. The closing string e'–f' completes the moment diagram for the case when the moving loads have the position on the beam shown in the figure.

Now when the beam is moved to other positions with respect to the moving loads, the dead uniform load moves with it, therefore the end strings of the funicular polygon cannot be extended to the left and to the right, as in Fig. 379. Additional divisions of uniform load must be represented by vectors laid off above A and below Y, and the funicular polygon continued to the left of e' and to the right of f', as shown in Fig. 383. The beam is now assumed to be shifted back and forth, and numerous closing strings are drawn, as in Fig. 379, and the curve m–n–o is drawn tangent to them. The maximum moment that can be produced is given by the maximum vertical intercept between m–n–o and the funicular polygon, times the pole distance H. The intercepts w and w' are found to be equal to each other but greater than any of the other intercepts. This means that when P_2 is at a certain position on the beam, the maximum moment that can be produced exists under it, and also that an equal maximum moment exists

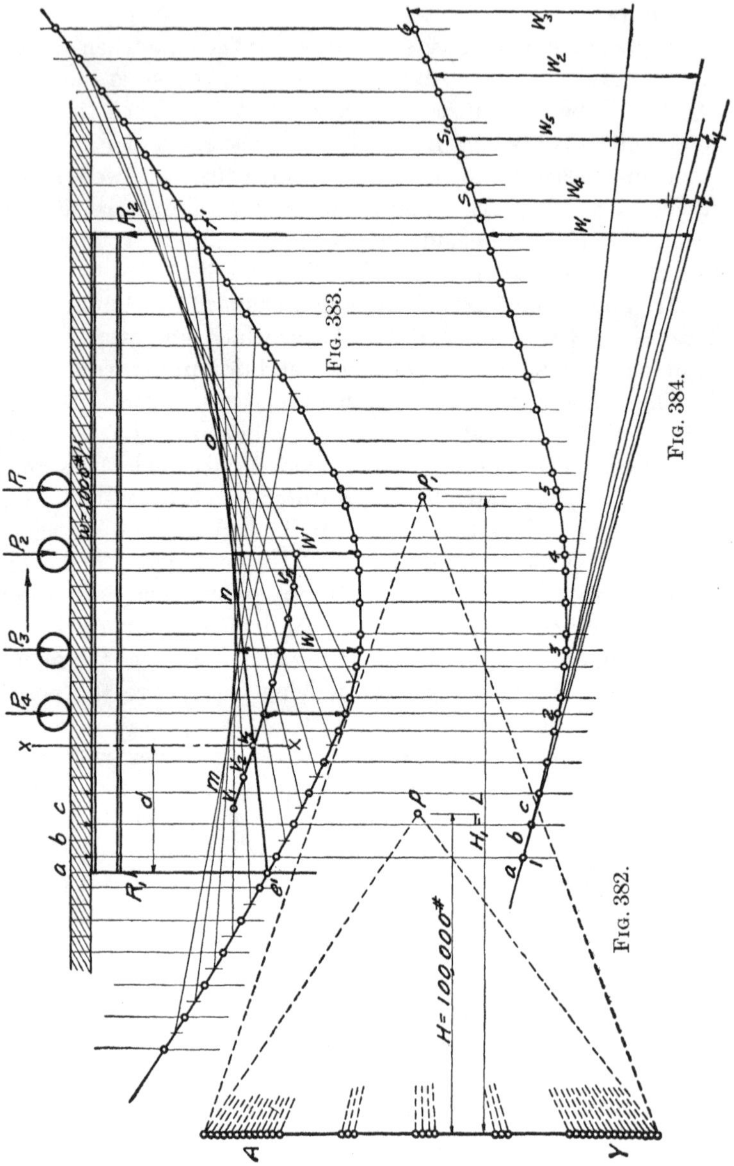

FIG. 381.

FIG. 382.

FIG. 383.

FIG. 384.

under P_3 when it is at some other section of the beam. The
position of the loads which will produce the maximum value of
M may be located by drawing tangents to the curve m–n–o at
the ends of the intercepts w and w', as explained in connection
with Fig. 379.

Suppose it is desired to find the maximum moment that can
be produced at any section x–x distant d from the left support.
The points $v_1, v_2 \ldots v_n$ are located on the various closing strings
so that each one of them is a distance d horizontally from the left
end of its closing string. The curve v_1–$v_2 \ldots v_n$ is now drawn,
and the maximum vertical intercept between this curve and the
funicular polygon, times the pole distance H gives the maximum
moment that can be produced at section x–x by the given loading.
This is true because every vertical intercept between the curve
v_1–$v_2 \ldots v_n$ and the funicular polygon, times H, gives the
moment at x–x for some particular position of the loads on the
beam, therefore the maximum intercept times H will give the
maximum moment that can be produced there.

Let it be required to find the maximum reaction that can be
produced at the left end of the beam. Any pole p_1 is chosen
with a pole distance H_1 equal to L to the scale at which the beam
was drawn, and the funicular polygon 1–2–3 . . . 6 is drawn
in Fig. 384. When the moving loads are at the position
on the beam shown in the figure, the load at the left end of
the beam is a–b, therefore the a string is extended back, and the
intercept w_1 measured to the scale at which the loads were laid
out in Fig. 382, gives R_1. The maximum value of R_1 and also
the maximum shear just to the right of R_1, will occur when one
of the concentrated loads is just to the right of R_1. When P_4
is just to the right of R_1, the string at the left of point 2 is extended
back to the right, and the intercept w_2 is taken at a horizontal
distance to the right of point 2 equal to the length of the beam.
This intercept w_2 gives R_1 for the case when P_4 is just to the right
of R_1.

When P_3 is just to the right of R_1 the string at the left of
point 3 is extended back to the right and the intercept w_3, which
is at a distance L to the right of P_3, gives the value of R_1. The
load P_2 might be placed just to the right of the left support and
R_1 found, if there were any reason to think that this position of
the loads would give a larger value for R_1. With the four moving

loads considered in this problem the maximum value of R_1 is found to be given by the intercept w_2, and is produced when P_4 is just to the right of the left support. Also the maximum shear that can be produced near the left support is equal to the maximum value of R_1.

Let it be required to find the maximum shear that can be produced at any section x–x distant d from the left support. The shear at this section is always equal to R_1 less the load between R_1 and the section, and the maximum shear will of course occur when one of the moving loads is just to its right. This is true because, as any one of the concentrated loads approaches the section x–x from the right, the value R_1 less the loads between R_1 and x–x increases; but when the load crosses to the left of section x–x, this value suddenly decreases. When P_4 is just to the right of x–x the load b–c is the last load at the left end of the beam, therefore the string b will be extended back to the right and the left reaction R_1 is given by the intercept s–t, since the horizontal distance between this intercept and P_4 is equal to L less d. The load between R_1 and the section amounts to 8000 pounds of uniform load, so the shear at section x–x when P_4 is just to its right is equal to the intercept s–t less 8000 pounds, and is given by the intercept w_4. When P_3 is just to the right of x–x, R_1 is given by the intercept s_1–t_1, while the loading between R_1 and x–x is 8000 pounds of uniform load plus the 20,000 pound concentrated load P_4. Therefore the shear at section x–x when P_3 is just to its right is given by the intercept w_5. Since w_5 is much smaller than w_4, and since P_2 and P_1 are no larger than P_3, it is evident that the maximum shear that can be produced at x–x is given by the intercept w_4 and is produced when P_4 is just to the right of the section.

98. A Large Number of Concentrated Moving Loads.—A railroad engine crossing a plate girder bridge is a typical example of a large number of concentrated moving loads. Fig. 385 shows one of the girders of a 70-foot bridge. The loads from the various engine wheels are laid off to some convenient scale along the load line of Fig. 386, the pole p is chosen, and the funicular polygon a–b–c . . . k is drawn in Fig. 387. Numerous closing strings are then drawn, and the curve m–n–o obtained as already explained. The maximum vertical intercept w is found to be under the third drive wheel, as shown, and its length measured in feet times the pole distance H gives the maximum moment that can be produced. This moment exists under P_4 when P_4 is to the right of R_1 a distance equal to the horizontal distance between s and the intercept w, the line s–t being drawn tangent to the curve m–n–o at the upper end of w.

In order to find the maximum moment that can be produced at any section x–x, the points v_1–v_2 . . . v_n are located on the various closing strings so that the horizontal distance between each one of these points and the right end of its closing string is just equal to the distance between the section x–x and the right reaction. The curve v_1–v_2–v_3 . . . v_n is then drawn, and the maximum vertical intercept between this curve and the funicular polygon a–b–c . . . k times the pole distance H gives the maximum moment that can be produced at x–x. The intercepts w_1 and w_2 are here found to be practically the same, therefore the maximum moment that can be produced at x–x exists when either P_2 or P_3 is at the section, and it is equal to either w_1 or w_2 times H.

In order to find the maximum shear that can be produced just to the left of the right support, the pole p_1 is chosen, in Fig. 386, with the pole distance H_1 equal to the span of the girder, and the funicular polygon in Fig. 388 is drawn, the last string at the right end being extended back to the left. The maximum value of R_2 and therefore of the shear just to its left will occur when one of the loads is just to its left. First let P_1 be placed just to the left of R_2. The value of R_2 for this position of the loads is given by the vertical intercept w_3, which is to the left of P_1 a horizontal distance of 70 feet, the span of the girder. When P_2 is just to the left of R_2, the load P_1 is off of the girder, therefore the string just to the right of P_2 is extended back to the left, and the value of R_2 is given by the intercept w_4. When P_3 is just to the left of R_2 the value of R_2 is given by the intercept w_5. Other intercepts should be obtained by placing other loads just to the left of R_2, unless it is evident that the maximum intercept has been passed. In this problem the maximum intercept is found to be w_4. This means that R_2 has its maximum value when P_2 is just to its left, and this value, which is also the maximum shear that can be produced just to the left of R_2, is found by measuring the intercept w_4 to the scale at which the loads were laid off along the load line of Fig. 386.

Fig. 389 was drawn from Fig. 386 in the same way that Fig.

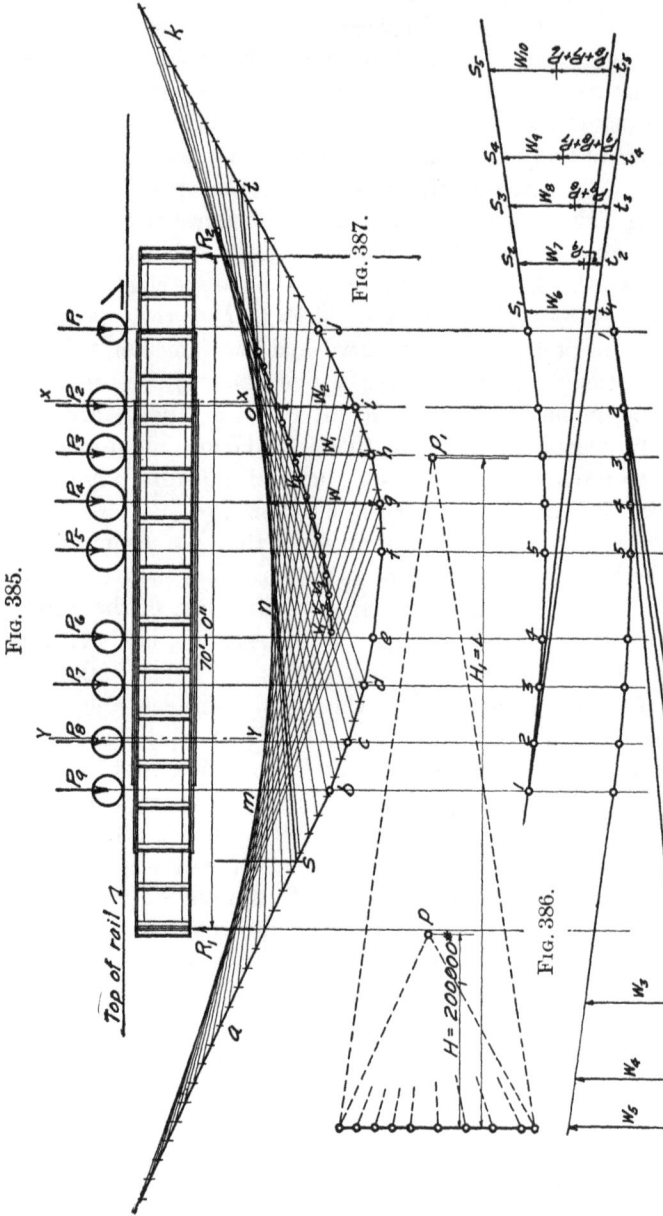

FIG. 385.

FIG. 386.

FIG. 387.

FIG. 388.

FIG. 389.

388 was drawn, the difference between the two figures is that in Fig. 388 the string at the right end is projected back to the left, while in Fig. 389 the string at the left end is projected back to the right.   Fig. 389 might be used to find the maximum value for R_1, which would give the maximum shear for the left end, but only the intercepts for finding the maximum shear at section $Y-Y$ are shown.   The maximum shear at $Y-Y$ will, of course, occur when one of the loads is just to its right.   When P_9 is just to the right of $Y-Y$, the value of R_1 is given by the intercept w_6, which is to the right of P_9 a horizontal distance equal to the distance between $Y-Y$ and R_2.   Since there are no loads between R_1 and $Y-Y$, the shear at section $Y-Y$ when P_9 is just to its right is equal to the intercept w_6.   When P_8 is just to the right of $Y-Y$, the value of R_1 is given by the intercept s_2-t_2 which is to the right of P_8 a horizontal distance equal to the distance between $Y-Y$ and R_2.   For this position of the loads the shear at $Y-Y$ is equal to R_1 less P_9, and is given by the intercept w_7.   In the same way the shear at $Y-Y$ when P_7, P_6, and P_5 are just to its right, is found to be w_8, w_9 and w_{10}, respectively.   When P_5 is just to the right of $Y-Y$, P_9 has passed off of the beam, therefore the string just to the left of P_8 is projected back to the right and the intercept s_5-t_5 taken to it.   The maximum of all of these intercepts is found to be w_6.   Therefore the maximum shear that can be produced at section $Y-Y$ is given by the intercept w_6, and it is produced when P_9 is just to the right of the section.

99. Maximum Shears and Moments in a Turntable.—An interesting problem is to find the maximum shears and the maximum moments that can be produced in a turntable when it is crossed by a locomotive.   A diagram of the turntable with its three supports is shown in Fig. 390, and the engine loads are laid off to scale along the load line of Fig. 391.   Using any pole p, the funicular polygon $K-L-M \ldots U$ is drawn in Fig. 392, and the outside strings are extended back towards the center until they meet at I, locating the resultant of all of the loads.   When this resultant is between R_2 and R_3 these two supports will take bearing, and the left end will be an overhanging end; but when this resultant is between R_1 and R_2 the support at the left end, and the one in the middle are supposed to take bearing, while the right end is an overhanging end.   It is reasonable to consider that the turntable acts in this way with only two supports taking bearing at one time, because the center support is usually of such an elevation that the turntable cannot be in contact with both end supports at the same time.

When the loads are in the position on the turntable shown in Fig. 390, the moment diagram is L–M–N ... T–H–J–L, the closing string being J–H. This is the moment diagram for an overhanging beam, the intercepts below the polygon K–L–M–N ... U giving negative moments, and the intercepts above the polygon giving positive moments. The largest negative moment occurs at R_2. If the turntable is moved a distance x to the left, the closing string becomes H_1–J_1 and the moment diagram is L–M–N–O ... T–H_1–J_1–L. The turntable is now moved to other positions with respect to the loads and the other closing strings shown are drawn, the resultant of all of the loads being kept between R_2 and R_3. The maximum negative moment is found to be equal to the intercept w times the pole distance H, and it occurs when the center support R_2 is directly under the resultant of all of the loads. The maximum positive moment that can be produced between R_2 and R_3 is equal to H times the maximum vertical intercept above the polygon K–L–M ... U. This maximum vertical intercept is found to be w_1 under the load P_4. The section of the turntable at which the maximum positive moment w_1 times H occurs, may be located in a way similar to that used in Fig. 387.

The funicular polygon in Fig. 393 is the same as the one in Fig. 392, except that the closing strings have been drawn for the case when the resultant of all of the loads falls between R_1 and R_2. The maximum positive moment that can be produced between R_1 and R_2 is equal to H times w_2, the maximum vertical intercept above the polygon, while the maximum negative moment is found to be H times w_3, the length w_3 being the same as that of w in Fig. 392.

Let it be required to find the maximum moment that can be produced at any section Y–Y, distant d to the left of R_2. Various points v_1–v_2–v_3 ... v_n are located on the closing strings in Fig. 393, so that each of these points is to the left of the right end of its closing string a horizontal distance equal to d. The curve v_1–v_2–v_3 ... v_n is now drawn and the maximum negative moment that can be produced at Y–Y is found to be w_4 times H,

while the maximum positive moment that can be produced there
is w_5 times H.

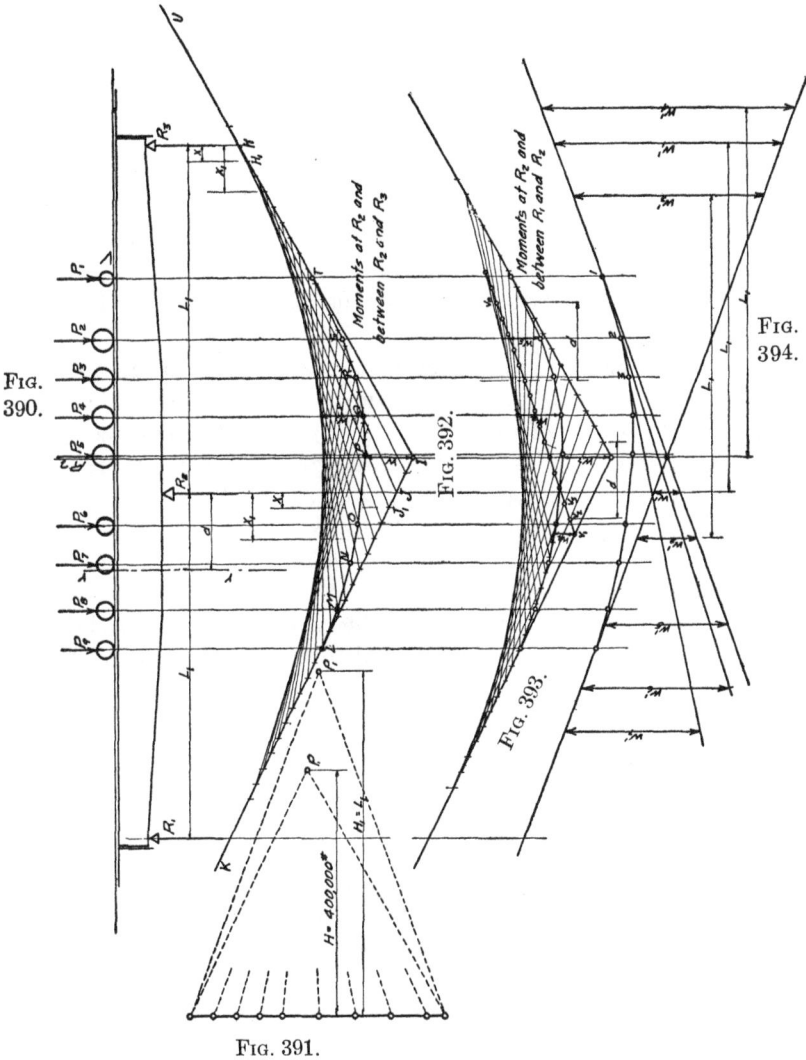

FIG. 390.

FIG. 391.

FIG. 392.

FIG. 393.

FIG. 394.

The reactions may be conveniently found by the use of Fig.
394. In Fig. 391 the pole p_1 is chosen with the pole distance H_1
made equal to L_1, the distance between R_1 and R_2 and between

R_2 and R_3. Using this pole p_1 the funicular polygon in Fig. 394 is drawn with the end strings extended towards the center and crossed as shown.

When the loads are in the position on the turntable shown in Fig. 390, R_3 is given by the intercept w' and R_2 is given by the intercept w'_1, both measured to the scale at which the loads were laid out along the load line of Fig. 391. The horizontal distance between the two intercepts w' and w'_1 is L_1, w' being directly under R_2 and w'_1 under R_3. If the turntable is moved to the left a distance x_1, R_3 is given by the intercept w'_2 and R_2 is equal to the intercept w'_3. The largest possible value of R_2 exists when R_2 is directly under the resultant of all of the loads. It is, of course, equal to the sum of all of the load, and is given by the intercept w'_4.

Suppose the maximum shear that can be produced just to the left of R_3 is desired. This maximum will exist when one of the loads is just to the left of R_3. When P_1 is just to the left of R_3, the value of R_3 is given by the intercept w'_5; when P_2 is just to the left of R_3 the value of R_3 is given by the intercept w'_6; and when P_3 is just to the left of R_3 its value is given by the intercept w'_7. It is now evident that the maximum value of R_3 and therefore of the shear just to the left of R_3 exists when P_2 is just to the left of R_3, this maximum value being given by the intercept w'_6. Fig. 394 may also be used to find the maximum shear that can be produced at any section of the turntable.

100. Moving Loads on Trusses.—Graphical construction can be used for finding the stresses that are produced in a bridge truss, of the type shown in Fig. 395, by engine and train loads as they move across it. In Fig. 395 the engines are shown moving across towards the right followed by a uniform train load. The engine loads, and also the train loads, are laid off along the load line of Fig. 396, and using any pole p, the funicular polygon a–b–c–d–e in Fig. 397 is constructed. Then the various closing strings shown in the figure are drawn.

It is evident that the stress in the member U_1–U_2 will be equal to the stress in the member L_2–L_3, but of different sign, and that the maximum value of this stress will be equal to the maximum moment that can be produced at section x–x divided by h, the vertical distance between chords. The points v–v_1–$v_2 \ldots v_n$ are located on the various closing strings so that each

one of these points is to the right of the left end of its closing
string a horizontal distance equal to the same distance that the
section x–x is to the right of R_1. The curve v–v_1–v_2 . . . v_n is now
drawn, and the maximum vertical intercept between this curve and
the funicular polygon, measured to the scale at which the beam
was drawn and multiplied by the pole distance H, gives the
maximum moment that can be produced at the section x–x.
This moment divided by the distance between chords, which
is 25 feet for this truss, gives the magnitude of the maximum
stress that can be produced in U_1–U_2 and in L_2–L_3. Now the
truss was drawn to a scale of 1 in. = 20 ft., and H in Fig. 396 was
made equal to 500,000 pounds, therefore, if the maximum intercept
between v–v_1 . . . v_n and the funicular polygon is measured to
the scale of 1 in. = 10,000,000 foot pounds, the maximum moment
will be obtained direct. Also, if this intercept were measured
to a scale of 1 in. = $\dfrac{10,000,000}{25}$ = 400,000 pounds, the stress in U_1–U_2
would be obtained direct. The maximum stress that can be
produced in U_2–U_3 can be found in a similar way by the use of
the curve u_1–u_2 . . . u_n, which is drawn for the section x_1–x_1
in the same way that v_1–v_2 . . . v_n was drawn for section x–x.
The maximum stress is found by scaling the maximum intercept between the curve u–u_1 . . . u_n and the funicular polygon,
to the scale of 1 in. = 400,000 pounds. This scale is obtained by
multiplying the scale at which the truss was drawn by H divided
by h. The maximum stress that can be produced in other chord
members can be found in a similar way.

The maximum stress that can be produced in the end post
L_9–U_8 and in the two end panels L_7–L_9 of the lower chord will
exist when R_2, the reaction at the right end of the truss, has its
maximum value. The pole p_1 is chosen with a pole distance
H_1 equal to L, and the funicular polygon of Fig. 398 is drawn
with the last string at the right extended back to the left. When
the load P_1 is at L_8, R_2 is given by the intercept f–f'. When
the load P_1 moves to the right, a part of its weight will be carried
to the pier by the floor beams, and this part will not affect the
reaction at the end of the truss or the stresses in the truss. In
Fig. 399, a little force polygon is drawn for the first five loads
using the scale of Fig. 396 and a pole distance H equal to the
length of the floor beams in the panel L_8–L_9. From this force

FIG. 395.
FIG. 396.
FIG. 397.
FIG. 398.
FIG. 399.
FIG. 400.
FIG. 401.
FIG. 402.

polygon the funicular polygon in Fig. 400 is drawn, the right string being extended back. Now when P_2 is at L_8, the portion of P_1 going to the pier through the floor beams is given by the intercept w. Therefore when P_2 is at L_8 the reaction at the right end of the truss is given by the intercept $g-g'$. When P_3 is at L_8, R_2 is given by the intercept $h-h'$, and when P_4 is at L_8 it is given by the intercept $i-i'$. When P_1 is at R_2, R_2 is given by the intercept $j-j'$, and just after P_1 has passed off of the truss R_2 is given by the intercept $k-k'$. The intercepts $l-l'$ and $m-m'$ give R_2 for the case when P_5 is at L_8 and when P_2 is at R_2 respectively. The maximum of all of these intercepts which is here found to be $i-i'$ is the maximum value of R_2, and the maximum stress for U_8-L_9 is given by the line $i'-n$, while the maximum stress for L_7-L_9 is given by $i-n$, both lines to be measured to the scale used in Fig. 396.

The maximum stress in U_8-L_7 will occur when the maximum shear exists in this panel. The shear which exists in this panel for any position of the loads is equal to R_2 less all of the loads to the right of L_8 and less that part of the loads between L_7 and L_8 which is brought to L_8 by the floor beams. When P_1 is at L_7 this shear is given by the intercept $f_1-f'_1$, and when P_2 is at L_7 it is given by $g_1-g'_1$. Other points h_1, j_1, i_1 and k_1 may be located until it is evident that the maximum shear is given in this case, by the intercept $h_1-h'_1$. Then the maximum stress in U_8-L_7 is given by the line h'_1-n_1 measured to the scale of Fig. 396. In a similar way the maximum stress for L_6-U_7 is found to be given by the line h'_2-n_2. The stress in U_7-L_7 is, of course, just equal to the vertical component of the stress in U_7-L_6. Therefore the maximum stress U_7-L_7 occurs when L_6-U_7 is maximum, and is given by the line h'_2-h_2. The maximum stress for any other diagonal and for any other vertical, except for U_8-L_8 and for U_1-L_1, can be found in a way similar to what is given above.

The maximum stress that can be produced in U_8-L_8 is found by the use of Figs. 401 to 403. The stress in U_8-L_8 is just equal to the reactions at L_8 from the floor beams between L_8 and L_9 and between L_7 and L_8. The force polygon for a number of the heavy engine loads is drawn in Fig. 401, from which the funicular polygon of Fig. 402 is drawn. These loads are assumed to be placed at various positions with respect to L_8, and the closing strings drawn in Fig. 402, parallel to which the corresponding closing rays are drawn in Fig. 401. The distance marked off along the load line in Fig. 401 between the two closing rays for any given position of the loads, gives the total reaction at L_8, and therefore the stress in U_8–L_8 for that position of the loads. The maximum of these distances gives the maximum stress in U_8–L_8.

The variation of these distances, as the position of the loading changes, is shown by the intercepts in Fig. 403, the maximum of which is Y directly under P_4. Then the maximum stress for U_8–L_8 occurs when P_4 is at L_8, and is equal to the intercept Y.

Since the truss is symmetrical, it is necessary to find the maximum stress for the members in only one-half of the truss, unless there is reason to believe that greater stresses might be produced in some of the members if the engine and train were turned end for end. In such a case it may be desirable to find the maximum stresses that can be produced in each member by the engine and train crossing in one direction. Then if the maximum for any one of two corresponding members, such as U_1–U_2 and U_7–U_8, or U_2–L_3 and U_7–L_6, is greater than that of the other, this greater value may be taken for the maximum of each one.

CHAPTER VII

MASONRY

It is very interesting to study the stresses which exist in various parts of a masonry structure, and to consider how they may vary under different loads. In some cases the determination of the maximum unit stresses and the distribution of stress is rather simple, but there are other problems in which it is decidedly difficult. In the solution of these more difficult problems the graphical methods are of special value and they can often be used to advantage in the solution of the simplest problems also.

Ordinary masonry, not reinforced, is usually weak and unreliable in tension, so it is customary when analyzing the stresses in masonry to assume that it can take no tension, that the joints would open up rather than take tension.

101. Stresses in Rectangular Piers.—Fig. 404 shows a masonry pier of rectangular cross-section, with P, the resultant of all the loads, acting at the centroid of the cross-section. In this case the distribution of stress over the cross-section will be uniform and the magnitude of the unit stress S will be $P \div A$, A being the area of the cross-section.

In Fig. 405, the same pier is shown, but P the resultant of all the loads has moved a small distance e away from the centroid. The average stress S_a is of course equal to $P \div A$, but since P acts at the left of the centroid the distribution of stress will not be uniform. The unit stress on the left side of the pier will be greater than the average, and there will be a uniform decrease towards the right until at the right side the unit stress will be less than the average. The moment produced by the eccentricity of P is P times e. This moment is responsible for the increase of stress on the left side and the decrease on the right side. The amount of this increase S_1 may be found by the use of the formula $\dfrac{M}{S} = \dfrac{I}{c}$, which may here be written $S_1 = \dfrac{Pec}{I}$, in which I is the moment of inertia about the axis x-x.

The maximum unit stress S_M is therefore equal to the average unit stress S_a plus the increase S_1 caused by the eccentricity of P, and it may be expressed by the following formula:

$$S_M = \frac{P}{A} + \frac{Pec}{I}.$$

For a rectangular section, the decrease S_2 is equal to the increase S_1; therefore the minimum unit stress S_m is given by the formula $S_m = \frac{P}{A} - \frac{Pec}{I}$. The above formulas should be used only when S_2 is less than S_a, as will soon be evident.

In Fig. 406, the eccentricity has increased until the unit stress on the right side has been reduced to zero. For finding the maximum and minimum stresses, S_M and S_m, the formulas given

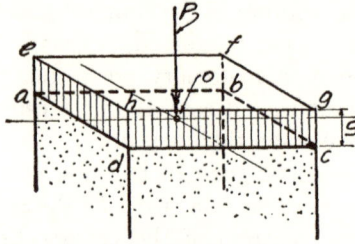

Fɪɢ. 404.

above may be used, but there is an easier way for this particular case; since $S_1 = S_2 = S_a$ the maximum stress S_M is just twice the average S_a and the minimum is zero.

Suppose the value of e is desired for the case when the stress at the right side of the pier is zero. The expression for S_m is $\frac{P}{A} - \frac{Pec}{I}$, and when this is equal to zero $\frac{P}{A} = \frac{Pec}{I}$, and it follows that $e = \frac{I}{Ac}$.

In Fig. 407 the eccentricity e is still greater, and if the material could take tension, a small portion of the right side would have tension as shown. If this tension could really be carried, S_M would be equal to $\frac{P}{A} + \frac{Pec}{I}$ and S_m would equal $\frac{P}{A} - \frac{Pec}{I}$. But since masonry is weak and unreliable in tension, and it is customary to assume that it can take no tension, the distribution of stress shown in Fig. 407 is not generally used for masonry.

Fig. 408 shows the distribution of stress which is assumed to exist when the eccentricity is large. A part of the section carries no stress, the joints tending to open up. When P acts on the neutral axis $Y-Y$, the stress volume is wedge shaped, as shown,

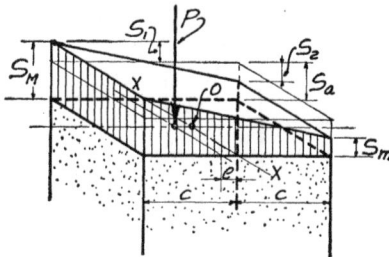

Fig. 405.

and of course the resultant load P passes through the centroid of this stress volume.

The centroid of a wedge is at the third point, therefore if the distance from the left face of the pier to the action line of P is d, then the distance from the left face to the axis of zero stress

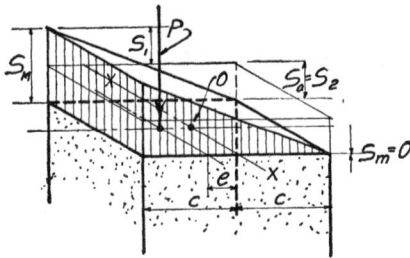

Fig. 406.

$Z-Z$ is $3\,d$. The area carrying stress is equal to the width of the pier times $3\,d$, equals A_1. The average stress is $\dfrac{P}{A_1}$, and, since the maximum stress, S_M, is twice the average, $S_M = \dfrac{2P}{A_1}$. The student should remember that A_1 is not the cross-sectional area of the pier, but rather that part of the cross-sectional area which carries stress.

102. Stress Volumes.—The term stress volume is here applied to a volume used to represent the distribution and intensity of

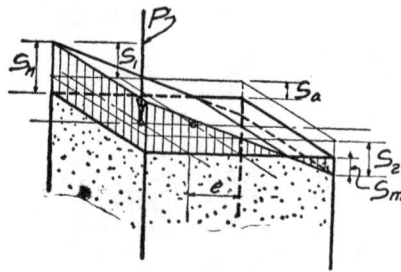

Fig. 407.

stress over a certain cross-section. In Fig. 404 the cross-section of the pier a–b–c–d is the base of the stress volume a–b–c–d–e–f–g–h. The base of a stress volume is always measured in square inches or square feet, but the dimension normal to the base may be measured in pounds per square inch or per square foot, or it may be measured in inches or feet. When the base is measured in square inches and multiplied by the average dimension normal

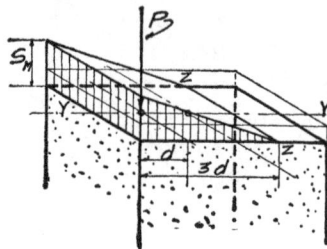

Fig. 408.

to it, measured in pounds per square inch, the content of the volume is obtained in pounds; and the content should be numerically equal to P, since the stress volume represents the stresses which just balance P. When the base is measured in square inches and multiplied by the average normal dimension measured in inches, the content of the volume is obtained in cubic inches. Each one of these cubic inches represents a certain number of pounds such that, when the content in cubic inches is multiplied by this number of pounds per cubic inch, a value numerically equal to P will be obtained.

The action line of P always passes through the centroid of the stress volume representing its balancing stresses because the resultant of all of these balancing stresses must have the same action line as P in order to produce equilibrium. It also follows that at any place over the section the dimension of the stress volume normal to its base, measured in pounds per square inch to the proper scale, gives the unit stress at that place.

In Fig. 404 the vertical dimension of the stress volume is constant, and is equal to S, which in turn is equal to $\dfrac{P}{A}$. In Fig. 405 the vertical dimension on one side of the stress volume is greater than on the other with intermediate values between. In Fig. 406 the stress volume is wedge shaped, and in Fig. 407 there are two wedge-shaped stress volumes, the one below the base line being negative. The content of the large positive stress volume less the content of the small negative stress volume should be equal to P, and P should pass through the centroid of the total volume, the smaller volume being considered negative in locating this centroid.

103. Problems, Rectangular Piers.—Consider the rectangular pier shown in Fig. 410 with a cross-section 15 by 24 and with $P = 27,000$ pounds. When P acts at the centroid o, there is a uniform distribution of stress and the unit stress equals $\dfrac{P}{A} = \dfrac{27,000}{15\times24} = 75$ pounds per square inch. When P moves toward the left along the axis Y–Y to the point o_1 its eccentricity is $2''$ and the maximum unit stress S_M is given by the formula

$$S_M = \frac{P}{A} + \frac{Pec}{I} = \frac{27,000}{15\times24} + \frac{27,000 \times 2 \times 12}{\frac{1}{12}\times15\times24^3} = 75 + 37.5$$
$$= 112.5 \text{ pounds per square inch.}$$

In like manner, the minimum unit stress S_m may be obtained by the use of the formula

$$S_m = \frac{P}{A} - \frac{Pec}{I} = \frac{27,000}{15\times24} - \frac{27,000 \times 2 \times 12}{\frac{1}{12}\times15\times24^3} = 75 - 37.5$$
$$= 37.5 \text{ pounds per square inch.}$$

When P acts at o_2 the eccentricity is 4 in. and $S_M = \dfrac{P}{A} + \dfrac{Pec}{I}$
$= 75 + 75 = 150$ pounds per square inch, and S_m is found to be zero.　In this case, since the stress along the edge a–b is zero, the maximum unit stress S_M could be found by using the formula

$$S_M = \frac{2\,P}{A} = \frac{2 \times 27{,}000}{15 \times 24} = 150 \text{ pounds per square inch.}$$

When P acts at o_3 the eccentricity is $6''$, also the distance from the f–d edge of the pier to o_3 is $6''$.　Therefore the stress volume is a wedge, and the distance over to the line of zero stress z–z is $3 \times 6 = 18$ inches.　The maximum unit stress S_M is given by the formula $\dfrac{2\,P}{A_1} = \dfrac{2 \times 27{,}000}{15 \times 18} = 200$ pounds per square inch.

104. Problems, Irregular Piers.—A cross-shaped pier is shown in Figs. 411 and 412, the dimensions are given and the resultant load P is 80,000 pounds.　When this resultant load acts at o, the centroid of the section, there is a uniform distribution of stress over the pier and the unit pressure is equal to $\dfrac{P}{A} = \dfrac{80{,}000}{500}$ $= 160$ pounds per square inch.

When P moves toward the right over to o_1 the eccentricity is $2''$, and there will be an unequal distribution of stress.　The maximum stress S_M will be given by the formula

$$\frac{P}{A} + \frac{Pec}{I} = S_M = \frac{80{,}000}{500} + \frac{80{,}000 \times 2 \times 15}{24{,}166} = 160 + 99 =$$
$$259 \text{ pounds per square inch.}$$

The minimum unit stress S_m equals $\dfrac{P}{A} - \dfrac{Pec}{I} = 160 - 99 = 61$ pounds per square inch.

When P has moved over to the point o_2 the eccentricity is 7 in.　In this case the eccentricity is so great that there will be compression over only a part of the pier section.　But the location of the axis of zero stress is not known and it cannot be found as in Figs. 408 and 410, because the cross-section of the pier considered is not a rectangle.　The highest part of the stress volume will be at the right side of the pier, and there will be a uniform decrease towards the left until the axis of zero stress z–z is reached.

In Fig. 413, draw the vertical line a–b directly below the right edge of the pier and from a the inclined line a–d. Now the plan of the stress volume is that portion of the cross-section of the pier to the right of the line z–z, and its elevation is the triangle a–x_1–x_2 in Fig. 413. The problem resolves itself into locating the base line x_1–x_2 and from it the line z–z. To start with, we know that when the location of x_1–x_2 is found the vertical line

FIG. 409.

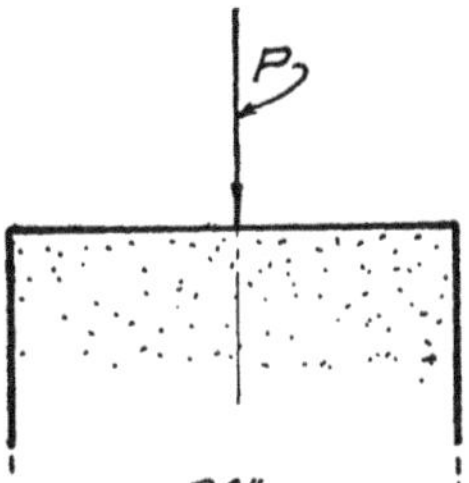

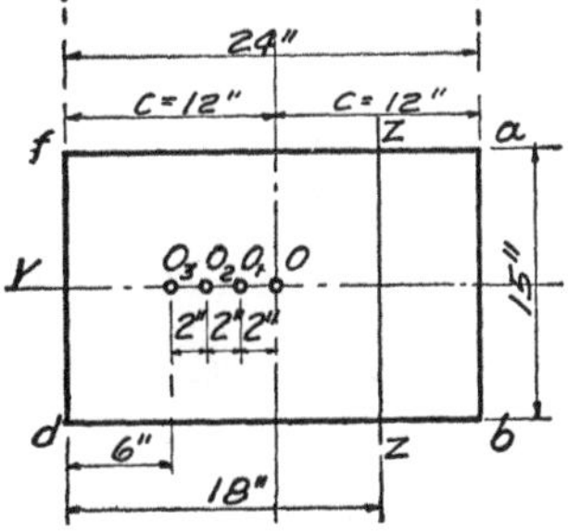

FIG. 410.

from o_2 will pass through the centroid of the volume above x_1–x_2. The vertical line passing through the centroid of the volume above the trial base line e–f intersects this base line at c, and the content of this volume above e–f, measured in cubic inches to the scale of Fig. 412, is found to be 500 cubic inches. The line f–f' is made equal to this value to some convenient scale.

When the trial base is taken at e_1–f_1 a peculiar shaped volume is obtained. With the trial base line at e–f the volume above is wedge shaped and c is at the third point between e and f, but the point c_1 cannot be located so easily. It will be convenient to divide the volume above e_1–f_1 into two parts by a vertical

Fig. 411.

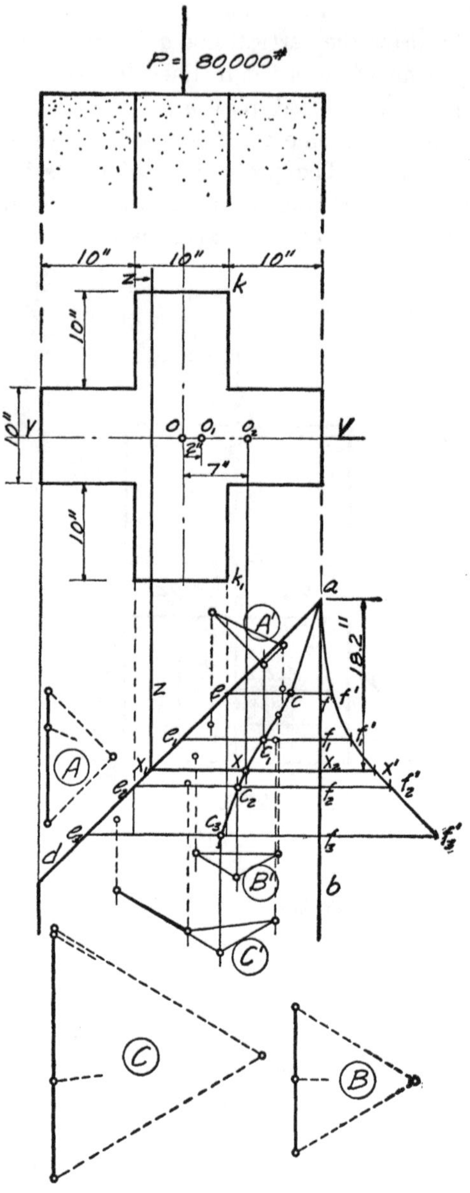

Fig. 412.

Fig. 413.

plane k–k_1. The left division will be a wedge, while the right will be a square in plan and a trapezoid in elevation and its centroid will be directly back of the centroid of its elevation. The centroids of these divisions are found, and then from them, using the polygons Ⓐ and Ⓐ' in Fig. 413, a vertical line containing the centroid of the volume above e_1–f_1 is located and the intersection c_1 obtained. The line f_1–f_1' is now drawn representing the number of cubic inches above e_1–f_1 to the same scale that f–f' represents the number of cubic inches above e–f.

When the base line is at e_2–f_2, the vertical line containing the centroid and also c_2, is located by the use of polygons Ⓑ and Ⓑ'. The point c_3 is located in a similar way by the use of polygons Ⓒ and Ⓒ'. The curve a–c–c_1–c_2–c_3 is now drawn, and when a vertical line is drawn from o_2 it is found to intersect this curve at x, thus locating the base line x_1–x_2 and from it the line of zero stress z–z. Also the line x_2–x', measured to the scale used for lines f–f' and f_1–f_1', will give the content of the volume above the base line x_1–x_2 in cubic inches. This content is found to be 2330 cubic inches. Since this volume measured in pounds must be just equal to P, each cubic inch represents $\dfrac{80,000}{2330} = 34.3$ pounds. The maximum height of the stress volume is found, by measuring the vertical x_2–a, to be 18.2 inches, and since each cubic inch represents 34.3 pounds, the maximum unit pressure along the right edge of the pier is $18.2 \times 34.3 = 625$ pounds per square inch.

105. Kerns.—The kern of the cross-section of a pier or compression member is an area such that, if the resultant load P remains within it, there will be compression over the entire cross-section. But when the eccentricity of P is so large that it acts outside of the kern, a part of the cross-section will not have compression. It therefore follows that when P acts at the edge of the kern, the pressure over the section will decrease down to zero at one edge or one corner. Therefore, if a stress volume is assumed, which has for its base the entire cross-section, but has its normal dimension decreasing down to zero at one edge or corner, and a vertical line passing through its centroid is drawn, this vertical line locates a point on the edge of the kern.

Fig. 414 shows the cross-section of a pier 30 inches square. Assume that the eccentricity of P is such that there is zero stress along the edge a–b. When this is the case the stress volume is a wedge shown in plan as a–b–c–d and in elevation by the triangle of Fig. 415. The centroid of this wedge is at the third point, and a line passing through this centroid and normal to the section gives the point o_1 which locates P. The point o_1 is therefore a point on the edge of the kern. Since the section is a square,

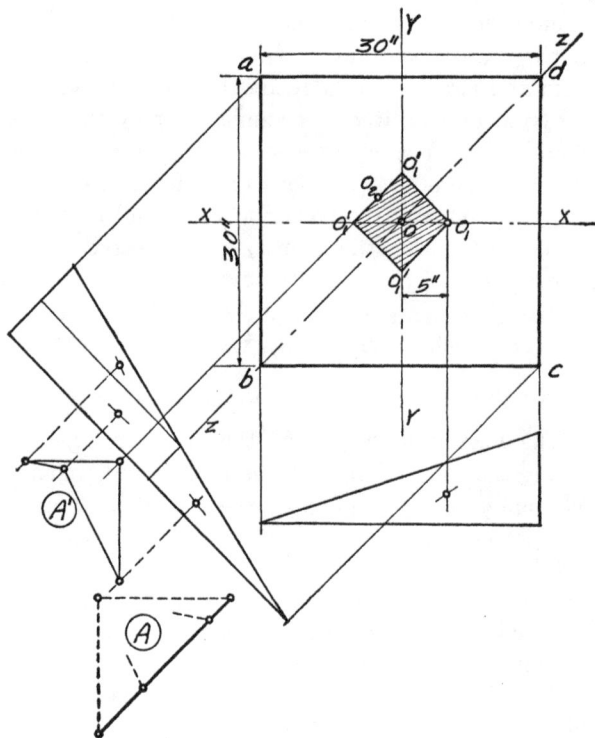

FIG 414.

FIG. 416.   FIG. 415.

the points o'_1 on the edge of the kern may be located, as shown, at the same distance from o as o_1.

Four points on the edge of the kern have now been located, but are the lines connecting these points straight lines?

Now consider the case when the stress at the corner c is zero and increases up to a maximum at a, the stress at b being the same as at d. The stress volume for this distribution of stress is of peculiar shape, its plan is the cross-section of the pier a–b–c–d, and its elevation is shown in Fig. 416. This volume is divided into three divisions, two pyramids and a triangular prism. By the use of polygons Ⓐ and Ⓐ', a plane containing the centroid of the whole volume is located. The centroid is found to be directly above o_2 which is found to lie on a straight line connecting the two adjoining o'_1 points. It therefore seems evident that the kern is a square, the diagonal of which is found to be 10 in. or one-third the length of a side of the pier.

For an explanation of the methods of locating centroids the student is referred to Chapter II.

If there is a question in the student's mind as to whether

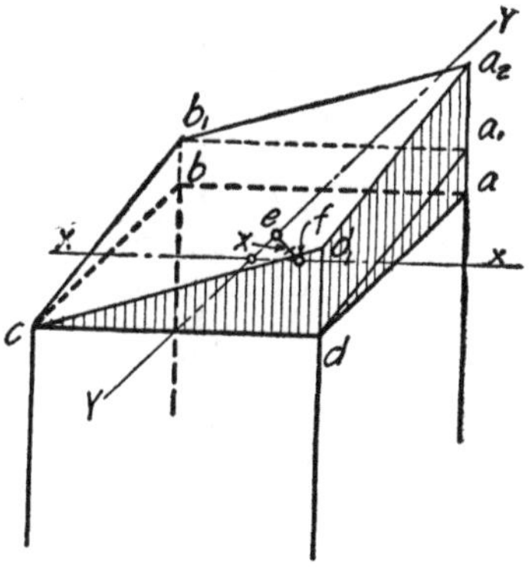

FIG. 417.

the lines o'_1–o_2 in Fig. 414 are straight or curved, it may be answered by reference to Fig. 417. The cross-section of a square pier is shown by the parallelogram a–b–c–d. The stress varies from zero at c to a maximum at a, and the stress at b is different than the stress at d. The plane c–b_1–a_1–d divides the stress volume into two wedge-shaped divisions. The centroids of these wedges are at the third points, that of the upper division being directly above f and that for the lower division directly above e. Therefore the centroid of the whole stress volume must be directly above some point x on the straight line e–f. The location of x varies as the ratio of d–d_1 to b–b_1, but it will always be on the line e–f. Therefore the side of the kern for a square pier must be a straight line.

Fig. 418 shows the cross-section of a hollow pier, the kern

for which is desired. When the stress is zero along the edge
a–d, and increases up to a maximum along the edge c–b, the stress
volume has for its base the cross-section of the pier, and for its
elevation the triangle shown in Fig. 419. By the use of polygons Ⓐ and Ⓐ′, a plane containing the centroid is found and the
point o_1 located in Fig. 418.

Fig. 420 shows the elevation of the stress volume for the case

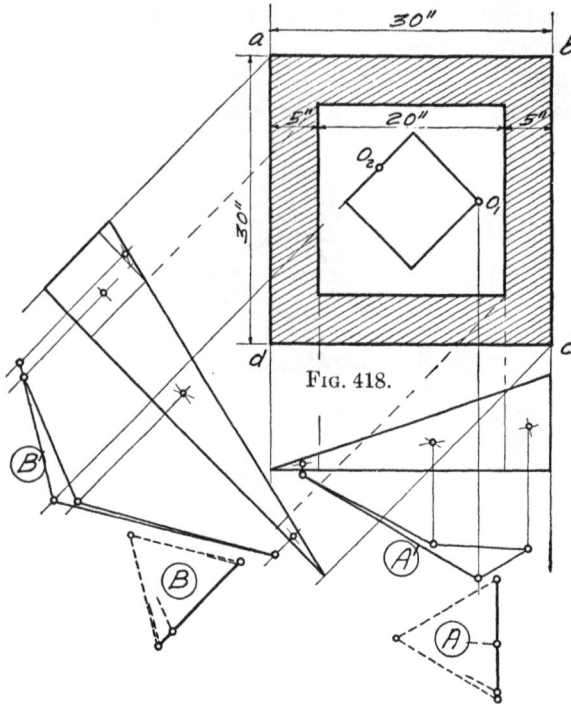

FIG. 418.

FIG. 420.   FIG. 419.

when the stress is zero at c, increasing up to a maximum at a,
the stresses at b and d being equal. Now by the use of polygons
Ⓑ and Ⓑ′ the point o_2 on the edge of the kern is located. It
is now evident that the kern is a square. Attention is called to
the fact that the kern for the hollow pier of Fig. 418 is larger
than the kern for the solid pier of the same size shown in Fig.
414. This does not mean that the hollow pier is the stronger
but simply that the eccentricity of the load on the hollow pier

may be greater and still have pressure over the entire section. It could be proved that the sides of the kern in Fig. 418 are straight lines by drawing a figure similar to Fig. 417.

The cross-section of a triangular shaped pier is shown in Fig. 421. When the stress is zero along the edge a–c, increasing up to a maximum at b, the stress volume is a pyramid, shown in elevation by Fig. 422. Making use of the fact that the centroid of a pyramid is at the quarter point of the line connecting the centroid of the base with the vertex, the point o_1, on the edge

Fig. 421.

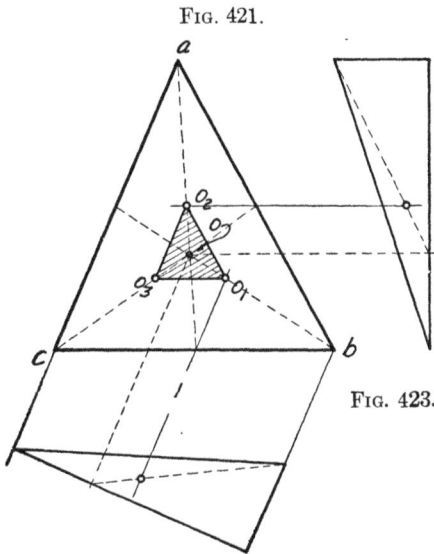

Fig. 423.

Fig. 422.

of the kern is located. Likewise the point o_2 is located by the use of Fig. 423, also o_3 may be located in a similar way. It can be proved that the lines o_1–o_2, o_2–o_3 and o_3–o_1 are straight lines by the use of Fig. 424. Consider the case when the distribution of stress is similar to that shown in the figure. The stress volume may be divided by the plane a_1–b–c into two pyramids, the one on top having its centroid directly above f, while the lower one has its centroid directly above e. Therefore the centroid of the whole stress volume must be directly above some point x on the straight line e–f, the location of x along this line depending upon the relative lengths of a–a_1 and b–b_1.

106. Location of Points on the Edge of the Kern Analytically.—
In some cases it is very easy to determine the size of the kern
analytically. The minimum stress S_m is given by the formula
$\dfrac{P}{A} - \dfrac{Pec}{I}$, and when the resultant load P is at the edge of the kern

this minimum stress is zero. That is $\dfrac{P}{A} - \dfrac{Pec}{I} = o$ and $\dfrac{P}{A} = \dfrac{Pec}{I}$.
It follows that the e, which is the distance from the centroid to
the edge of the kern and which we will call e_1, is given by the

expression $e_1 = \dfrac{I}{Ac}$. In order to illustrate how this formula

may be employed let it be used to locate o_1 and o_2 in Fig. 414.
For locating o_1 consider the axis Y–Y. The distance from o

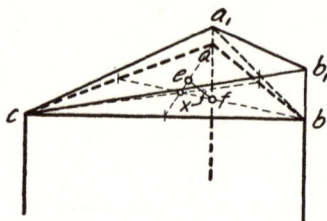

Fig. 424.

to o_1 equals $\dfrac{I_y}{Ac} = \dfrac{67,500}{900 \times 15} = 5$ inches, which checks with the

graphical result. For locating o_2 the axis z–z is considered and

the distance from o to o_2 is $\dfrac{I_z}{Ac} = \dfrac{67,500}{900 \times 21.21} = 3.54$, which

also checks with the graphical results.

Figs. 425 to 428 show four different cross-sections and their
kerns. The dimensions of these kerns were found analytically
using the formula given above. The student should remember

that c in the formula $e_1 = \dfrac{I}{Ac}$ is the distance from the neutral

axis to the edge or corner of the pier which has zero stress, and
e_1 is always measured on the opposite side of the neutral axis.
The neutral axis considered is the axis about which the bending
due to eccentricity occurs.

In Fig. 414 the over-all dimension of the kern along the x–x
axis, and also along the Y–Y axis, is equal to one-third the length

of a side of the pier, hence the expression that the kern of a square or a rectangle is the middle third.

A word regarding maximum unit pressure may not be out of place at this time. The formula $S_M = \dfrac{P}{A} + \dfrac{Mc}{I}$ may be used when the resultant load falls within the kern, but should not be

FIG. 425.   FIG. 426.

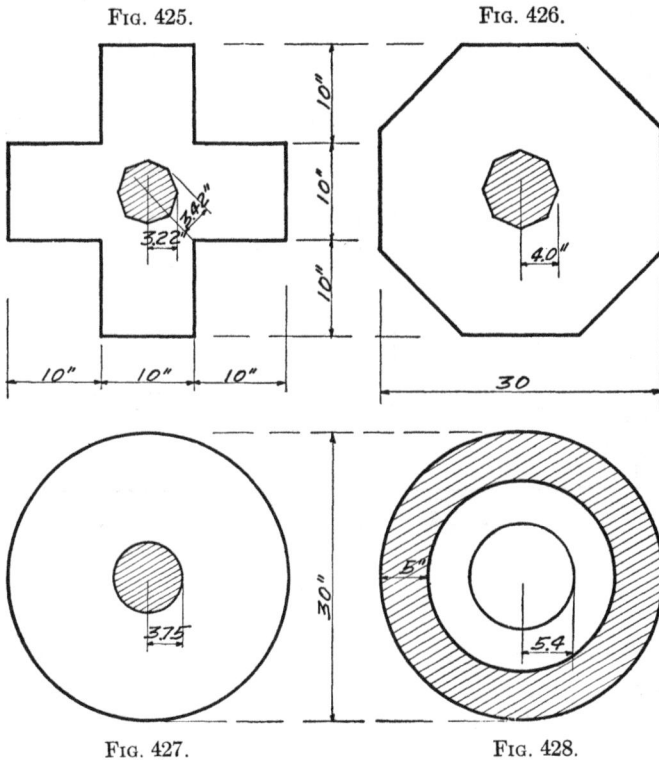

FIG. 427.   FIG. 428.

used when P falls without the kern.   When the section is a square or a rectangle and P is without the kern but on a neutral axis parallel to a side, the formula $S_M = \dfrac{2P}{A_1}$ may be used, but for more complicated cases the construction of Figs. 412 and 413 should be employed.

The kern for the cross-section shown in Fig. 429 may be found without much difficulty.   Consider the neutral axis x-x, and using the formula $e_1 = \dfrac{I}{A\,c}$, the points o_1 and o_2 are located. In like manner o_3 and o_4 are located by considering the neutral axis Y-Y. In order to locate o_5, z-z is assumed to be the axis of zero stress with the stress increasing directly as the distance from it. The stress volume is divided into four wedges and the centroid of the whole volume located as shown. The point o_5 is directly below the centroid. It can easily be proved that the lines o_1-o_4; o_4-o_5, etc., are straight lines, therefore the six-sided area indicated is the kern,

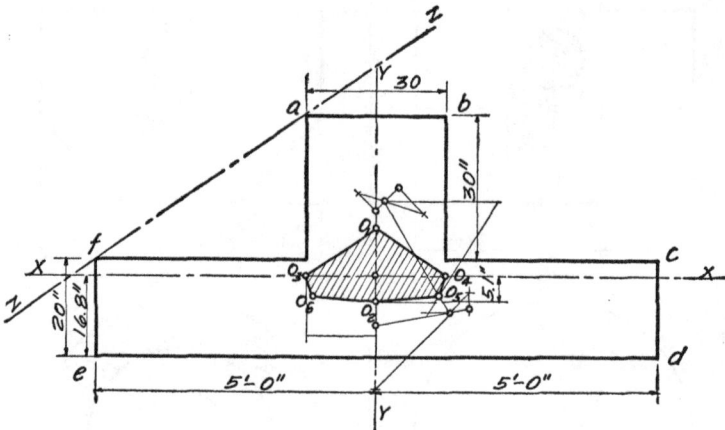

Fig. 429.

107. Pressure on Wall Footings.—Fig. 430 shows a vertical section through a brick wall and concrete footing. The load P at the top of the wall is supposed to come from the roof and P_1 from the floor. On the right side there is an earth fill, while on the left there is a basement; the wall then acts somewhat as a retaining wall. By the use of polygon Ⓐ and Fig. 431, R the resultant of P, P_1 and the weight of the wall above the footing, is found in action line and magnitude. In computing P, P_1, W, and W_1, it will be found convenient to consider a one-foot length of wall. Now there is a horizontal thrust from the earth fill, the unit intensity of which is assumed to increase directly as the depth. A rate of increase of 20 pounds per square foot per foot was assumed in this case. The resultant H of the horizontal

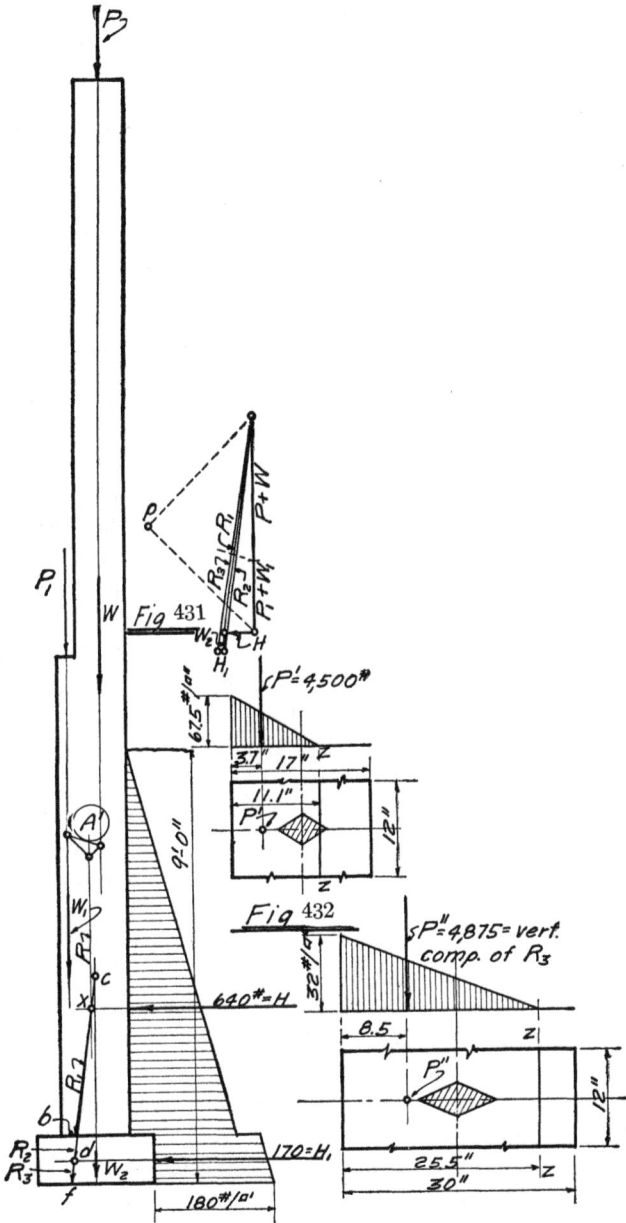

FIG. 430.

FIG. 433.

thrust above the footing is found to be 640 pounds for a one-foot length of wall, and of course it acts at the third point since its distribution is in the form of a triangle.

The action line of H intersects that of R at x. Then x is a point on the action line of their resultant R_1 which is given in magnitude and direction in Fig. 431. The action line of R_1 intersects the bottom of the wall at b, 3.7 inches from its left side. The plan of the base of a one-foot length of wall is shown in Fig. 432. The upper part of the figure shows the distribution of pressure between the base of the wall and the top of the footing. The maximum unit stress is given by the formula $S_M = \dfrac{2\,P'}{A_1}$ in which P' is the vertical component of R_1. $S_M = \dfrac{2 \times 4500}{11.1 \times 12}$ = 67.5 pounds per square inch, the 11.1 being three times 3.7.

In order to find the maximum unit pressure and distribution of pressure between the footing and the soil, the resultant R_3 must be found in action line and magnitude. The action line for the weight of the footing W_2 intersects the action line of R_1 at c which is, therefore, a point on the action line of R_2. R_2 is given in magnitude and direction by Fig. 431. The resultant of R_2 and H_2 is now found, and its action line intersects the bottom of the footing at f, 8.5 inches from the left edge of the footing. A plan of the base of the footing for a one-foot length of wall is shown in Fig. 433, the upper part of which shows the distribution of pressure. The maximum pressure on the soil is found to be 32 pounds per square inch or about 4600 pounds per square foot.

It would be better if the eccentricity of R_1 and of R_3 were small enough so that there would be compression over the entire cross-section at the base of the wall and also at the bottom of the footing. However, the wall is safe as long as the maximum unit pressure is not excessive.

108. Retaining Walls.—It was not considered advisable to include in this volume a discussion of earth pressures. The student is, therefore, referred to Ketchum's book on Walls, Bins, and Grain Elevators for the methods of computing the thrust or pressure exerted by fills against retaining walls. One of the standard formulas is

$$P = \tfrac{1}{2}\,w \cdot h^2 \frac{1 - \sin \phi}{1 + \sin \phi},$$

in which P is the total horizontal thrust against the wall when the fill is level with its top, h is the height of the wall, w is the weight of the fill per cubic foot, and ϕ is the angle of repose for the fill. The unit thrust p at any distance h below the top of the fill is given by the formula $p = h \cdot w \dfrac{1 - \sin \phi}{1 + \sin \phi}$.

The drawing in Fig. 434 illustrates a reinforced concrete retaining wall with an earth fill which is assumed to weigh 100 pounds per cubic foot. Assuming the angle of repose ϕ to be about 42°, the unit thrust is found to vary uniformly from zero at the top to 440 pounds per square foot at the bottom. This horizontal thrust which is distributed over the back of the wall is divided into a number of small divisions and an equivalent concentrated load substituted for each one of the divisions. For convenience, a portion of the wall 1 foot long is considered in computing these horizontal loads which are laid off to scale in Fig. 435. From Fig. 435 the funicular polygon of Fig. 436 is drawn, giving the moment diagram shown by the shaded area. The main part of the wall acts as a large vertical cantilever and the moment which should be provided for at any section x–x is given by the horizontal intercept z, in Fig. 436, times the pole distance H of Fig. 435.

In order to find the maximum unit pressure on the bottom of the footing, the weight of the wall must be found, and since the fill directly above the right part of the footing rests upon the footing and helps to resist overturning, its weight must also be considered. The total vertical load carried by the soil just below the bottom of the footing is therefore equal to the resultant of the following forces:

> The weight of the footing, P_2, the weight of the wall, P_1 plus P_3, and the weight of the fill above the right part of the footing, P_4.

The resultant R of these four loads is found by drawing the force polygon of Fig. 437 and the funicular polygon Ⓐ. The resultant of R and the horizontal thrust H acts through the point b, and its direction is found in Fig. 437. The action line of R_1 intersects the bottom of the footing at c which is 6 inches to the left of the center. The vertical component of R_1 is resisted by vertical pressure from the soil, while the horizontal component may be

Fig. 437.

Fig. 436.

Fig. 434.

Fig. 438.

Fig. 440.

Fig. 435.

Fig. 439.

resisted by friction along the bottom of the footing, by bearing against the surface g–h, or by a portion of the footing projecting down below the rest, as shown by the dotted lines in Fig. 434. If part of the horizontal component is taken up by friction along the bottom, and the rest by a lower projection such as that shown dotted, the vertical component will act at c with an eccentricity of 6 inches. The maximum and minimum unit pressures are found by the use of the formulas,

$$S_M = \frac{P}{A} + \frac{Pec}{I} \quad \text{and} \quad S_m = \frac{P}{A} - \frac{Pec}{I},$$

in which

$$I = \frac{1}{12} bd^3 = \frac{1}{12} 10.5^3$$

in this case.

$$S_M = \frac{P}{10.5} + \frac{P \cdot \frac{1}{2} \cdot 5\frac{1}{4}}{\frac{1}{12} \cdot (10\frac{1}{2})^3} = 2174 \text{ pounds per square foot.}$$

The minimum unit pressure is found to be 1210 pounds per square foot, and the distribution of pressure is shown by Fig. 438. The distributed pressure is divided into small divisions and an equivalent concentrated force substituted for each. The right half of the footing has, in addition to the upward pressure from the soil, a downward pressure from the fill above. This downward pressure is divided into small divisions and equivalent concentrated loads substituted. From these loads are subtracted the corresponding concentrated loads substituted for the distributed pressure below, and the results obtained are the forces which tend to produce moment in the right part of the footing. The concentrated loads which have been substituted for the distributed pressures on the footing are laid off along the load line in Fig. 439, from which the moment diagram, shown in Fig. 440, is drawn. Now at any section Y–Y the bending moment in the footing is given by the intercept in the moment diagram times the pole distance H of Fig. 439. Attention should be called to the fact that the scale has been changed in Fig. 439 below x, but H_1, measured to the proper scale, has the same value as H.

The moment diagrams, Figs. 436 and 440, show how the bending moment varies in the various parts of the retaining wall and its footing, and may be used to advantage in designing the reinforcement.

109. Line of Pressure in a Pier.—Consider the pier shown in Fig. 441 with the thrusts P_1 and P_2 acting upon it as indicated. The pier is divided into a number of divisions by horizontal planes x_1-x_1, x_2-x_2, etc. The resultant force acting on section x_1-x_1 is the resultant of P_1, P_2 and P_3, the force P_3 being the weight of that portion of the pier which is above section x_1-x_1. In Fig. 442, the resultant of P_1 and P_2 is found in magnitude

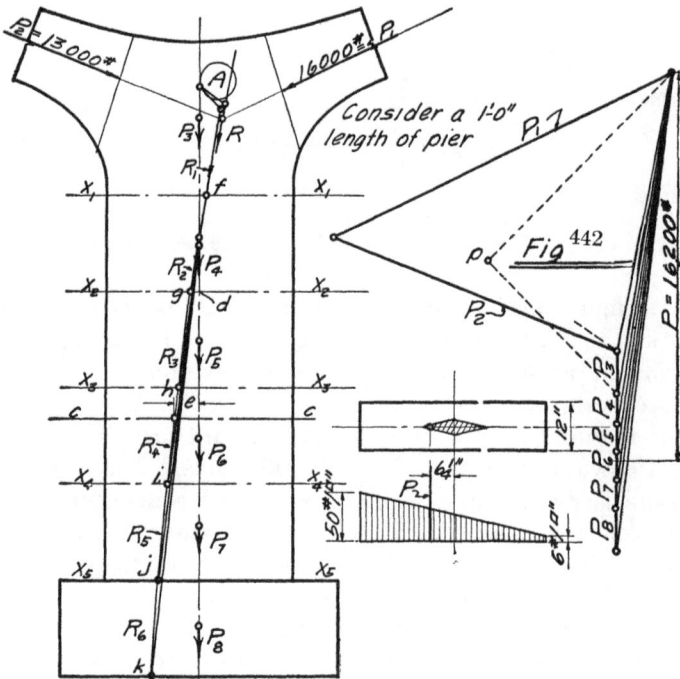

Fig. 441.   Fig. 443.

and direction, and of course, its action line passes through the intersection of the action lines of P_1 and P_2. The resultant of this resultant and P_3 is found in magnitude and direction in Fig. 442, and its action line is located by using the pole p and drawing the funicular polygon Ⓐ.

The total load which the section x_2-x_2 must carry is the resultant of R_1 and the weight of the masonry between x_1-x_1 and x_2-x_2, or in other words R_2 is the resultant of R_1 and P_4. R_2 is found in magnitude and direction from Fig. 442, and its

action line must pass through the intersection of the action lines
of R_1 and P_4. The action line of R_2 intersects the section x_2–x_2
at g, and the distance from the center d over to g shows the eccentricity of the loading at this section. In like manner R_3, R_4, etc.
are found and the points h, i, j, etc., located. Then the smooth
curve f–g–h–i–j–k is called the line of pressure. If a section
c–c is taken at any level, the intersection of this section by the
line of pressure locates the point where the action line of the
resultant of all the loads above would cut the section. The
distance from the center of the pier over to this intersection
gives the eccentricity of the loading. The eccentricity for the
section c–c shown in the figure is found to be $6\frac{1}{4}$ inches; Fig. 443
shows the maximum unit pressure and the distribution of stress
over this section.

In Fig. 444 a pier is shown which has a number of horizontal
loads applied at different elevations. The resultant load that
must be carried by section x–x is the resultant of the three forces
W_1, P_1 and P_2. This resultant is shown in magnitude and
direction as R_1 in Fig. 445, and its action line passes through the
intersection of the action line of W_1 with that of P_1 plus P_2, and
intersects x–x at b. Point b is therefore a point on the line of
pressure. The resultant force that must be carried by section
x_1–x_1 is the resultant of the three forces R_1, W_2, and P_3. First
the resultant of R_1 and P_3 will be found and then the resultant
of their resultant and W_2. The resultant of R_1 and P_3 is shown
as R_2 in Fig. 445, and its action line, passing through a, the intersection of R_1 and P_3, intersects the action line of W_2 at point c.
Point c is therefore a point on the action line of R_3, which is the
resultant of R_2 and W_2, or in other words the resultant force
that must be carried by the section x_1–x_1. The magnitude and
direction of R_3 is found in Fig. 445, and its action line intersects
the section x_1–x_1 at d, which is therefore a point on the line of
pressure or line of resistance.

The resultant force which must be carried by section x_2–x_2
is the resultant of the three forces R_3, P_4, and W_3 and will be called
R_5. The action line of R_3 intersects the action line of P_4 at e,
and the action line of this resultant intersects that of W_3 at f,
which is therefore a point on the action line of R_5. R_5 intersects section x_2–x_2 at g, thus locating another point on the line

of pressure.   In a similar way points j, and L are located and the line of pressure a–b–d–g–j–L drawn.

At any section m–n, the intersection o with the line of pressure locates a point on the action line of the resultant force carried by the section.   The distance from the intersection o to the center of the section is the eccentricity.

110. Masonry Chimneys.—It is interesting to study the stresses and the distribution of stress in a masonry chimney

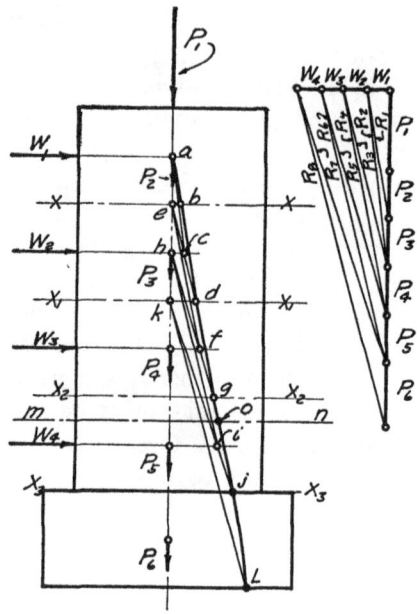

Fig. 444.   Fig. 445.

under wind load.   When a chimney is subjected to a wind pressure there are two sets of forces acting, the horizontal forces due to the wind and the vertical forces due to the weight of the chimney.   The weight of the chimney above any section can be found by computing the volume of masonry above the section and multiplying by the weight per unit volume.   In order to get an idea how the forces due to the wind may be computed, let Fig. 446 be an enlarged cross-section of the chimney shown in Fig. 448.   Consider the straight-line formula which was used to find the unit wind pressure for roof surfaces of various slopes.

The curve a–b–c–d–e shows how the unit normal pressure varies
on the windward side of the chimney when the unit pressure on
a vertical surface normal to the wind is assumed to be 30 pounds
per square foot. The distributed pressure from the wind is
divided into small divisions and an equivalent concentrated load
substituted for each division, considering a portion of the chimney
one foot in height. These forces are laid off in Fig. 447 and their
resultant R obtained, which is found to act normal to the diameter a–e and to have a magnitude of 243 pounds. If R is divided
by the length of the diameter a–e, 27 pounds per square foot
is obtained, and the magnitude of R could be obtained by multiplying a–e by 27. It therefore follows that the resultant hori-

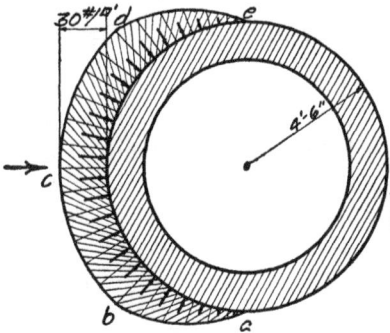
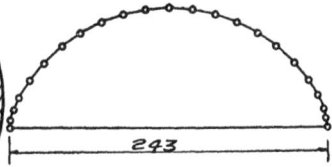

FIG. 446.   FIG. 447.

zontal force acting on any portion of the chimney is given by the
mean diameter times the vertical dimension of the division times
a certain number of pounds per square foot. The number of
pounds per square foot that should be used is, of course, proportional to the assumed pressure on a vertical surface normal to the
wind. The value of 27 pounds per square foot found above is
for an assumed pressure of 30 pounds per square foot on the vertical. The reduction from 30 to 27 is very conservative; many
designers make a reduction of one-third which means that 30
would reduce to 20, and one authority even says that the reduction should be one-half. Experiments have not been complete
enough to give very definite results.

The chimney shown in Fig. 448 is divided into a number of
divisions by sections x–x, x_1–x_1, x_2–x_2, etc. The respective
weights P_1, P_2, P_3, etc., of these divisions are computed, also

the respective resultant forces w_1, w_2, w_3, etc., from the wind. These forces are used in the force polygon of Fig. 449, and the points a, b, c, etc., on the line of pressure, are located in Fig. 448 in the same way that points b, d, g, etc., were located in Fig. 444. Fig. 449 corresponds to Fig. 445. The line of pressure is now obtained by drawing the smooth curve a–b–c . . . i–j.

The maximum unit pressure for all sections having the line of pressure passing through their kern may be found by using the formula $S_M = \dfrac{P}{A} + \dfrac{Pec}{I}$. But when the line of pressure is outside of the kern there will be compression over only a part of the section and the construction of Fig. 413 should be used. Of course if the resultant force is just a little outside of the kern, the error produced by using the formula $S_M = \dfrac{P}{A} + \dfrac{Pec}{I}$ will be small.

In order to illustrate the method, the maximum unit pressure on section Y–Y will be found. The section is shown to an enlarged scale in Fig. 450. The kern is drawn and the resultant load is found to act outside of it, as shown. In order to find the maximum unit stress correctly, a method similar to that shown in Fig. 413 will be used. The line a–c in Fig. 451 may be drawn at any angle, but an angle of 45° will be found convenient. The trial base line e–f is drawn and the stress volume above it divided into small slices which are shown in plan in Fig. 450 and in elevation in Fig. 451. The content of each of these divisions is found in cubic feet, using the scale to which Fig. 450 was drawn. The values thus obtained are laid off along the load line of Fig. 452 to some convenient scale. The funicular polygon Ⓐ is drawn from Fig. 452, and locates the vertical line containing the centroid of the volume above the base line e–f. This vertical line intersects e–f, locating point c, which is therefore directly below the centroid. P, the vertical component of the resultant loading, acts at o some distance to the left of c, therefore the base line e–f is too high. The base line e_1–f_1 is taken at a position somewhat lower than e–f and the point c_1 located by the use of Fig. 453 and the funicular polygon Ⓑ. The point c_1 is also found to be a short distance to the right of the vertical line from o, showing that the assumed base line is still too high. Point c_2 is now located on the base line e_2–f_2 by the use of Fig. 454 and the

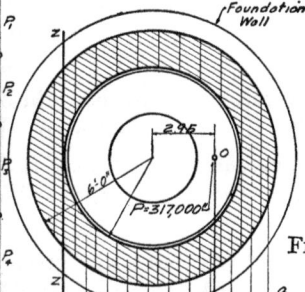

FIG. 450.

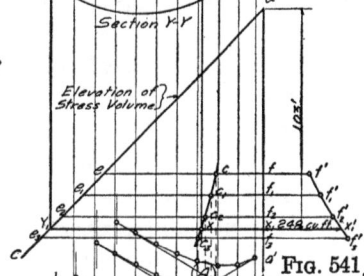

FIG. 541.

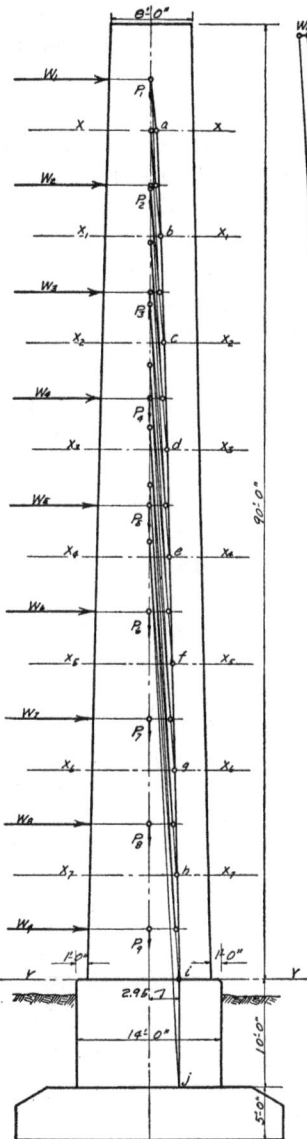

FIG. 452.

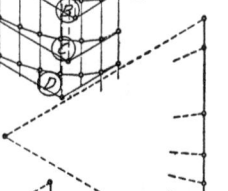

FIG. 454.

FIG. 453.

Vertical component
of resultant load on
section Y-Y= 317,000#

$\frac{317,000}{248} = 1280^{\#}$

$1280 \times 103 = 13200^{\#}$ per sq. ft.
or $916^{\#}$ per sq. in. = max. unit pressure.

FIG. 448.   FIG. 449.   FIG. 455.

polygon ©. Also point c_3 is located on the base line e_3–f_3. A smooth curve is now drawn through the points c, c_1, c_2 and c_3, and the vertical line from o gives the intersection x which locates the true base line y_1–x_1 from which the axis of zero stress z–z is located.

Now measuring to the right of the line a–b, the line f–f' is made equal to the total content of the volume above the base e–f to some convenient scale, also the line f_1–f'_1 is made equal to the content of the volume above the base line e_1–f_1 to the same scale. The points f'_2 and f'_3 are located in a similar way and the line x_1–x'_1, measured to scale, gives the content of the volume above the true base line y_1–x_1. This content is found to be about 248 cubic feet.

Now P, the vertical component of the resultant load carried by the section is found to have a magnitude of 317,000 pounds. $317,000 \div 248 = 1280$ pounds per cubic foot. That is, each cubic foot of the volume above the base line y_1–x_1 represents 1280 pounds, and, since the height of this volume is 10.3 feet, the maximum unit pressure is $1280 \times 10.3 = 13,200$ pounds per square foot or 91.6 pounds per square inch.

Usually only two trial base lines in Fig. 451 will be necessary if they are carefully chosen, since the curve c–c_1–c_2–c_3 for short distances is practically a straight line.

It is interesting to know what the shape of the curve c–c_1–c_2–c_3 would be if it were extended on up to a and also down below c_3. Fig. 456 shows the plan of a chimney cross-section similar to the one shown in Fig. 450. Various base lines are taken in Fig. 457, and by the use of the force polygons shown in Figs. 459 and 460, and the funicular polygons Ⓐ, Ⓑ, Ⓒ . . . Ⓘ the points a, b, c, . . . i are located and the desired curve drawn. When the resultant load acts at any point o, the intersection x locates the base line y_1–x_1 of the stress volume from which the axis of zero stress z–z can be found in Fig. 456, also the line x_1–x'_1 at the right of Fig. 457 measured to scale gives the content of the volume above the base line y_1–x_1.

Fig. 458 is a drawing illustrating the appearance of the stress volume above the base line y_1–x_1.

111. Line of Pressure in an Arch.—In Fig. 461 a small portion of an arch is shown. There are three forces acting on this part of the arch, P_3, P_4, and P. P_4 is the resultant force exerted

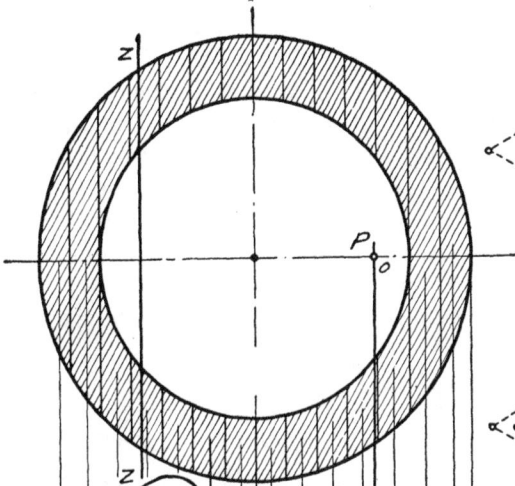

FIG. 456.

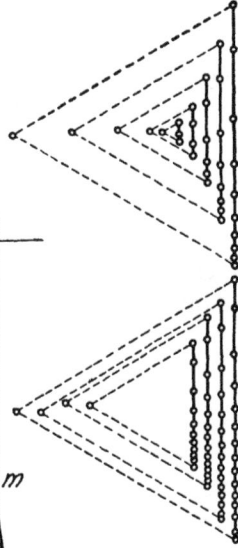

FIG. 460.

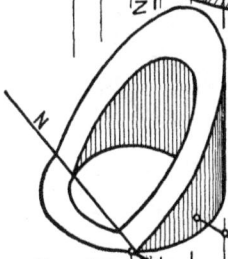

FIG. 458.

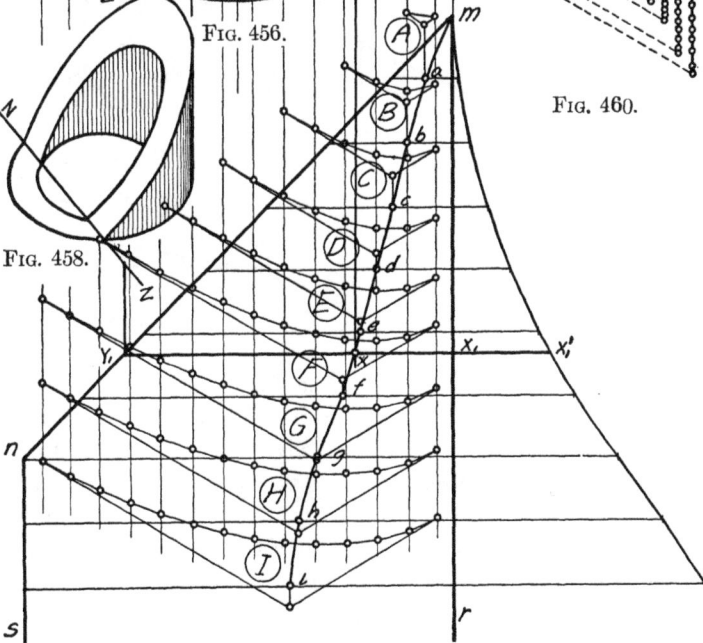

FIG. 457.

on the portion shown in the figure by the part of the arch at its
right, while P_3 is the resultant force exerted by the part of the
arch at its left. P is the resultant of the load P_1 and the weight
P_2.   These three forces are in equilibrium, therefore they must
be concurrent.   If the action lines of any two of them are known,
their intersection locates a point on the action line of the third,
and if the magnitude, direction and sense of any two be known,
the magnitude, direction, and sense of the third can be found
by drawing the force polygon ⓑ.   Also, since P, P_3 and P_4
are in equilibrium, any one of them is the anti-resultant of the
others.

In Fig. 462, a larger portion of an arch is shown divided into

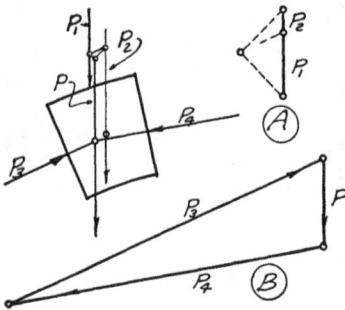

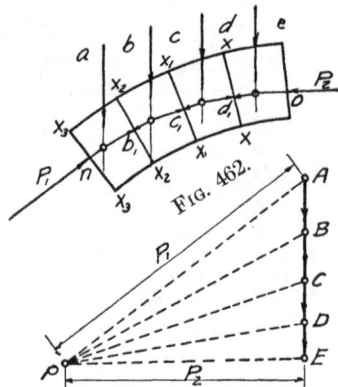

Fig. 462.

Fig. 461.   Fig. 463.

a number of small divisions.   The forces $A–B$, $B–C$, etc., are
supposed to represent the resultant vertical loading carried by
each division including its own weight.   Suppose the force P_2
has the magnitude, direction, and sense shown in Fig. 463 and
action line passing through o in Fig. 462.   The resultant of the
two forces $D–E$ and P_2 is found in Fig. 463 to be $D–p$, and its
action line passes through the intersection of the action lines
of P_2 and $D–E$.   This resultant is the resultant force acting on
the section $x–x$ from the right, and of course there must be an
equal and opposite reaction from the left which holds it in equilibrium.   This reaction is the anti-resultant of the forces P_2 and
$D–E$.   The resultant pressure on the section $x–x$ is therefore

given by the line D–p in Fig. 463, and it acts at d_1. The result-ant of C–D and the pressure on x–x is given in magnitude and direction by the line C–p, and is the resultant pressure on section x_1–x_1 acting at c_1. Also the line B–p gives the resultant pres-sure on the section x_2–x_2 acting at b_1 and the line A–p gives the resultant force acting on the section x_3–x_3 from the right, this resultant being balanced by the reaction P_1 which must have the same action line and magnitude but opposite sense. The broken line n–b_1–c_1–d_1–o may be called the line of pressure for the portion of the arch shown in the figure, but the student should keep in mind that the true line of pressure for a distributed load is a smooth curve tangent to the various lengths of the broken line, on the inside. Attention should be called to the fact that the broken line n–b_1–c_1–d_1–o is nothing more than a funicular polygon for the forces A–B, B–C, etc., passing through the points o and n, with p of Fig. 463 used as the pole.

The resultant force acting on any section like x_1–x_1 or x_2–x_2 is given in magnitude and direction by the corresponding ray in the force polygon, and it acts at the place where the section cuts the line of pressure. Having located the point where the line of pressure intersects the section, the eccentricity may be measured, and since the magnitude of the force is known, the maximum unit pressure on the section can be found and also the distribution of stress.

112. Three-hinged Arches.—A small three-hinged arch carry-ing a number of unequal unsymmetrically placed loads is shown in Fig. 464. At the hinges X, Y, and Z there can be no moment or eccentricity, and the line of pressure must therefore pass through these hinges. In other words, the hinges locate the line of pressure, or we might say that they locate three points on the funicular polygon which represents the line of pressure. The line of pressure for the arch may be drawn conveniently by either of two methods: The pole p in Fig. 465 may be located by finding the reactions R_1 and R_2, using the regular construc-tion for the three-hinged arch as explained in Chapter V, or the construction for passing a funicular polygon through three given points as explained in Chapter I might be used.

In this particular case the construction for finding the reactions of a three-hinged arch will be used, and the student should refer to the discussion on three-hinged arches in Chapter V for a more

detailed explanation. The loads are laid off in Fig. 465, taking
them in order beginning with A–B, and any convenient poles
p_1 and p_2 are chosen. From pole p_1 the funicular polygon Ⓐ
is drawn for the loads between hinges X and Y, and the action
line of their resultant R_L located. By the use of pole p_2 and the
funicular polygon Ⓑ the action line of R_r is located.

First we will find the reactions at X and Z due to R_L, and then

Fig. 464.

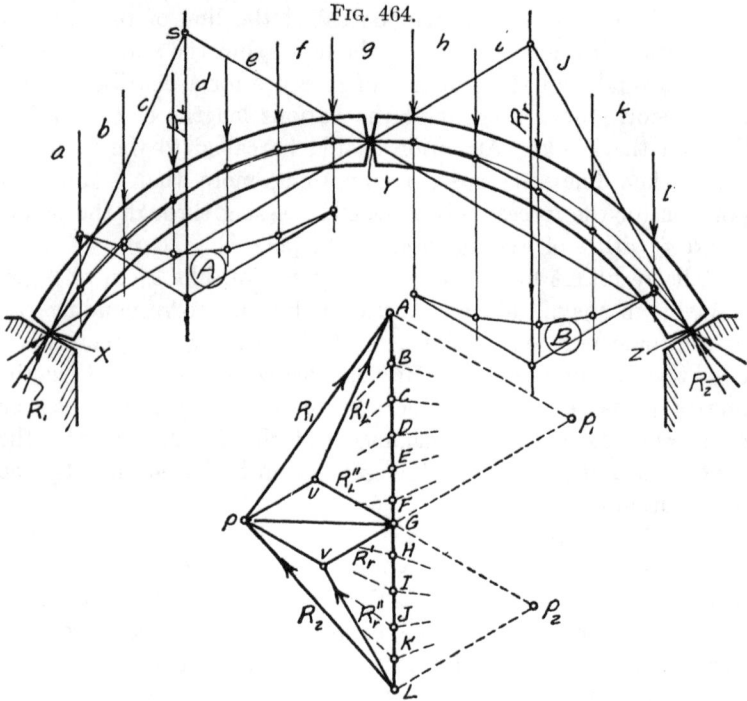

Fig. 465.

those due to R_r. The loads on the left of the arch produce just
two forces on the right half, a reaction at Y and also one at Z.
Since these two reactions are in equilibrium they must be equal
in magnitude and must have the same line of action. The line
of action must therefore be Y–Z, which is extended until the action
line of R_L is intersected at S. There are three forces acting on
the left half of the arch, R_L, the reaction at X, and the reaction
at Y. Three forces in equilibrium must be concurrent, there-

fore the reaction at X due to R_L must have X–S as its action
line. In Fig. 465, the line A–U is drawn parallel to S–X. and G–U
parallel to S–Z. R'_L is the reaction at X due to R_L, and R_L'' is
the reaction at Z due to R_L. In a similar way R'_r is found to be
the reaction at X due to R_r, and R''_r the reaction at Z due to R_r.
R_1 the total reaction at X, is the resultant of R'_L and R'_r, and is
shown in magnitude and direction in Fig. 465; also R_2 is the
resultant of R''_L and R''_r. Now with p as a pole the funicular
polygon in Fig. 464, representing the line of resistance, is drawn
starting at either X, Y, or Z. This polygon if correctly and
carefully drawn will pass through the three points X, Y, and Z.

In Fig. 466 a three-hinged arch somewhat similar to the
one in Fig. 464 is shown. The general construction for passing a
funicular polygon through three points will be used for drawing
the line of pressure of this arch. The loads are laid off in order
along the load line of Fig. 467 starting with A–B, and using any
pole p', the funicular polygon Fig. 468 is drawn. Lines are now
drawn down from X, Y, and Z parallel to the action lines of the
loads and extended until the polygon, Fig. 468, is intersected and
the points r, s and t located. In Fig. 467, the lines p'–w and p'–w_1
are drawn parallel to r–s and s–t, respectively, and the points w
and w_1 located. Then from w the line w–p is drawn parallel to
X–Y, and from w_1 the line w_1–p parallel to Y–Z, and the pole
p located. Using the pole p, the funicular polygon representing
the line of pressure is drawn in Fig. 466, starting at either X,
Y or Z. The polygon will be found to pass through each of the
three points X, Y, and Z, if the work is correctly and carefully
done. For a proof of the construction just used the student
is referred to Chapter I. If the loading on the arch consists of
a number of concentrated loads as shown, the broken line is the
true line of pressure; but, if the loading is a distributed load,
for small parts of which the concentrated loads have been substituted, the true line of pressure is a smooth curve tangent
on the inside to the various parts of the broken line. The smaller
the divisions into which the distributed load is divided, or in
other words the closer the substituted concentrated loads are
together, the closer will the broken line representing the line
of pressure approach the smooth curve.

The reaction at the left end of the arch is given in magnitude
and direction by the line A–P in Fig. 467, while the reaction at

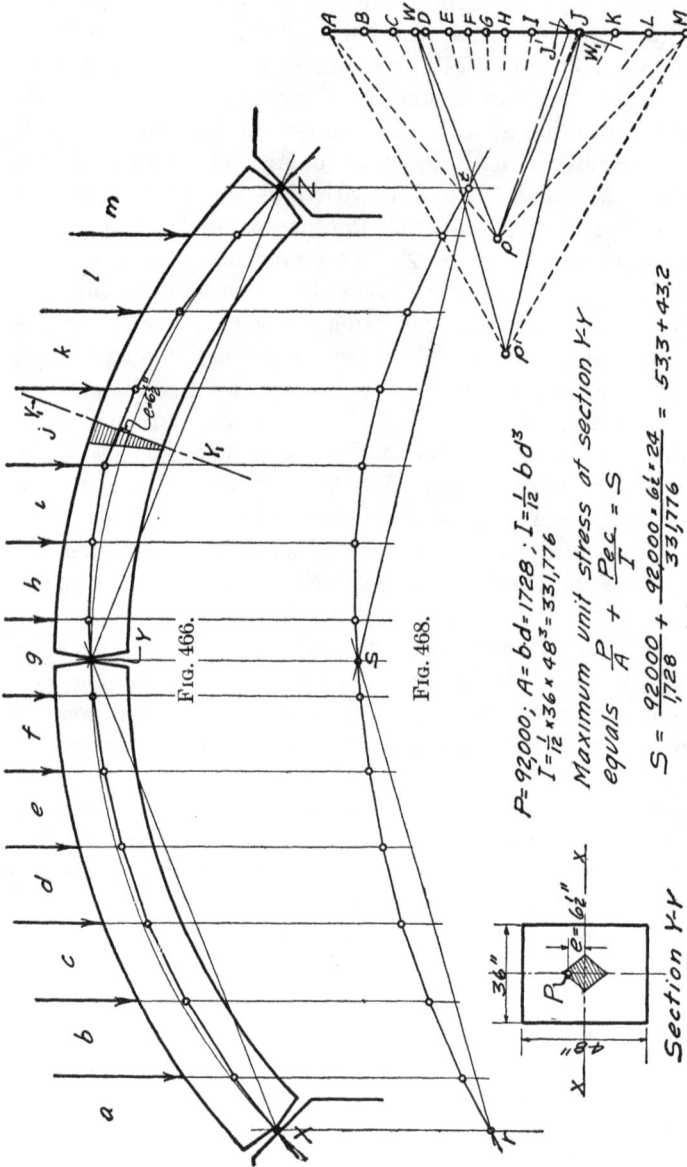

FIG. 467.

FIG. 466.

FIG. 468.

$P = 92000; A = bd = 1728; I = \frac{1}{12} bd^3$

$I = \frac{1}{12} \times 36 \times 48^3 = 331,776$

Maximum unit stress of section Y-Y

equals $\frac{P}{A} + \frac{Pec}{I} = S$

$S = \frac{92000}{1728} + \frac{92000 \times 6\frac{1}{4} \times 24}{331,776} = 53.3 + 43.2$

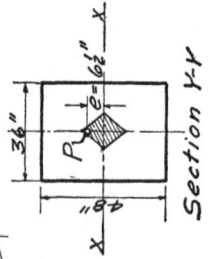

$S = 96.5 \#/\square''$

Section Y-Y

FIG. 469.

the right end is given by the line $M-p$. The resultant force carried by any section Y_1-Y_1, normal to the axis of the arch, is given in magnitude by the corresponding ray in Fig. 467. The j string is cut by the section Y_1-Y_1, therefore the magnitude of the resultant acting on the section is given by the line $p-J$ in Fig. 467. The eccentricity of the resultant load is given by the distance from the center line of the arch to the point where the line of pressure intersects the section, and is here found to be $6\frac{1}{2}$ inches. The component of the resultant force normal to the section is found, after scaling the line $p-J'$, to be 92,000 pounds. Fig. 469 shows the section Y_1-Y_1, assuming the width of the arch to be 36 inches, and since the resultant loading falls within the kern, the maximum unit pressure on Y_1-Y_1 may be found by using the formula $S_M = \dfrac{P}{A} + \dfrac{Pec}{I}$. The maximum unit pressure is found to be 96.5 pounds per square inch, and the distribution of stress is shown in Fig. 466.

The maximum unit pressure at any other section may be computed in a similar way. There is, of course, one section which will have a greater maximum than any other section, and the maximum unit pressure on this section will therefore be the maximum unit pressure in the arch. It is sometimes rather difficult to locate by inspection the section at which the maximum pressure will occur, since both P and e vary along the length of the arch and both have an effect upon the unit pressure. It is therefore often desirable to compute the pressure at several different sections in order to be sure that the maximum has been found. The maximum pressure will not always occur where the eccentricity is greatest, because some other section, which has almost as large an eccentricity, may have a larger P. In case the loading is distributed, the eccentricity should be measured to the smooth curve tangent to the various parts of the broken line, rather than to the broken line. However, in many cases the difference would be small.

Fig. 470 shows a section of a three-hinged arch masonry bridge. The line of pressure and the maximum unit pressure are desired when the left half of the arch carries a live load assumed to be 200 pounds per square foot. A portion of the arch one foot wide will be considered in computing the loads. The live load fill, and the arch itself are divided into small vertical slices and

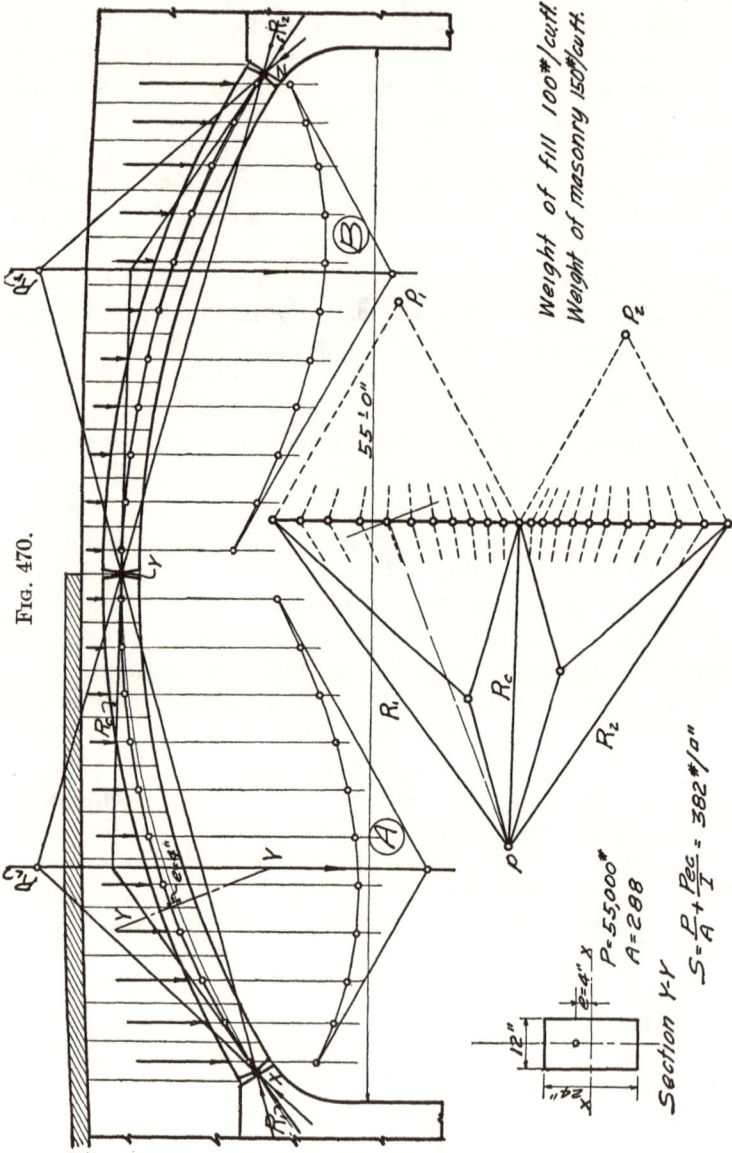

FIG. 470.

FIG. 471.

Weight of fill 100#/cu.ft.
Weight of masonry 150#/cu.ft.

$55'-0''$

Section Y-Y

$P = 55,000\#$

$A = 288$

$S = \dfrac{P}{A} + \dfrac{Pec}{I} = 382\#/\square''$

FIG. 472.

an equivalent concentrated load substituted for the weight of each slice. These loads are laid off in order along the load line in Fig. 471, the poles p_1 and p_2 are chosen, and the polygons Ⓐ and Ⓑ are drawn, locating the action lines R_L and R_r. Using the construction already explained, R_1, R_2 and R_c are found and the pole p located. With p as a pole, the funicular polygon representing the line of pressure is drawn in Fig. 470, starting at any one of the three hinges. At the section Y–Y the eccentricity is found to be 4 inches, and by referring to Fig. 471, the normal component of the resultant force acting on Y–Y is found to be 55,000 pounds. The maximum unit pressure on the section Y–Y is therefore 382 pounds per square inch. If there is reason to suppose that any other section might have a higher unit pressure, its maximum unit pressure should be computed and compared with the maximum for Y–Y. The greatest maximum that can be found will be the maximum unit pressure for the arch. After the student has had a little experience finding the maximum unit pressure in arches, computations at two or three carefully chosen sections should be sufficient for determining the maximum.

113. Three-hinged Arch Symmetrically Loaded.—When a three-hinged arch is symmetrical and symmetrically loaded, the resultant pressure at the crown must be horizontal, that is, a tangent to the line of pressure at the middle hinge or center of the arch must be horizontal. The construction of the true line of pressure for such a three-hinged arch is very simple, as may be seen by reference to Figs. 473 and 474. The loads are laid off in order along the load line in Fig. 474, and using any pole p_1 for the forces on the left half of the arch, the funicular polygon is drawn, locating the resultant. Now there are three forces acting on the left half of the arch, the reaction R_1, the crown pressure R_c, and the resultant of the loads. From Y a horizontal line is drawn which is the action line of R_c. This horizontal line intersects the resultant of the loads at o, and therefore o–X must be the action line of R_1. In Fig. 474 a line is drawn from A parallel to the action line of R_1 and another line from J parallel to the action line of R_c. These two lines by their intersection locate the point p. Now if a polygon is drawn, using this pole and starting at any one of the three hinges, it will pass through the other two hinges and represent the true line of pressure for the arch.

114. Two-hinged Arches.—The advantage of the three-hinged arch is the fact that it has three hinges or points through which the line of pressure must pass, and of course only one funicular polygon can be drawn for a given loading when it is necessary for it to pass through three fixed points. Therefore the three hinges of a three-hinged arch make it easy to locate the true line of pressure. When an arch having only two hinges, as shown in Fig. 475, is considered, the problem becomes much more difficult because the two hinges locate only two points on

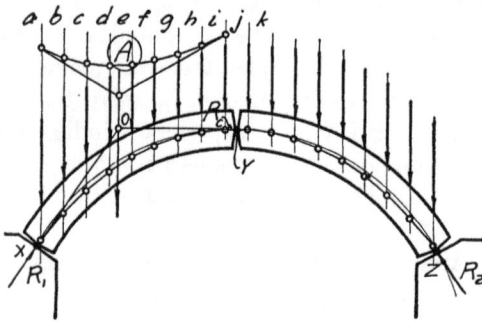
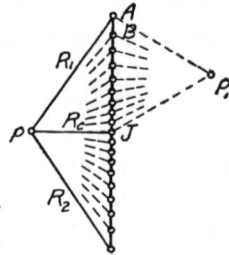

FIG. 473.   FIG. 474.

the line of pressure, and for a given system of forces any number of polygons can be drawn passing through two points.

The loads are laid off along the load line of Fig. 476, and using the pole p', the funicular polygon of Fig. 477 is drawn. Then taking some point t_1 near the center of the arch, the pole p_1 is located in the same way that p was located in Fig. 467. Using the pole p_1, a funicular polygon is drawn passing through r, t_1, and s. The pole p_1 is next moved to p_2, then to p_3, etc., and a number of other polygons drawn in Fig. 475 passing through r and s. Many more might have been drawn. Now the question arises as to which one of these many polygons represents the true line of pressure.

115. Two-hinged Arch, Method of Least Work.—An article by W. M. Smith entitled " True Pressure Line in a Masonry Arch " is given in the *Engineering Record*, Vol. 70, page 401.

The arch shown in Fig. 475 will be solved by the method of least work which is given in this article.

The total work of resistance in the arch is given by the formula,

W = summation of $\dfrac{M^2 \cdot dc}{2\,E \cdot I}$ plus summation of $\dfrac{P^2 \cdot dc}{2\,A \cdot E}$ for the length of the arch L; in which M is the bending moment at any assumed section, P is the total force normal to the section, A is the area of the cross-section, E is the modulus of elasticity of the material, I is the moment of inertia of the section, L is the length of the arch measured along its axis, and dc is a short length of the arch. If b equals the width of the arch and d is its depth then $I = \frac{1}{12}\,b \cdot d^3$, and $A = b \cdot d$, also $P \cdot e$ may be substituted for M. Making these substitutions and taking a width of one foot for b, the formula becomes:

$$W = \text{summation of } \frac{P^2 \cdot e^2\,dc}{2\,E\,\frac{1}{12}d^3} \text{ plus summation of } \frac{P^2 \cdot dc}{2\,E \cdot d}.$$

In this formula e and d are to be measured in feet.

Let the arch shown in Fig. 475 be divided into small divisions, all of equal length, by planes normal to its axis, and let the length of each one of these divisions be called dc. Then for the given arch, dc will be constant as well as E, and the above equation may be written

$$W = \frac{dc}{2\,E}\left[\text{summation of } \frac{P^2 \cdot e^2}{\frac{1}{12}\,d^3} \text{ plus summation of } \frac{P^2}{d}\right].$$

Since $\dfrac{dc}{2\,E}$ is a constant, the value of W will be a minimum when summation of $\dfrac{P^2 \cdot e^2}{\frac{1}{12}\,d^3}$ + summation of $\dfrac{P^2}{d}$ has its minimum value. The value of this expression will, therefore, be found for various funicular polygons, or trial lines of pressure, and the polygon which will give the minimum value is assumed, by the theory of least work, to be the true line of pressure.

With the pole p_1 a polygon was drawn passing through r, t_1, and s. At the center of the first length dc, the section 1 is taken. The eccentricity is measured, also d, and P is found by reference to Fig. 476. P^2 and e^2 are computed, also the value of $\dfrac{P^2 \cdot e^2}{d^3 \div 12}$ and of $\dfrac{P^2}{d}$, and the results recorded in the table of **Fig.**

478, as shown in the horizontal line marked 1. *P* was measured in thousands of pounds and not in pounds, thus reducing the size of the numbers in the table. The same computations are made for the second length or division of the arch, and the results recorded in line 2 of the table, Fig. 478. Similar computations are made for all the other *dc* divisions of the arch, and by adding up the last two vertical columns in Fig. 478, summation of $\dfrac{P^2 \cdot e^2}{d^3 \div 12}$ and summation of $\dfrac{P^2}{d}$ are obtained. These two values are added together and the result, which is equal to summation of $\dfrac{P^2 \cdot e^2}{d^3 \div 12}$ plus summation of $\dfrac{P^2}{d}$, is used to locate point 1 in Fig. 479.

Using the pole p_2 of Fig. 476, the second polygon was drawn in Fig. 475. It passes through *r* and *s* and also some point t_2 a short distance above t_1. Using this second polygon, the average *e* and *P* are found for the various lengths *dc* of the arch the same as was done in connection with the first polygon, and a table similar to Fig. 478 is made. By summing up the last two vertical columns of this second table, which is not shown, and adding them together the value, summation of $\dfrac{P^2 \cdot e^2}{d^3 \div 12}$ plus summation of $\dfrac{P^2}{d}$ is obtained. Using this value the point 2 is located in Fig. 479.

Using the poles p_3, p_4, and p_5 of Fig. 476, three more polygons were drawn in Fig. 475, as shown. Then a table corresponding to Fig. 478 is made for each of these polygons and the value, summation of $\dfrac{P^2 \cdot e^2}{d^3 \div 12}$ plus summation of $\dfrac{P^2}{d}$ is obtained in each case.

These values are used in locating points 3, 4, and 5 in Fig. 479. A smooth curve is now drawn through the points 1, 2, 3, 4, and 5 in Fig. 479, and its lowest point locates the place where summation of $\dfrac{P^2 \cdot e^2}{d^3 \div 12}$ plus summation of $\dfrac{P^2}{d}$ has its minimum value. Or, in other words, when a polygon is drawn in Fig. 475, using a pole p_x which is found from Fig. 479 to be about midway between p_2 and p_3, the work of resistance of the arch is a minimum and the

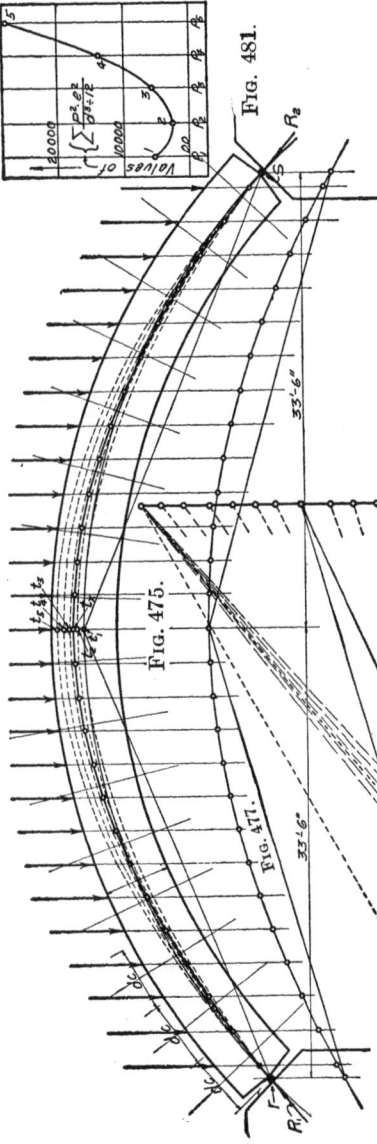

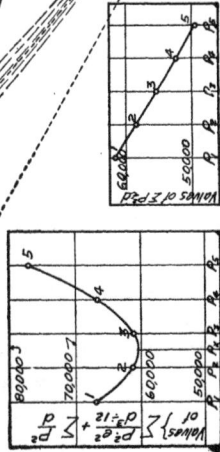

d	$\frac{d''}{12}$	e	e^2	P	P^2	$\frac{P^2 e^2}{d^2 \div 12}$	$\frac{P^2}{d}$
1	10.4	.20	.040	156.0	24,336	93.0	4867
2	5 feet	.50	.250	152.0	23,104	556.0	4621
3		.65	.422	144.0	20,736	8400	4147
4		.60	.360	140.0	19,600	680.0	3920
5		.50	.250	135.0	18,225	438.0	3645
6		.35	.122	129.0	16,641	1950	3328
7		.15	.022	126.0	16,876	34.0	3175
8		.05	.003	125.0	15,625	4.0	3125
9		.00	.00	125.0	15,625		3125
10		.05	.003	126.0	15,876	4.0	3175
11		.20	.040	129.0	16,641	64.0	3328
12		.40	.160	134.0	17,956	1760	3541
13		.50	.250	141.0	19,881	477.0	3976
14		.60	.360	145.0	21,025	7280	4205
15		.50	.250	152.0	23,104	6550	4621
16		.25	.062	157.0	24,649	147.0	4930

$$\Sigma \frac{P^2 e^2}{d^2 \div 12} + \Sigma (P^2) \frac{d}{65770} \quad \frac{499}{61779}, \frac{499}{65770}$$

FIG. 478.

FIG. 481.

FIG. 475.

FIG. 477.

FIG. 476.

FIG. 480.

FIG. 479.

polygon represents the true line of pressure. The correct polygon is shown in Fig. 475 by the full heavy line.

It is interesting to note that as the pole distance in Fig. 476 becomes shorter and the polygon in Fig. 475 rises, the value of summation of $\dfrac{P^2}{d}$ decreases as shown in Fig. 480, also the value of summation of $\dfrac{P^2 \cdot e^2}{d^3 \div 12}$ varies as shown in Fig. 481.

116. Hingeless Arches, Method of Least Work.—The hingeless arch presents a more difficult problem than the two-hinged arch. In the two-hinged arch two points on the line of pressure are known, but in the hingeless arch there is no known point on the line of pressure.

The method of least work may be applied to such a problem and a solution obtained, but when the student has finished he will probably think it was the method of most work for him. The solution would be similar to that for the two-hinged arch, except that more polygons and computation would be necessary because the location of the trial line of pressure would have to be varied at the supports as well as at the center. Of course, by carefully choosing the location of the trial polygons, the amount of work required for the solution of both two-hinged and hingeless arches can be reduced. For the former, three or four carefully chosen polygons will be sufficient. A number more will, of course, be required for the hingeless arch. It may be desirable to draw several polygons from each pole, one above the other. The values of summation of $\dfrac{P^2 \cdot e^2}{d^3 \div 12}$ plus summation of $\dfrac{P^2}{d}$ for the polygons from each pole would be plotted and the low point of the curve found. This low point would be the value plotted for that particular pole. In case the loading is unsymmetrical, it may even be necessary to vary the location of the poles, in the Fig. corresponding to Fig. 476, vertically as well as horizontally.

117. Hingeless Arches, General Discussion.—The method just given for the solution of the two-hinged arches and for hingeless arches is rather long and requires a considerable amount of labor. While it is probably true that no method requiring much less work will give a solution which is theoretically as accurate, yet there are often cases when a shorter but more approximate method will give satisfactory results. In many small arches a rather approximate location of the line of pressure is considered satis-factory, especially if there is a wall above the arch, as is often the case when small arches are used in buildings. It is therefore usually considered that a small arch is of satisfactory shape if a polygon for the loading can be drawn which will follow along the arch, always remaining within the middle third, or more strictly speaking, within the kern of the arch. Also the cross-section is considered of sufficient size if the computed unit stress, using this polygon as the line of pressure, is well within the allow-able pressure for the material of which the arch is constructed. If, for the given loading, a polygon can be drawn which will follow closely the axis of the arch, the shape of the arch may be said to be exceptionally good. In other words, if, for the given loading, a polygon is drawn which follows the axis of the arch as closely as possible, the distance which this polygon deviates from the axis gives an indication of how well the shape of the arch is adapted to the loading which it must carry. When a polygon cannot be drawn which will remain within the middle third, the shape of the arch is unsatisfactory. The student should remember that a polygon which follows the axis of the arch as closely as pos-sible may not be the true line of pressure, but it indicates whether or not the arch is of proper proportions. The true line of pres-sure will probably give a somewhat greater eccentricity than the polygon which follows the axis as closely as possible; it cannot give less, and therefore the actual stress will be somewhat greater than that computed from this polygon.

When an arch is well designed, and is of such a shape as to permit the drawing of a polygon, for the given loading, which will follow very close to the axis, the true line of pressure would be expected to pass a short distance above the axis at the center and a short distance below the axis at or near the ends. The theory of least crown pressure, so called, is perhaps a develop-ment from this fact. This theory assumes that the polygon which represents the true line of pressure is at the top of the middle third in the central part of the arch and at the bottom of the middle third at or near the ends. This polygon will give a smaller crown pressure, or resultant force acting at the center of the arch, than any other polygon that can be drawn within the middle third, hence the expression "least crown pressure." This method is only approximate, for the polygon obtained does not locate

the true line of pressure; in many cases it will give a greater
eccentricity than the true line of pressure.

Satisfactory designs can be made for small arches by using
an approximate method, if the results obtained from the con-

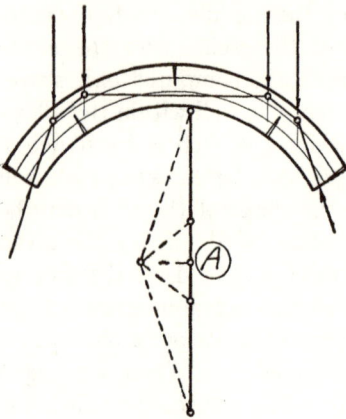

Fig. 482.

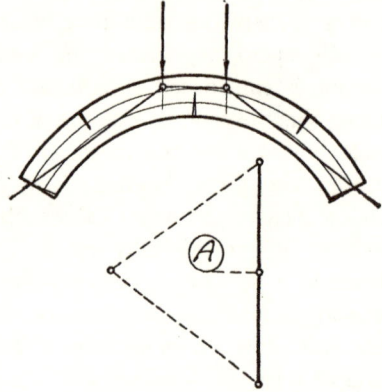

Fig. 483.

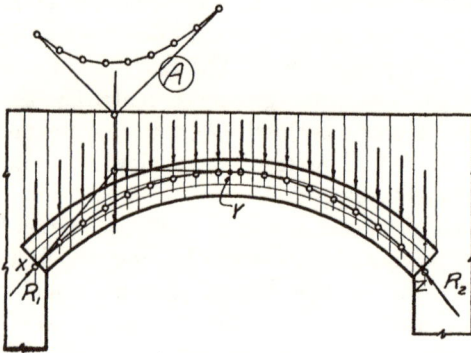

Fig. 484.

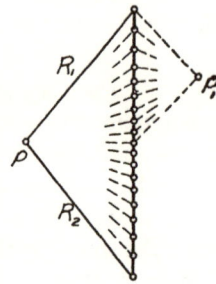

Fig. 485.

structions and computation are modified by good judgment.
The designer's judgment will be greatly improved by a study
of the more exact methods.

Large and heavily loaded arches should be very carefully
investigated.

It is of interest to note that heavy loads on the central part
of an arch cause the line of pressure to rise. An arch heavily

loaded at the crown is shown in Fig. 483, the polygon illustrating the shape of the line of pressure being drawn from the force polygon Ⓐ. Under such a loading an arch of this shape would tend to fail by the formation of cracks, as shown, and by the crushing of the material opposite the cracks. In Fig. 482 an arch is shown heavily loaded at the haunches. The loads are laid off in the force polygon Ⓐ from which the funicular polygon illustrating the shape of the line of pressure is drawn. Notice that under this loading the cracks would tend to form on the opposite side of the arch from which they would under a loading like that shown in Fig. 483.

118. Solution of an Arch Using Theory of Least Crown Pressure.—Consider the symmetrically loaded arch shown in Fig. 484. Since the theory of least crown pressure assumes that the line of pressure is at the top of the middle third in the center of the arch, and at the bottom at or near the ends, the points X and Z are located at the bottom of the middle third, while Y is located at the top. The loads are laid off along the load line of Fig. 485, p_1 is chosen, polygon Ⓐ drawn, and the pole p located as in Fig. 485. Using this pole, the desired polygon may be drawn passing through points X, Y, and Z . If the loading were not symmetrical, the points X, Y, and Z could be assumed and the polygon drawn using the general construction for passing a funicular polygon through three points. Or the construction used for an unsymmetrically loaded three-hinged arch might be used, the points X, Y, and Z being used in place of the three hinges.

119. Investigation of a Gothic Vault.—An investigation of a Gothic vault will be found very instructive. Fig. 486 shows a section of the vault that will be investigated, while Fig. 487 shows a plan looking down from above. The two diagonal ribs are the supporting ribs and divide the vault into four equal parts which are filled in by the thin portions of vault arching over from one rib to the other. Fig. 488 shows a half section through the center of a diagonal rib ending at column 1. Sections through the diagonal ribs ending at column 2 would of course look the same. The thin portions of the vault which arch over from rib to rib are divided into narrow slices, as shown in Fig. 487, each of which acts somewhat as an arch supported by the ribs. Half sections through these slices are shown in Fig. 489. Each slice is divided into small divisions, and the weight of the masonry in each division is computed and an equivalent concentrated load substituted for each slice. These loads are laid off in a force polygon, and the action line of their resultant located by choosing any pole and drawing a funicular polygon. Since the arches are symmetrical and are symmetrically loaded, the lines of pressure must be horizontal at the center. With this known, each half section is investigated by seeing if a polygon can be drawn which will follow reasonably close to the axis of the arch, or at least remain within the middle third. All of the arches or slices are found to be of satisfactory shape, and the approximate magnitudes of the thrusts R'_1, R'_2, R'_3, etc., which each slice delivers to the rib may be found by scaling the rays R'_1, R'_2, R'_3, etc. in the force polygons Ⓐ, Ⓑ, Ⓒ, etc. For the investigation of the ribs, consider the half rib extending from the center of the vault down to column and buttress 1, a longitudinal section of which is shown in Fig. 488. This half rib is divided into six divisions, as shown in Fig. 487, divisions 2, 3, etc., being that part of the rib which is in contact with the ends of slices 2, 3, etc., of the thin portion. Consider the division of the rib marked 2. There are three forces acting on this division, the weight of the division itself, and a thrust R'_2 from the slice 2 on each side. Let these thrusts R'_2 be broken into vertical and horizontal components, V_2 and H_2, at the edge of the rib. The two vertical components thus obtained will be equal and parallel to each other, therefore their resultant V'_2 will act vertically down through the axis of the rib and will be numerically equal to 2 V_2. The weight W_2 of this division of the rib will also act down and may be added to V'_2, The two horizontal components of the thrusts R'_2 are equal and act in the same plane, but their action lines make an angle of 90 degrees with each other. Their resultant H'_2, which acts horizontally and through the axis of the arch, is found in magnitude in Fig. 490 and laid off to scale as shown in Fig. 488. P_2, the resultant of W_2, V'_2 and H'_2, is found by constructing a parallelogram of forces in Fig. 488. This is the resultant loading which acts on division 2 of the rib. P_1, P_3, P_4, etc., the resultant loads on the other divisions of the rib, are found in a similar way.

These loads are laid off to scale along the load line of Fig. 491. Using any convenient pole p_1, the funicular polygon Ⓐ in Fig. 488 is drawn and the action line of R, the resultant loading on

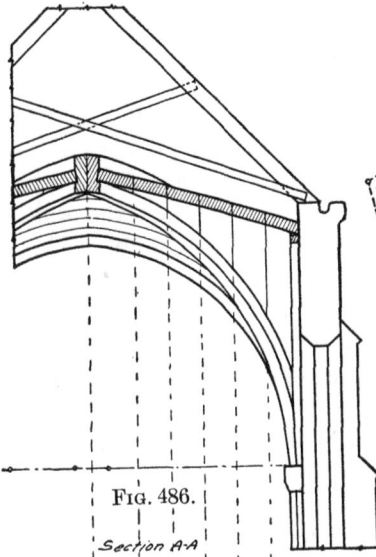

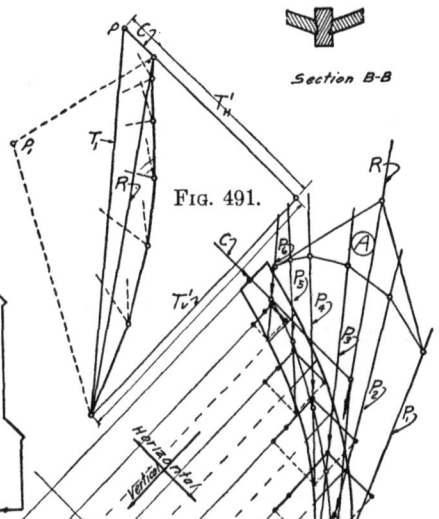

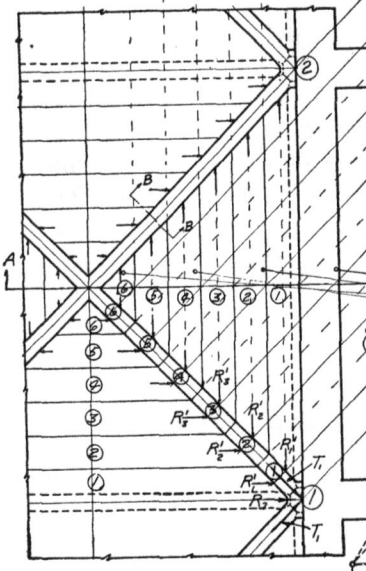

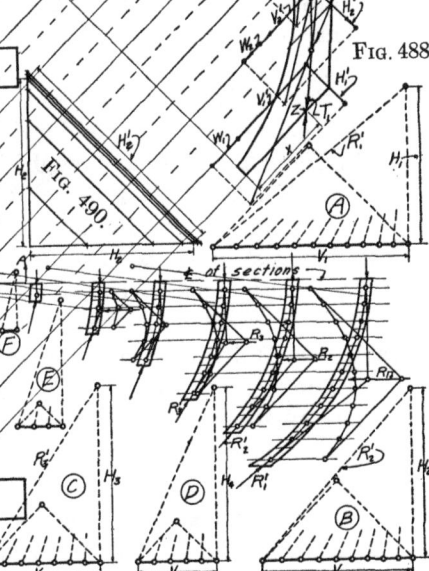

Section B-B

FIG. 491.

FIG. 486.

Section A-A

FIG. 488.

FIG. 490.

FIG. 487.

FIG. 489.

this half of the rib, is located. Since the rib is symmetrical and symmetrically loaded, the crown pressure must be horizontal. It is found by trial that a funicular polygon can be drawn which will remain within the middle third of the rib. Such a polygon is shown in the figure. From this polygon and Fig. 491, using the pole p from which the approximate line of pressure was drawn, values for the crown pressure C and the thrust T_1 may be obtained. The resultant thrust T_1 passes out into the wall and buttress at point Z, a distance X above the capitol. The horizontal and vertical components T'_H and T'_V, of the resultant thrust T_1 are found as shown in Fig. 491. Now by referring to Fig. 487, it will be seen that there are two diagonal ribs ending at each column and buttress. Vertical planes passed through the axes of these ribs make angles of 45 degrees with the wall and an angle of 90 degrees with each other. There is a false rib or molding projecting below the thin portion of the vault and running from one column directly across to the other. The part of this molding which projects below the thin portion was not considered in finding the thrust R'_1. The weight of this molding is therefore found and compared with the weight of slice 1, and the thrust R_7 found by proportion. The molding along the wall just under the thin portion of the vault is assumed to be fastened to and carried by the wall.

In Fig. 492 a section through the wall and buttress is shown. The magnitude of T_H is equal to the resultant of two forces T'_H, acting at 90 degrees to each other, plus the horizontal of R_7. The last force is of course very small. The magnitude of T_V is just equal to two T'_V plus the vertical component of R_7. T, the resultant of T_V and T_H, is the resultant thrust of the vault against the wall and buttress. Since the cross-section of the column is very small in comparison with the cross-section of the wall and buttress, and since the highest stress will be along the outside edge of the buttress, the column will be neglected.

The resultant force acting on section X_1–X_1 is found, also the resultant forces acting on the sections X_2–X_2, X_3–X_3, etc., Fig. 493 being used. The points a, b, c, d, and e are thus located and the line of pressure drawn. The roof was here assumed to be trussed or braced so that the reaction from the roof construction would be vertical. There could be no pushing out or overturning of the buttress without a large portion of the wall

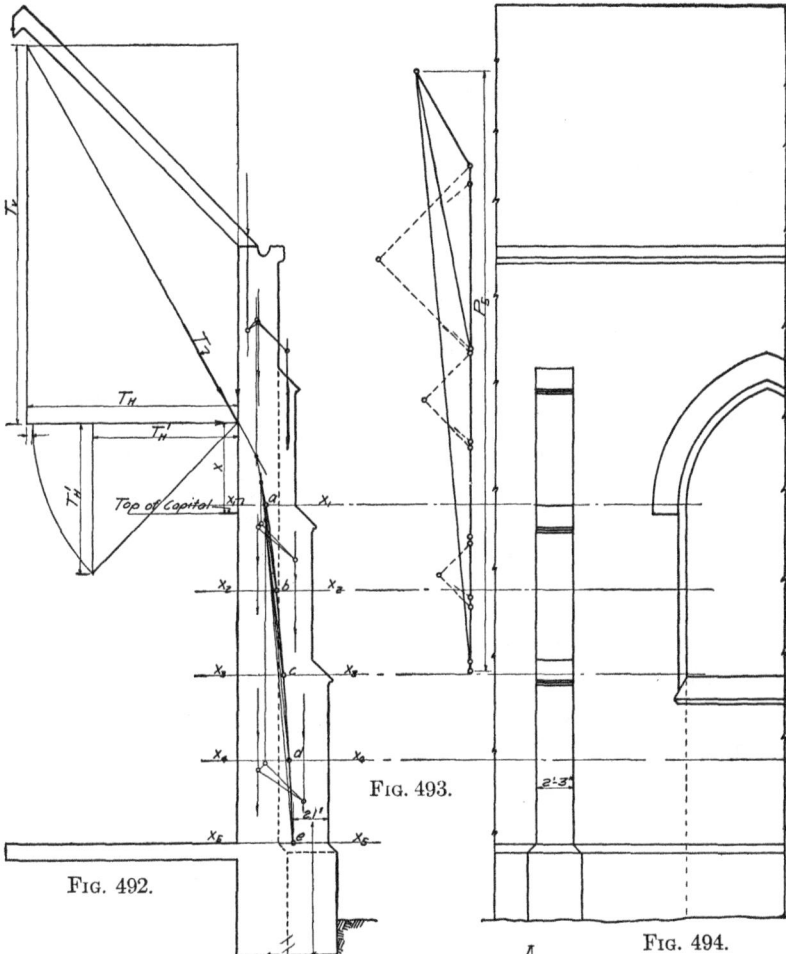

FIG. 493.

FIG. 492.

FIG. 494.

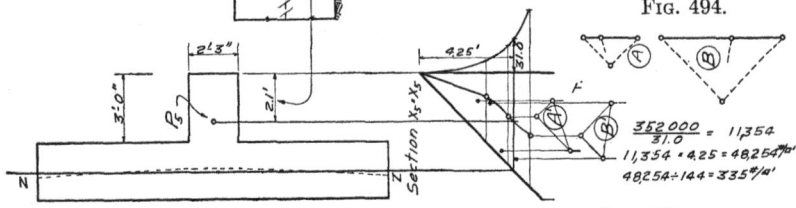

FIG. 495.

$$\frac{352,000}{31.0} = 11,354$$

$$11,354 \times 4.25 = 48,254^{\#/lo'}$$

$$48,254 \div 144 = 335^{\#/lo'}$$

FIG. 496.

being overturned also. In figuring the weights above sections X_1–X_1, X_2–X_2, etc., the wall and buttress were assumed to act together, and the wall clear over to the window was added in. However, the masonry under the window was neglected. It is a question how soon the weight of the masonry and roof above the window should be added in when locating the line of pressure for the wall and buttress. In this case the roof load above the window was added in when the resultant on X_1–X_1 was found; one-half the weight of the masonry above the window was added in when finding the resultant on X_2–X_2, and the other half was added in with the resultant on X_3–X_3, this would seem reasonably conservative.

Any section through the wall and buttress might be taken and its maximum unit pressure found. In order to illustrate the method, the maximum unit pressure on section X_5–X_5 will be found. This section is drawn as shown in Fig. 495, the wall over as far as the side of the window being considered. The vertical component P_5 of the resultant force acting on section X_5–X_5 is found from Fig. 493 to be about 352,000 pounds. The same general construction that was used in finding the maximum unit pressure in a chimney is now applied, and the maximum unit pressure is found to be 335 pounds per square inch. The length of wall that should be considered for section X_5–X_5, as shown in Fig. 495, is a matter of judgment. The masonry under the window no doubt helps some, but it is safer to neglect it. Actually, the axis of zero stress Z–Z probably varies slightly from the straight line, as indicated by the dotted line.

In this solution the ribs and the slices of the thin portion of the vault were solved, using an approximate method. If a more exact solution were desired and the necessary time available, the method of least work might be used.

Wind pressure may be considered without any serious difficulty developing.

120. A Study of Domes.—The following is a brief study of the stresses in domes, but is not an exact analysis. Consider a hemispherical dome such as is shown in plan by Fig. 497 and in section by Fig. 498. Let the dome be divided into a number of small equal slices, one of which is shown in plan to a large scale by Fig. 499, and in section by the left half of Fig. 500, the line X–X being the central axis of the dome. Now let the slice shown in Figs. 499 and 500 be divided into small divisions by the planes 1, 2, 3, etc. An estimate of the volume of each one of these divisions is made and its weight obtained, which weight will be represented by a concentrated force at the centroid of each division. The volume of division d_2 may be found approximately by multiplying together the dimensions v, u, and w; the volume of other divisions may be found in a similar way.

Consider the wedge-shaped division d_1. We may say there are four forces acting, the weight of the masonry, the pressure from the division d_2, and the pressure on each side from the adjoining slices. These two thrusts from the adjoining slices are equal and horizontal in a hemispherical dome of uniform thickness under its own dead weight, and they act in the same plane. Their resultant together with the other two forces form a system of three forces in equilibrium; therefore these three forces must be concurrent and in the same plane. Take any other division, say d_5, and there are five forces acting, the weight, the two forces from the adjoining slices, one on each side, and the two forces from the adjoining divisions, in this case the divisions d_4 and d_6.

Now in Fig. 501 the loads A–B, B–C, etc., are laid off in order as shown, and in Fig. 500 a polygon is drawn following along the center line. Parallel to the strings of this polygon, the rays from A, B, etc., are drawn in Fig. 501 and extended until they intersect the horizontal from A. The intercepts marked off on this horizontal line are the resultants of the pressure from the adjoining slices. The magnitudes of the thrusts P_2 and P'_2 are found by drawing a line from 1, in Fig. 501, parallel to the action line of P_2 and another line from 2 parallel to P'_2; their intersection gives point b' and determines the magnitude of P_2 and P'_2. In a similar way the magnitudes of P_1, P_3, P_4, P'_1, P'_3, etc., may be found. It should be noted that the forces P_1 to P_{10} are compression while those from P_{11} to P_{18} are tension. This means that above section Y–Y there is circumferential compression, while below this section there is circumferential tension. If the material were unable to take tension the circumferential stresses below Y–Y would be zero and the line of pressure would move out to the position of the dotted line, the rays K to R all radiating from point 10. This dotted line passes outside of the masonry shown in section by the left half of Fig. 500, and the dome would therefore be unstable in case the masonry could not take tension. It would then be necessary to increase the thickness near the base, as shown by the right half of Fig. 500. For the portion from section X–X to Y_1–Y_1, the polygon

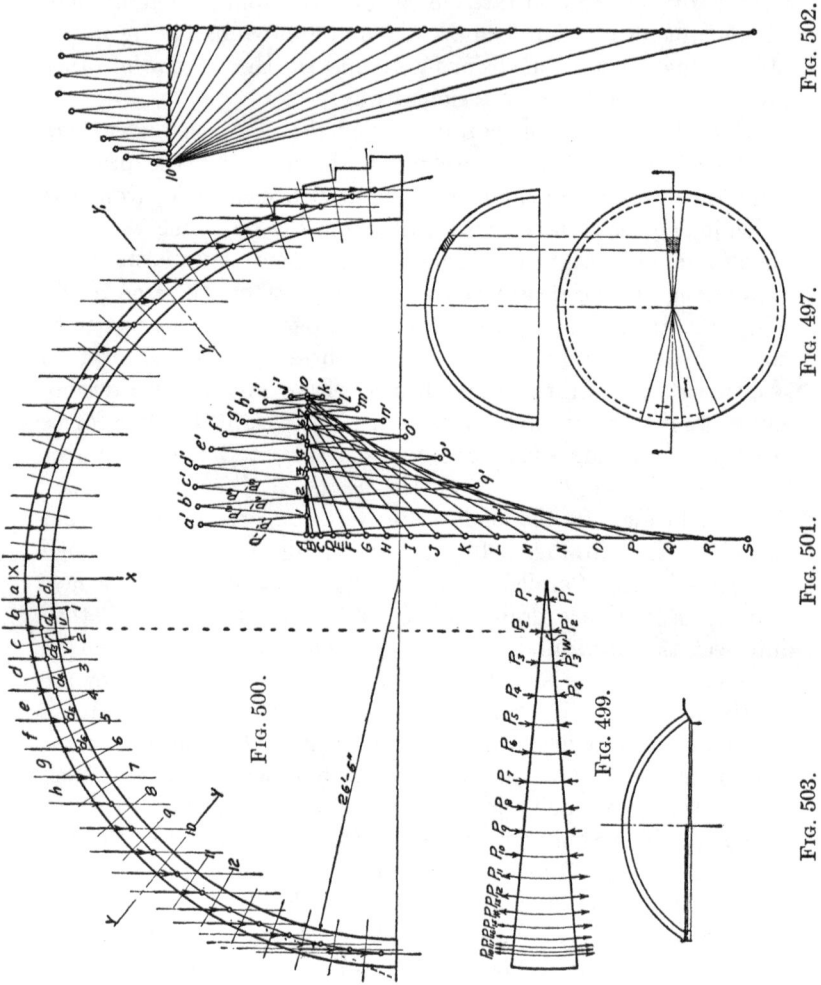

Fig. 502.

Fig. 497.

Fig. 501.

Fig. 500.

Fig. 499.

Fig. 503.

was drawn the same as for the left half of Fig. 500, but for the portion below section Y_1–Y_1, point 10 in Fig. 502 is used as a pole, and the polygon representing the line of pressure drawn as shown.

When a reinforced concrete dome is used the analysis of Fig.

501 and the left half of Fig. 500 may be used, reinforcing steel taking the circumferential tension. The reactions at the base of the dome are vertical in this case.

When an ordinary masonry dome is used, the analysis indicated by Fig. 502 and the right half of Fig. 500 may be used. In this case the reaction at the base of the dome has a horizontal component which must be taken up in some way by the structure below.

A dome of the shape shown in Fig. 503 might be analyzed in a way similar to the methods used for the hemispherical dome. The thrust at the base might be taken up by a steel ring, as shown at the left of the figure, or by the structure below.

It will be interesting for the student to study ring domes and masonry domes together, since the distribution of stress is somewhat similar.

The method given above is approximate, corresponding somewhat to the method of investigating an arch by drawing a polygon which will follow the center line as nearly as possible. However the results obtained from this method will be instructive, and when modified by good judgment, should be satisfactory for designing.

The theory of least work might be used in developing a more exact method.

The stresses produced by wind would be relatively small. When studying the effect of the wind the student should keep in mind the solution of the ring dome under wind loads.

CHAPTER VIII

REINFORCED CONCRETE

No attempt has been made to make this chapter a complete study of the theory of reinforced concrete. Its purpose is to show how a number of the problems in reinforced concrete can be solved to advantage by graphical construction. These graphical constructions are of special value when applied to complicated and unusual problems, such as eccentrically loaded columns, or beams with many layers of steel.

It is taken for granted that the student is familiar with some of the more important assumptions that are usually made in connection with reinforced concrete. We usually consider that the concrete on the tension side of the neutral axis takes no direct stress, but that the reinforcing steel takes all of the direct tension. On the compression side of the beam the compression in the concrete is assumed to vary directly as the distance from the neutral axis. The ratio of the modulus of elasticity of steel to that of concrete, which is usually denoted by n is taken as 15 for the common $1 : 2 : 4$ mixture of concrete; for different mixtures the value n should be varied from this value in accordance with standard practice. This means that for a given deformation, a given area of steel will take n times as much stress as the same area of concrete, or as much stress as an area of concrete n times that of the steel. If the area of the steel reinforcement is multiplied by 15, assuming n to have this value, and the area thus obtained substituted for that of the steel, we have what is sometimes called the transformed section. It should be observed that this new area obtained by multiplying the area of the steel by 15, can be considered just the same as so much concrete area, except that it is able to take either tension or compression.

121. Simple Rectangular Beams.—There are perhaps three types of problems which might arise. First, the beam and the allowable stresses are given and the allowable moment is required. Second, the beam and the moment may be given, and the stresses produced required. Third, the bending moment and the allowable stresses may be given, and a design of the beam required.

Consider the beam illustrated in Fig. 504, having dimensions as shown and being reinforced with four $\frac{3}{4}$-inch square bars. It will be assumed that the allowable f_c is 650 pounds per square inch and that the allowable f_s is 16,000 pounds per square inch.

The upper or compression side of the beam is divided into small slices. These slices are taken parallel to the axis about which bending occurs, and the dividing into slices is continued down until it is evident that the neutral axis has been passed. The area of the top slice will be called A_1, that of the second slice A_2, etc. Lines are now drawn from the centroids of the various slices and parallel to the axis about which bending occurs. A similar line is also drawn from the center of the reinforcing steel. The area of the steel is called A_s, and this area times n (n is here taken equal to 15) will be called A_{sn}. From the point X in Fig. 505, a vector is laid off to the right, equal to A_{sn} to some convenient scale, and to the left of point X another vector is laid off equal to A_1; then A_2, A_3, etc., follow in order, after which the pole p is chosen. It is usually found convenient to make the line X–p vertical, although this is by no means necessary. From Fig. 505 the funicular polygon Fig. 506 is drawn, observing the rules for the construction of funicular polygons. The intersection O locates the neutral axis as will now be shown. Take any intercept Z, in Fig. 506, above the intersection O and parallel to the lines drawn from the centroids of the areas. This intercept times the pole distance H gives the first moment of the portion of the beam above the intercept Z about the line Z extended. Also take any similar intercept Z_1 below the intersection O, and this intercept times the pole distance H will give the first moment of the stress-taking portion of the beam below Z_1 about the line Z_1 extended. Now let the intercept Z be moved down to the position of the intercept W, and the intercept Z_1 moved up to the same position. The intercepts Z and Z_1 will now be equal, which means that about the line W extended the moment of the portion of the beam above is numerically equal to the moment of the stress-taking portion below, but is of different sign. Therefore, the line Y–O–j contains the centroid of the stress-taking portion of the beam, and is the neutral axis. It should be noted that the areas of concrete below the intersection

O have not been used in obtaining the intersection O, thus conforming with the assumption that the concrete on the tension side of the neutral axis does not take direct tension.

At a convenient point along the neutral axis, Y is chosen, and the base line C–F is drawn in Fig. 507. The line C–D is measured off equal to the allowable f_c to some convenient scale, and the line D–Y drawn and extended to G. The line G–F is now measured to the same scale and multiplied by 15; this gives the stress in the steel when the concrete is stressed up to the

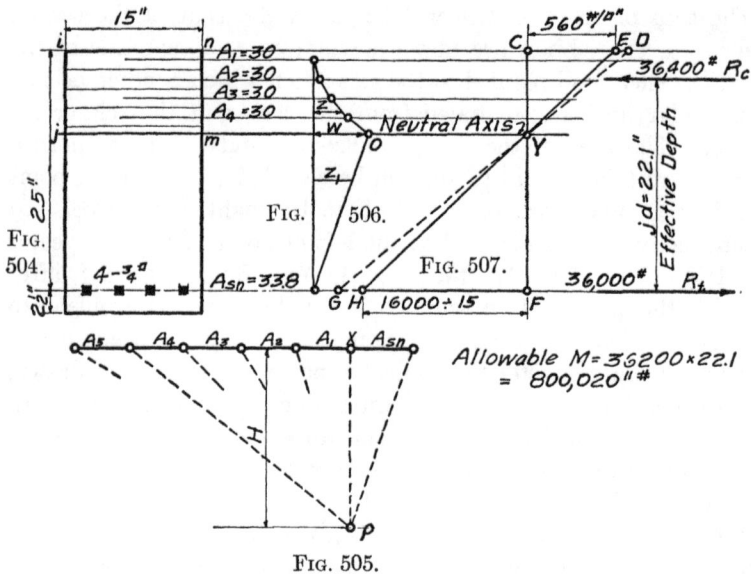

FIG. 504.

FIG. 506.

FIG. 507.

Allowable $M = 36200 \times 22.1$
$= 800,020''^{\#}$

FIG. 505.

allowable. In this case G–F times 15 is more than 16,000, which shows that the steel governs. The line H–F is, therefore, measured off equal to $16,000 \div 15$ and the line H–Y–E drawn. C–E is measured and found to be 560; this means that when the steel is stressed up to 16,000 pounds per square inch, the maximum compression in the concrete is 560 pounds per square inch.

There are compressive stresses in the concrete above the neutral axis which are assumed to increase uniformly from zero at the neutral axis up to a maximum of 560 pounds per square inch at the top of the beam. The stress volume of the compressive stresses is, therefore, wedge-shaped, and is shown in plan by the area i–j–m–n, and in elevation by the triangle C–E–Y. The average stress is, of course, one-half of the maximum, and it follows that R_c, the summation of all of the compressive stresses, is equal to the area i–j–m–n times one-half of C–E. This is found to be 36,400, and it acts at the centroid of the stress volume, which is $\frac{1}{3}$ of C–Y below the top of the beam. The summation of all of the tensile stresses is equal to F–H times 15 times the area of the steel and is here found to be 36,000. The summation of the compressive stresses should, of course, be equal to the summation of all of the tensile stresses; in the above problem a variation of 400 pounds or about 1 per cent is found. The allowable bending moment is equal to the effective depth times R_c, or times R_t, or better yet, times the mean between R_c and R_t. The allowable moment for the given beam is found to be about 800,000 inch-pounds.

Problems of the second type may be solved in a very similar way. Let it be required to find the stresses produced in the beam shown in Fig. 504 by a bending moment of 500,000 inch-pounds, n to be taken as 15. Figs. 505 and 506 would be drawn as already described, also Fig. 507, except that any convenient length might be taken for the stress in the concrete, and the corresponding length for F–H. For these stresses, the resisting moment would be found. Then, knowing the bending moment which would produce the assumed stresses, the stresses produced by the given moment can be easily found by proportion. In the given beam it has already been found that a bending moment of 800,000 inch-pounds produces a stress of 16,000 pounds per square inch in the steel, and 560 pounds per square inch in the concrete. By proportion it is found that a bending moment of 500,000 inch-pounds will produce stresses of 10,000 pounds per square inch in the steel, and 350 pounds per square inch in the concrete.

The third type of problem requires constructions which are slightly different than those already described. Let it be required to design a beam to carry a moment of 1,500,000 inch-pounds, the allowable f_c being 650 pounds per square inch, f_s, 16,000 pounds per square inch, and n is to be taken as 15. Not only must the size of the beam be found, but also the amount of steel; and it should be remembered that the proportion of steel is supposed to be such that the allowable stress in both materials will be approximately reached under the given moment. To start with,

a beam of any convenient size may be chosen. In this problem
a beam 15 inches wide and $27\frac{1}{2}$ inches deep, down to the steel,
is assumed; the cross-section is shown in Fig. 508. In Fig. 509
the line C–D is drawn equal to the allowable f_c to some convenient
scale, and G–F is made equal to the allowable $f_s \div 15$ to the same
scale. Then Y, the intersection of C–F with D–G, locates the
neutral axis for the given stresses. Of course, if too much or

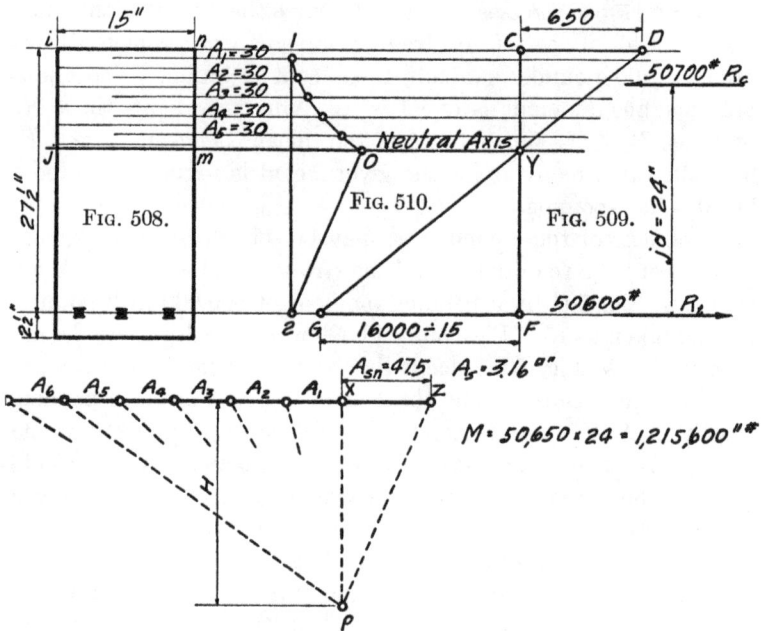

Fig. 508.   Fig. 510.   Fig. 509.

Fig. 511.

too little steel were used, the location of the neutral axis would
be changed, but the ratio of f_c to f_s would also be changed.

Having found the location of the neutral axis for the given
stresses, the next step is to find the proper amount of steel. The
compression side of the beam is divided into slices and lines are
drawn from the centroids of these slices and from the steel as in
Fig. 504. The areas A_1, A_2, etc., are laid off in Fig. 511, and
the funicular polygon of Fig. 510 is drawn from 2 to 1 and then
down to point O, where the neutral axis is intersected. The
closing string of this funicular polygon is the line 0–2 parallel

to which a line is drawn from p intersecting the load line at Z. The line Z–X measured to the proper scale gives A_{sn}, which, divided by 15, gives the proper area of steel, A_s, in this case found to be 3.16 square inches. The resultant of all of the compressive forces, R_c, is found and located as in Fig. 507, also the resultant of the tensile forces R_t. The resisting moment of the assumed beam is then found to be about $24 \times 50,650 = 1,215,600$ inch-pounds. This moment is less than the moment which must be carried. Now the strength or allowable resisting moment of a reinforced concrete beam varies directly as the width and also as the square of the depth, the percentage of steel remaining the same. Therefore, if the beam is to carry a moment of 1,500,000 inch-pounds and is to be $27\frac{1}{2}$ inches deep, from proportion we find that the width should be about $18\frac{1}{2}$ inches and the steel area should be about 3.9 square inches. Or if it is desired to keep the beam 15 inches wide, the proper depth is found by proportion to be 30.6 inches and the steel area 3.52 square inches.

122. I for Rectangular Beams.—It is of interest to note how easily the moment of inertia of reinforced concrete beams can be found by graphical construction. Fig. 512 shows the cross-section of a beam, the moment of inertia of which is desired. The compression side of the beam is divided into slices, Fig. 513 is drawn, and from it the funicular polygon of Fig. 514, which locates the neutral axis by the intersection O. The pole p' is chosen, by the use of which the funicular polygon Fig. 515 is drawn. Thus the construction for the moment of inertia by Culmann's approximate method is completed, and I is given by the formula $I = H \cdot H' \cdot V'$, for the derivation of which see Chapter III. It will be found convenient to take even numbers for H and H' in order to make the multiplication more simple. In this problem I is found to equal $150 \times 10 \times 10.4 = 15,600$ inches to the fourth power. To obtain the area, A_{sn}, which was used in finding the moment of inertia, the area of the steel was multiplied by 15; therefore, I will be in terms of concrete.

The moment of inertia may be used to find the actual stresses or the allowable bending moment in the following way: From mechanics we have the formula $\dfrac{M}{S} = \dfrac{I}{c}$. When the bending moment is given and the maximum stress in the concrete is desired, the formula may be written $f_c = \dfrac{M \cdot c_c}{I}$, and if the stress in the steel

is desired the formula is $f_s = \dfrac{M \cdot c_s}{I} \cdot n$. When the allowable

bending moment is to be found, the two equations given above,

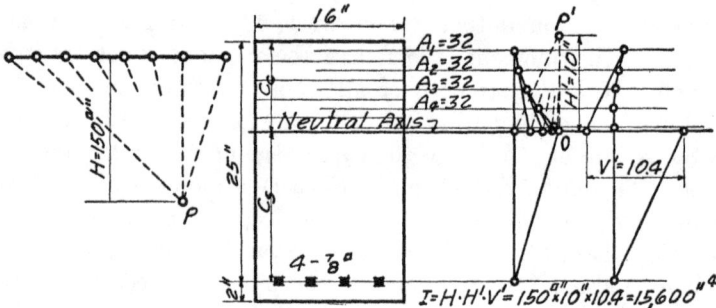

FIG. 513.　　　FIG. 512.　　　FIG. 514.　　　FIG. 515.

may be solved for M, using the allowable f_c and f_s. The smaller one of these values will be the allowable moment.

Figs. 516, 517, and 518 show the construction for the moment of inertia, using Mohr's method. I is obtained by multiplying the pole distance H by two times the area enclosed by the funic-

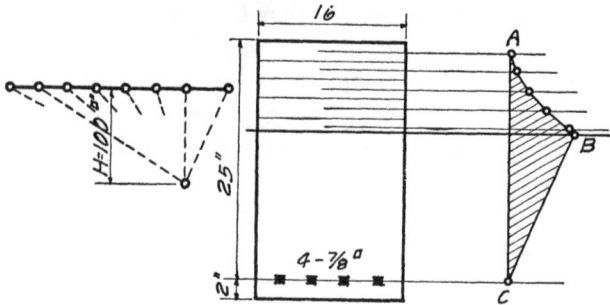

FIG. 517.　　　FIG. 516.　　　FIG. 518.

ular polygon in Fig. 518. This area to be measured to the scale of the space diagram.

123. T-Beams.—The common formulas used for designing and investigating T-beams neglect the compression in the stem below the flange. This approximation, of course, produces some error; when the stem is narrow and the lower side of the flange extends down almost to the neutral axis, the error is small, but when the lower side of the flange is a number of inches above the axis and the stem is large, the error may be considerable The graphical construction here given takes into account the compressive stresses that exist in the stem, thus avoiding the error which exists in the common formulas.

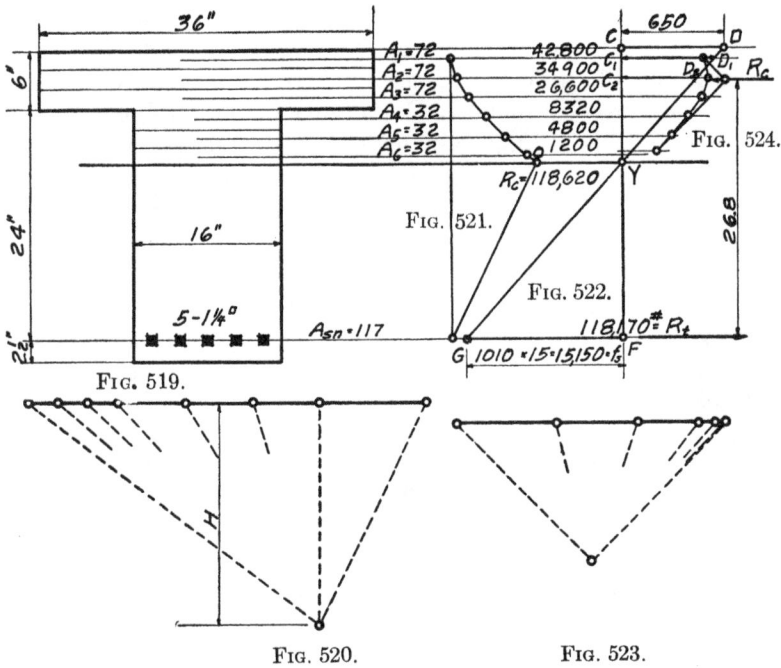

Fig. 519.
Fig. 520.
Fig. 521.
Fig. 522.
Fig. 523.
Fig. 524.

As with rectangular beams, three types of problems will be considered.

The allowable bending moment for the beam shown in Fig. 519 is desired. The allowable stresses are $f_c = 650$ pounds per square inch, and $f_s = 16,000$ pounds per square inch; $n = 15$. The compression side of the beam is divided into slices and, after A_{sn} and the concrete areas A_1, A_2, A_3, etc., have been found, Fig. 520 is drawn in the usual way. The neutral axis is now located by the construction of Fig. 521. The line C–Y–F is drawn and C–D is made equal to 650 to some convenient scale. The line D–Y is drawn and extended to G, thus marking off F–G, which is found to be approximately 1010. This shows that the stress in the steel is $1010 \times 15 = 15,150$ pounds per square inch when the concrete is stressed up to the allowable. When the maximum stress in the concrete is 650 pounds per square inch, the average stress over the top slice, A_1, is found by scaling the line C_1–D_1. This stress times the area A_1 gives the sum of the compressive stresses carried by the top slice, in this case found to be about 42,800 pounds. Similarly, the average stress over A_2 is found by scaling the line C_2–D_2, and the sum of all the compressive forces over this slice is found to be 34,900 pounds. The average stress for each of the other slices is found, and the sum of all the compressive stresses on each is obtained by multiplying its area by the average stress. When the slices are small, the resultant of all the compressive stresses on each slice will act approximately at the center of the slices. All of these resultants are laid off to scale along the load line of Fig. 523, from which Fig. 524 is drawn locating the action line of the resultant of all of the compressive stresses, R_c. The sum of all of the compressive stresses is found to be 118,620, which is slightly larger than 118,170, the sum of the tensile stresses. This checks the work which has been done, showing that the errors have been small. The mean between R_c and R_t is found and multiplied by 26.8 inches, which is the effective depth jd. The resulting 3,173,000 inch pounds is the allowable bending moment.

When the beam and moment are given, the actual stresses produced may be found by constructions similar to those shown in Figs. 519 to 523. The line C–D may be taken equal to any convenient stress, and when the resisting moment has been found for the assumed stresses, the actual stresses produced by the given moment can be found by proportion, as was illustrated for rectangular beams.

In order to illustrate how the third type of problem may be solved, let it be required to design a T-beam to carry a bending moment of 3,300,000 inch-pounds. To start with, the beam shown in Fig. 525 by the full lines will be considered, and the allowable stresses of the previous problem will be used. In Fig. 526 the intersection Y locates the neutral axis. Then by means of Figs. 527 and 528, using the intersection O, the proper steel area is found to be about 7.46 square inches. By the use of Figs.

529 and 530 the resultant of all of the compressive stresses is located, and using the effective depth thus found, the allowable moment for the assumed beam is found to be approximately 2,976,000 inch-pounds. This is smaller than the given moment, therefore the beam must be made stronger. This might be done by increasing the width of both flange and stem, in which case the proper width and steel area could be found by proportion, or by increasing the width of the flange only. In case the second

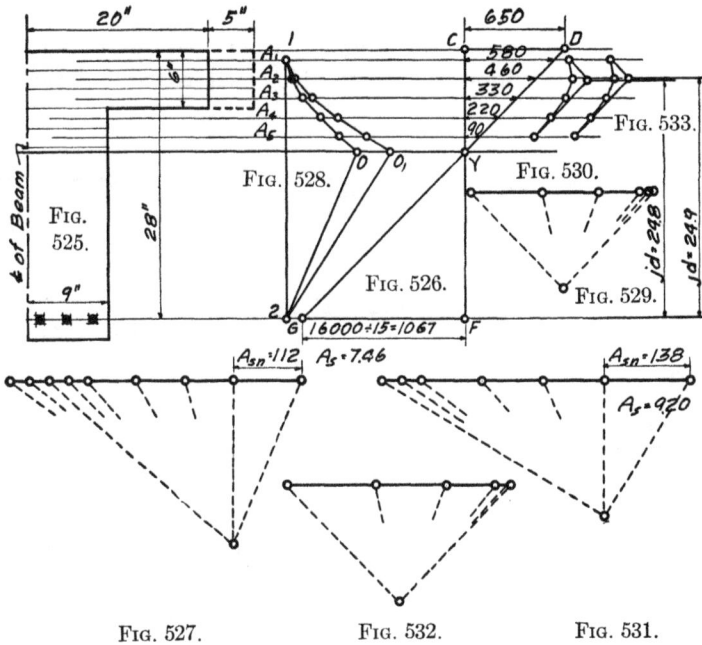

FIG. 527.   FIG. 532.   FIG. 531.

method is chosen, let the width of the flange be increased, say 10 inches, 5 inches on each side. By use of Fig. 531 and the intersection O_1, the proper steel area for this new beam is found to be 9.2 square inches. Then by use of Figs. 532 and 533 the action line of the resultant of the compressive stresses is found, and from it the effective depth. Using this new effective depth, the allowable moment for the second beam is found. Knowing the allowable moment for these two beams, the increase in allowable moment per inch of increase in width of flange can be found. With this known, the proper width of flange for the given moment

can easily be determined.   In a similar way the proper amount
of steel can be found.

124. I of T-Beams.—The moment of inertia of a T-beam may
be found in much the same way as the moment of inertia of
rectangular beams.   First, the cross-section of the beam is drawn
to scale as shown in Fig. 534.   Fig. 535 is drawn in the usual way
with some convenient pole distance H, and from it Fig. 536.
I may now be found by multiplying 2 times the pole distance H
by the area inclosed by the polygon o–q–r, or the pole p' may be
chosen and Fig. 537 drawn.   Then $I = H \cdot H' \cdot V'$.

125. Double Reinforced Concrete Beams.—By graphical construction, the solution of double reinforced concrete beams is

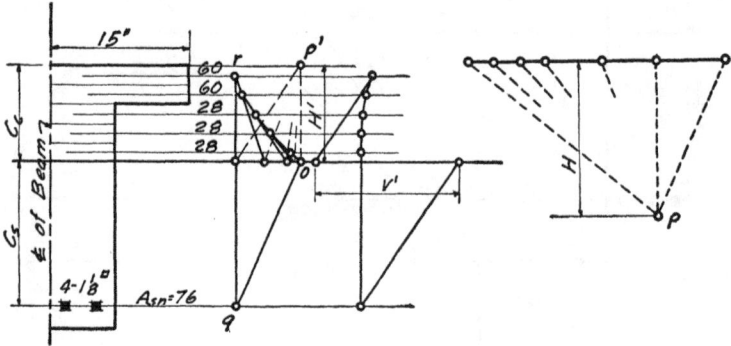

Fig. 534.   Fig. 536.   Fig. 537.   Fig. 535.

no more difficult than the solution of T-beams.   First consider
the case when the beam and the allowable stresses are given,
and the allowable moment is desired.   A cross-section of the
beam is shown in Fig. 538, the upper side or compression side
being divided into slices as usual.   For convenience, the slices
are so taken that the compression steel comes at the center of
one of them.   The area of the concrete, in this slice containing
the compression steel, is added to 15 times the area of the steel.
Attention should be called to the fact that the compression steel
displaces some of the compression concrete.   This should be
taken account of when the areas are computed.   Fig. 539 is drawn
in the usual way; and from it Fig. 540 which locates the neutral
axis by the intersection O.   Fig. 541 is next drawn and the resultant of all the compressive stresses is found and its action line

located by means of Figs. 542 and 543. Using the effective
depth, found by scaling to be 26.7 inches, the allowable resisting
moment is found to be about 3,042,500 inch-pounds.

When the beam and the moment are given and the actual
stresses are desired, they may be found by proportion as already
explained in connection with rectangular beams and T-beams.

Let it be required to design a double reinforced concrete beam
to carry a given moment. It will be assumed that the allowable

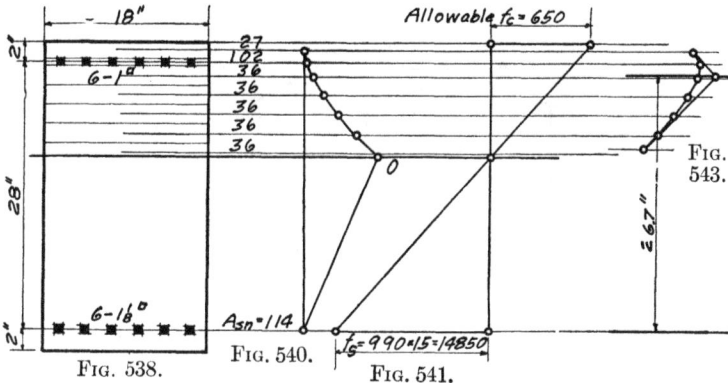

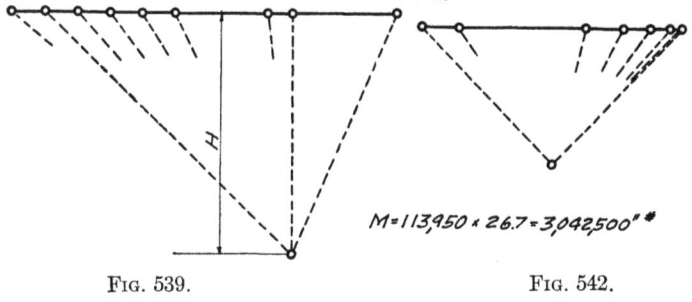

Fig. 538.   Fig. 540.   Fig. 541.
Fig. 543.

Fig. 539.   Fig. 542.

$M = 113,950 \times 26.7 = 3,042,500''^{\#}$

stresses are $f_c = 650$ pounds per square inch, and $f_s = 16,000$
pounds per square inch, 15 will be taken for n. The width of the
beam will be limited to 20 inches and the over-all depth to $32\frac{1}{2}$
inches, or 30 inches down to the steel. This beam must be
reinforced so that it can carry a bending moment of 3,200,000
inch-pounds. In Fig. 545 the line C–D is laid off equal to the
allowable stress in the concrete, while the line F–G is made equal
to $\frac{1}{15}$ the allowable stress in the steel to the same scale. Then
the intersection Y locates the neutral axis for these given stresses.

Figs. 546 and 547 are now drawn and the value of A''_{sn} is determined in the same way that A_{sn} was found by Figs. 510 and 511. The value of A''_{sn} thus found may be divided by 15 and A''_s obtained, which is the proper amount of tension reinforcement to develop the compressive strength of the concrete. With this amount of tension reinforcement and no compressive reinforcement, the maximum compression in the concrete would be 650 pounds per square inch when the tension in the steel was 16,000

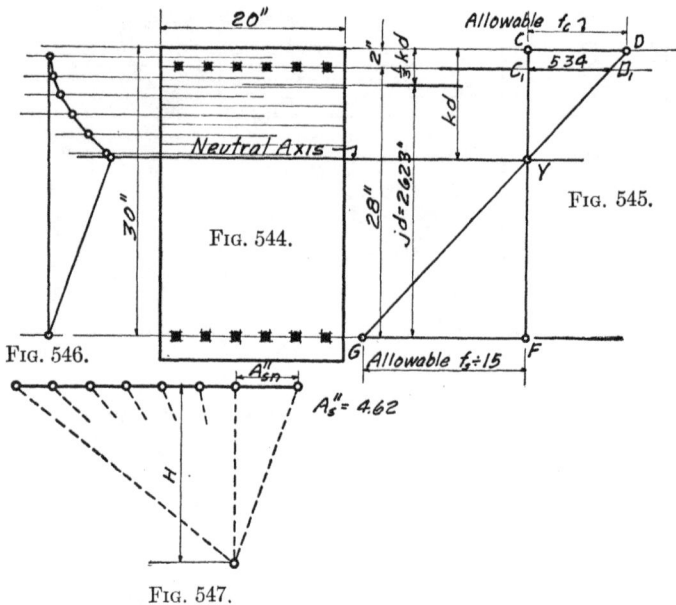

FIG. 544.

FIG. 545.

FIG. 546.

FIG. 547.

pounds per square inch. Also with this reinforcement the beam could carry a bending moment of $A''_s \cdot jd \cdot f_s = 4.62 \times 26.23 \times 16,000$ or 1,939,000 inch-pounds without exceeding these stresses. But the beam must support a moment of 3,200,000 inch-pounds. The remaining moment, 3,200,000 − 1,939,000 or 1,261,000, must be carried by compression steel and additional tension steel acting together as a couple. The arm of this couple is the distance between the tension steel and the compression steel, in this case 28 inches. 1,261,000 ÷ 28 = 45,000 pounds which is the load that must be carried by the compression steel and also by the additional tension steel. 45,000 ÷ 16,000 = 2.81

square inches, which is the additional area of tension steel required, and the total area of tension steel required is

$$2.81 + A''_s = 2.81 + 4.62 = 7.43 \text{ square inches} = A_s.$$

The stress in the concrete at the center of the compression steel is given by $C_1–D_1$ in Fig. 545, the stress in the compression steel is, therefore, 15 $C_1–D_1$. But the compression steel displaces its area of compression concrete. Therefore, the required area of compression steel A'_s, is 45,000 ÷ (14 × 534) = 6.02 square inches.

126. Bending Stresses in Complex Sections.—The graphical solution is especially well adapted to complex sections such as are shown in Figs. 548 and 556. Fig. 548 shows part of the cross-section of a concrete slab which is reinforced with a bar similar to that used in the " Columbian Floor-System." Although this type of reinforcing bar is not used very much at present, it has been used in the past, and occasionally an engineer is asked to determine the safe load for floors with such reinforcement. The constructions shown in Figs. 548 to 555 not only solve this particular problem but they show how other problems of the same general type can be solved.

In Fig. 548 a 7-inch slab is considered, reinforced with bars 12 inches on centers and of dimensions as shown. A 12-inch width of slab was used in finding the areas of the various slices, and the areas found are given at the right of the figure. In the left column compression values are given and in the right tension values. The compression value of a slice is the area of the concrete plus 15 times the area of its steel, while the tension value of any slice is just 15 times the area of its steel. Since the steel has a large vertical dimension, it was necessary to divide the tension side as well as the compression side into small slices. In Fig. 549 the compression areas are laid off in order to the left of x beginning with A_1, while to the right of x the tension values are laid off beginning with the lowest and 'taking the others in order. From Fig. 549, Fig. 550 is drawn in the usual way, the intersection O locating the neutral axis.

In Fig. 551 $C–D$ is made equal to the maximum allowable f_c, in this case assumed to be 500 pounds per square inch, and by drawing the line $D–Y–F$ the maximum corresponding stress in the steel is found to be 15 times $E–F$, which is about 10,500 pounds

per square inch.   The summation of all of the compressive stresses
is found and the resultant located in the usual way, Figs. 552
and 553 being used.   In this problem there are a number of
tension-taking slices.   The total tension taken by each of these
slices is computed and the sum of all of these totals gives the sum
or resultant of the tensile forces.   The action line of this resultant
is located by Figs. 554 and 555.   The effective depth, which
is the perpendicular distance between R_c and R_t, is found by
scaling to be 4.15 inches, and the allowable bending moment

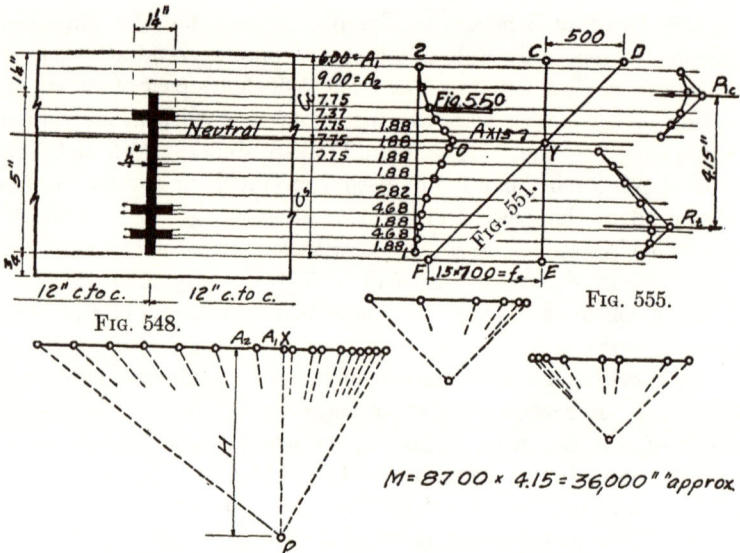

FIG. 548.   FIG. 555.

FIG. 549.   FIG. 552.   FIG. 554.

$$M = 8700 \times 4.15 = 36,000'' \text{ approx.}$$

per foot of slab is found to be 36,000 inch-pounds, for the case
when the allowable compression in the concrete is 500 pounds
per square inch.   The allowable bending moment for any other
allowable stresses could be computed by proportion, or for any
given moment the stresses could be found.

The moment of inertia for the above section is given as follows:
$I = 2 \cdot H$ times the area enclosed by Fig. 550.   With the moment
of inertia known, c_c and c_s could be measured, and the allowable
moment computed without drawing Figs. 551 to 555.

Fig. 556 shows a circular section with reinforcing rods
uniformly spaced around it near the outside.   In this case it is

found convenient to divide the compression side of the beam into
small slices as usual, from the centroids of which the lines parallel
to the axis about which bending occurs, are drawn. The areas
of these slices are computed by multiplying the width by the mean
length, and for convenience, the concrete displaced by the steel
is not deducted. Therefore, the areas thus obtained for the slices
give the area of the concrete plus one times the area of the steel.
Lines are now drawn from the steel rods on the compression side
of the beam and they are considered as representing areas $15 - 1$

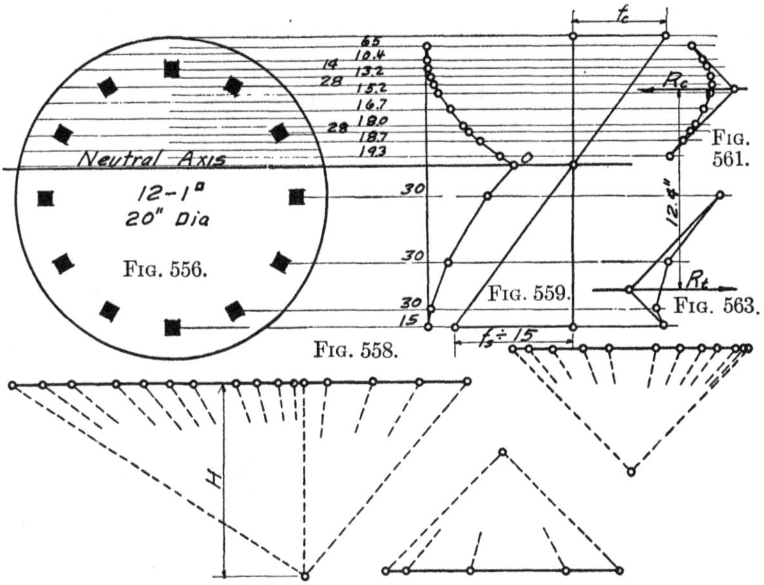

FIG. 556.

FIG. 561.

FIG. 559.

FIG. 563.

FIG. 558.

FIG. 557.   FIG. 562.   FIG. 560.

$= 14$ times the area of the steel. The steel on the tension side
of the beam is as usual multiplied by 15. Figs. 557 and 558 are
drawn as already explained, the intersection O locating the neutral axis. Figs. 559 to 563 are drawn in the same way as the corresponding figures in the previous problem, and the bending
moment may be obtained by multiplying the effective depth
by the mean between R_c and R_t.

In case the moment of inertia is desired, it may be obtained
by multiplying the pole distance H by two times the area enclosed
by Fig. 558, or Culmann's method might be used.

Other unusual beam sections could be handled in just the same way as the section shown in Fig. 556, the only difference would be in the areas of the slices.

127. Combined Stresses.—In reinforced concrete work it is not an uncommon thing to meet with structural members which carry both direct and bending stresses. That is, they must resist both an axial load and a bending moment. Some building laws require that wall columns supporting flat slabs be designed to carry a certain bending moment in addition to the direct load. Any combination of direct load and bending may be represented by an eccentric load P, P being equal to the direct load, and P times the eccentricity being the moment.

Problems dealing with combined stresses may be divided into two large classes. When the bending moment is relatively small, there will be compression over the entire cross-section. But as the moment increases with respect to the direct load, the compression on one side will decrease until zero is reached. If the relative value of the moment still increases, there will be tension over a part of the section and this tension is assumed to be taken by the steel.

Now the problems of the first class, that is, those in which there is compression over the entire cross-section, can be easily solved by analytical methods. The maximum stress in the concrete is given by the formula $f_c = \dfrac{P}{A} + \dfrac{Mc}{I}$, in which A is the area of the concrete plus 15 times the area of the steel, M is the moment about the center or neutral axis of the entire section, and I is the moment of inertia of the entire section, all of the concrete being considered and the area of the steel multiplied by 15. This I may be found either graphically or analytically, but it should not be confused with the moment of inertia as found for reinforced concrete beams where the concrete on the tension side of the neutral axis is neglected.

In the second class will be placed those problems in which there is tension over a part of the section. For problems of this kind the above formula cannot be used because it is assumed that the concrete cannot take tension. Of course, when there is tension only over a small part of the section, the error in using this formula would be small, but when the axis of zero stress moves farther in, the error would become considerable. Such problems can be solved by analytical methods when the sections are not too complicated; but when there are rods on all four sides, or when a hollow circular section such as is shown in Fig. 574 is considered, an analytical solution becomes more difficult.

128. Eccentrically Loaded Columns.—In order to illustrate how such problems can be solved by graphical construction, consider the column section shown in Fig. 564. This column is 24 inches square and is reinforced with 16 one-inch square bars placed as shown. The column carries a direct load of 100,000 pounds in addition to a bending moment of 1,150,000 inch pounds about the center. Both the bending moment and direct load can be represented by an eccentric load P of 100,000 pounds acting $11\frac{1}{2}$ inches off center, as shown in the figure. It will be assumed that the outside $1\frac{1}{2}$ inches of concrete is for fire-proofing and it will not be considered in computing the stresses.

The effect of the bending moment will first be found, and to it will be added the effect of the direct load. As usual, the compression side is divided into small slices parallel to the axis about which bending occurs. From the centroids of these slices the customary lines are drawn and parallel to them the lines from the tension steel. Figs. 565 and 566 are drawn and the neutral axis located by the intersection O in the same way as for beams. In Fig. 567 the line C–E is drawn and C–D is made equal to any convenient number of pounds per square inch, in this case 400. The line D–Y is next drawn and extended to F. R_c and R_t are both computed and their action lines located by Figs. 568 to 571. The mean between R_c and R_t times the effective depth gives 1,064,000 inch-pounds, which is the bending moment that must be exerted to produce a stress in the concrete of 400 pounds per square inch. Now for the eccentric load P we can substitute a load of the same magnitude P_1 acting at the neutral axis Y, and a bending moment of 900,000 inch-pounds. It should be noted that this moment is the moment about the neutral axis and not the moment about the center. Since a moment of 1,064,000 inch-pounds produces a stress of 400 pounds per square inch in the concrete, a moment of 900,000 inch-pounds will produce a stress of 338 pounds per square inch. The line C–H is measured off equal to 338 pounds per square inch, and the line H–Y–G is drawn. The effect of the moment about the neutral axis has

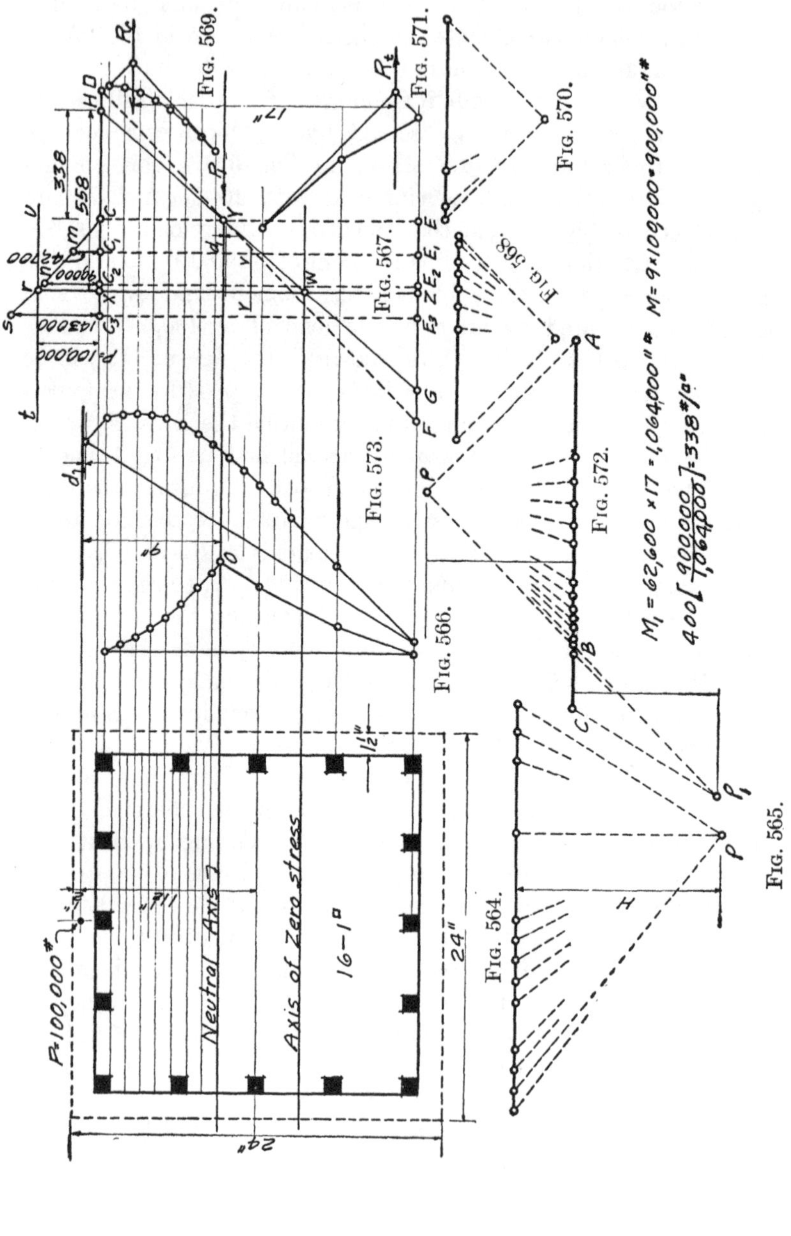

FIG. 569.

FIG. 570.

FIG. 571.

FIG. 567.

FIG. 568.

FIG. 573.

FIG. 572.

FIG. 566.

FIG. 565.

FIG. 564.

$$M_1 = 62600 \times 17 = 1,064,000 \text{"} \# \quad M = 9 \times 100,000 = 900,000 \text{"} \#$$

$$\frac{400}{900000}, \frac{900000}{142000} = 338 \#/\square \text{"}$$

Neutral Axis

Axis of Zero stress

16-1"

P=100,000#

24"

12½"

16-1"

now been obtained; it remains to find what will be the effect of P_1 acting at the neutral axis.

A direct load applied at the centroid of the stress-taking area, that is, at the neutral axis, will produce a uniform stress over the area, or in other words, the base line $C-E$ would be moved to the left. The compression over the area that now has compression would be increased, the tension in the tension steel would be decreased, and a small area of the concrete below the neutral axis would be subjected to compression. But how far to the left will the load of 100,000 pounds move the base line $C-E$? This is determined in the following way: The base line $C-E$ is moved 100 pounds per square inch to the left, to the position C_1-E_1, and the corresponding load required near the neutral axis is computed. This load is equal to the increase in the compressive stresses over the old compression area, plus the decrease in the tensile stresses, plus the new compressive stresses below the neutral axis. The increase in the compressive stresses is given by the area of the concrete above the neutral axis plus 15 times the area of the steel above the neutral axis, times the distance $C-C_1$. The new compressive stresses are given by the new compression area below the neutral axis times $\frac{1}{2}$ $C-C_1$, or times 50 pounds per square inch in this case. The decrease in the tensile stresses may be taken as 15 times the area of the steel below the neutral axis times $C-C_1$. The numerical sum of these three values will give the load which when applied near the neutral axis will move the base line $C-E$ 100 pounds per square inch to the left, to the position C_1-E_1. This sum is found to be 42,700 pounds, and it is measured off along the line C_1-m, to some convenient scale, and the point m is located.

In the same way it is found that when the base $C-E$ is moved 200 pounds per square inch to the left, to C_2-E_2, the corresponding load near the neutral axis is 90,000 pounds. The line C_2-n is made equal to this value using the same scale that was used for the line C_1-m. Also it would take 143,000 pounds to move the base line 300 pounds per square inch to the left. The curve $C-m-n-s$ is now drawn, and the vertical intercepts between this curve and the line $C-C_3$, measured to scale, give the necessary load near the neutral axis to move the base line to the position of the intercept. The line $t-u$ is now drawn parallel to $C-C_3$ and distant 100,000 pounds from it. This line intersects the curve *s–n–m–C* at *r*, and from *r* the line *r–X–Z* is drawn. This line *X–Z* is the base line for the given load and the intersection *w* locates the axis of zero stress, above which there is compression and below which there is tension. The direct load increases the compression in the concrete by *C–X*, which, added to the stress produced by the moment, gives a maximum stress in the concrete of 558 pounds per square inch. The maximum stress in the tension steel may be found by multiplying the line *G–Z* by 15.

The correctness of the results obtained may be checked in the following way: The resultant of all of the stresses in the column just balances the eccentric load *P*. That is, the sum of all of the compressive stresses measured to the base line *X–Z*, less the sum of all of the tensile stresses should equal *P* or 100,000 pounds, and the action line of the resultant of all of these stresses should coincide with the action line of the eccentric load *P*. The average stress over each of the various slices is obtained from intercepts between the lines *X–Z* and *H–G*, and these average stresses times the areas of the corresponding slices give the forces acting on the various slices. These forces are laid off to scale along the load line of Fig. 572. Beginning with the compressive force on the top slice and starting at *A*, the compressive forces are laid off in order, the last compressive force reaching the point *B*. From *B* to *C* the tensile forces are shown and the pole *p* is changed to p_1 with the same pole distance. *A–B* less *B–C* is found to be approximately 100,000 pounds. From Fig. 572, the funicular polygon in Fig. 573 is drawn, locating the action line of the resultant. This action line varies from the action line of the eccentric load *P* by a very small distance *d*, showing that the results obtained are only slightly in error.

It might be well to call attention to one place where an approximation has been made. When the base line is at *C–E*, the point *Y* is at the centroid of the stress-taking area. Therefore, a load applied at *Y* would tend to move the base line directly to the left parallel to its original position. But when the base line moves to the left, an ever-increasing area of concrete below the neutral axis begins to take compression. Therefore, the load ought to act a short distance from *Y* on the tension side of the column in order to keep the base line parallel to its original position. This distance, which increases as the base line moves farther to the left, is small and may be neglected when the distance $C-x$ is not too large; it is equal to d_1 when the base line has been moved to the position $X-Z$, as shown in Fig. 567. The fact that P_1 was assumed to act as Y, in place of a distance d_1 below Y, causes a slight error in the stress produced by the bending moment. In this case the error produced in f_c was about 8 pounds per square inch, or a little more than 1 per cent of the final result.

The distance d_1 and the amount of the error were found in the following way: It is only the compressive stresses in the concrete below the neutral axis that tend to pull the resultant of the resisting stresses below Y. When the base line is at the position shown in the figure by $X-Z$, the area of concrete below the neutral axis-taking compression is 110 square inches. The resultant of the compressive stresses on this area is found to be about 12,000 pounds, and it acts along the line $v-v$ distant from the neutral axis 1.76 inches. The remaining 88,000 acts along the neutral axis, and the resultant of these two forces is found to act .21 inch below the neutral axis; that is, d_1 in this case is .21 inch. When the distance d_1 is added to the arm used in finding the moment, the bending moment, and also the stresses produced by the bending moment, will be increased proportionally. This increase is here found to be 8 pounds, which is the error produced by the approximation we have been considering.

129. Reinforced Concrete Chimneys.—A reinforced concrete chimney under wind load gives an interesting problem, especially when the wind load is large enough to produce tension over a part of the section. Consider a cross-section of a chimney of dimensions as shown in Fig. 574, reinforced with 32 one-inch square bars and carrying an eccentric load of 500,000 pounds, as indicated. For the construction of the line of resistance of a chimney and for a general discussion of masonry chimneys, see Chapter VII.

The solution of this problem is very similar to that of the eccentrically loaded column. The compression side is divided into small slices and Figs. 575 and 576 are drawn as already explained, the neutral axis being located by the intersection O. Fig. 577 corresponds to Fig. 567, and the stress in the concrete produced by the bending moment about the neutral axis is found to be 330 pounds per square inch. In finding this value, the ordi-

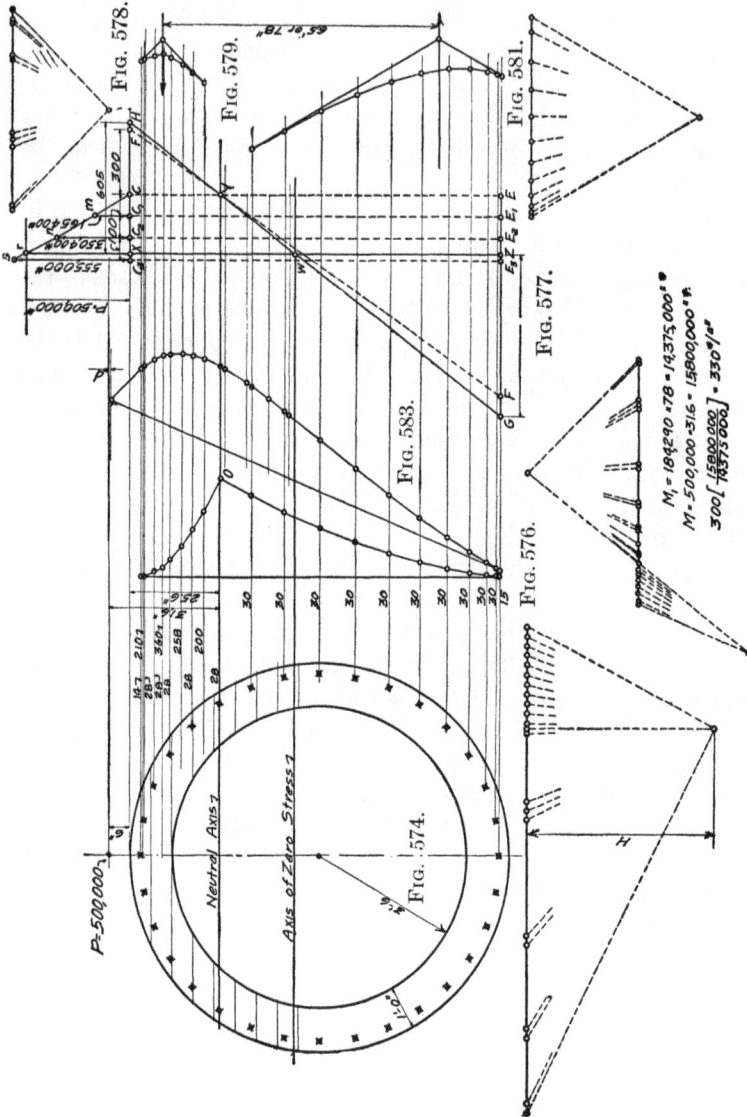

Fig. 578.

Fig. 579.

Fig. 581.

Fig. 580.

Fig. 577.

Fig. 583.

Fig. 582.

Fig. 576.

Fig. 575.

Fig. 574.

$M_1 = 184{,}290 \times 78 = 14{,}375{,}000$ #

$M = 500{,}000 \times 316 = 158{,}000{,}000$ #

$300 \left[\dfrac{158{,}000{,}000}{14{,}375{,}000} \right] = 330$ #/□″

Neutral Axis

Axis of Zero Stress

$P = 500{,}000$

nary method and Figs. 578 to 581 were used. The line C–H is made equal to 330 pounds per square inch and the line H–Y–G drawn.

The base line C–E is now shifted to the left, to the positions, C_1–E_1, C_2–E_2, and C_3–E_3. The values for C_1–m, C_2–n, and C_3–s are computed and the curve C–m–n–s drawn. The intersection r locates the base line X–Z and the maximum stress in the concrete H–X is found to be 605 pounds per square inch. Figs. 582 and 583 correspond to Figs. 572 and 573, and give a check.

In case the eccentricity of the loading were such as to produce compression over the entire section, the maximum compression in the concrete could be found by using the formula $f_c = \dfrac{P}{A} + \dfrac{Mc}{I}$, in which A is the area of the concrete plus 15 times the area of the steel, M is the moment about the neutral axis of the entire section, and I is the moment of inertia of the entire section, all of the concrete being included.

The moment of inertia of the section shown in Fig. 574, not considering the concrete on the tension side of the neutral axis, is equal to the pole distance H times twice the area enclosed by Fig. 576. This moment of inertia should not be confused with the I in the above formula.

It may be well to make clear the difference between the I in the above formula and the I as usually found for reinforced concrete beams. Fig. 584 shows the cross-section of a reinforced concrete column, the moment of inertia of which is desired when the outside inch and a half is neglected. By the use of Figs. 585 to 587 the I is found for the case when the concrete on the tension side of the neutral axis is not considered. This is the I which would be used to find the stresses produced by a bending moment before any direct load has been applied. Figs. 588 to 590 show the construction for finding I for the case when all of the concrete is considered. It should be used for finding the stresses produced by a bending moment only when there is, acting with the bending moment, a direct load large enough to keep most of the cross-section under compression.

Any column, no matter how complicated and irregular its cross-section or how eccentric its loading, can be solved by the methods given above.

130. Deflection of Reinforced Concrete Beams.—The same general constructions that were employed in Chapter IV to find

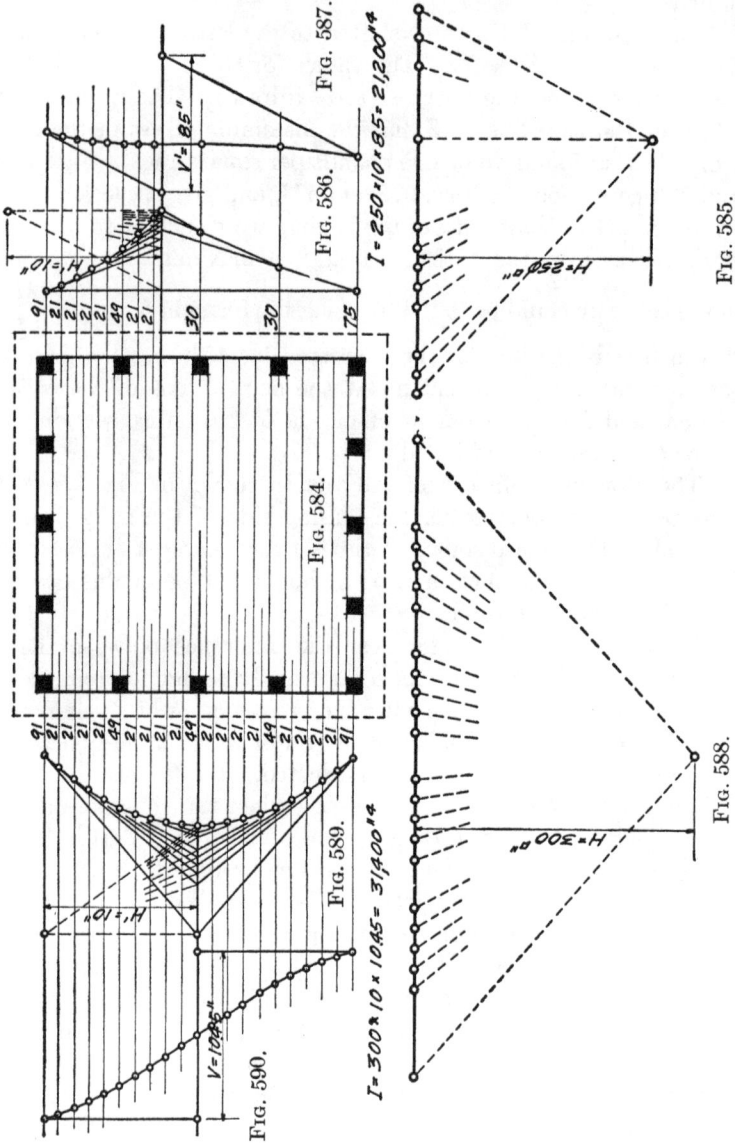

the deflections of steel and timber beams may be used for reinforced concrete beams.

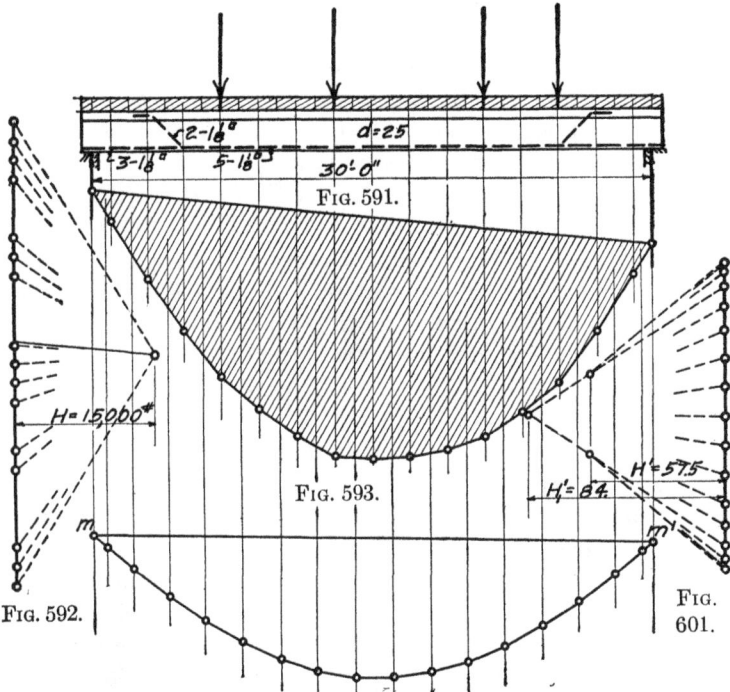

Fig. 591.

H=150,00#

Fig. 593.

Fig. 592.

Fig. 602.

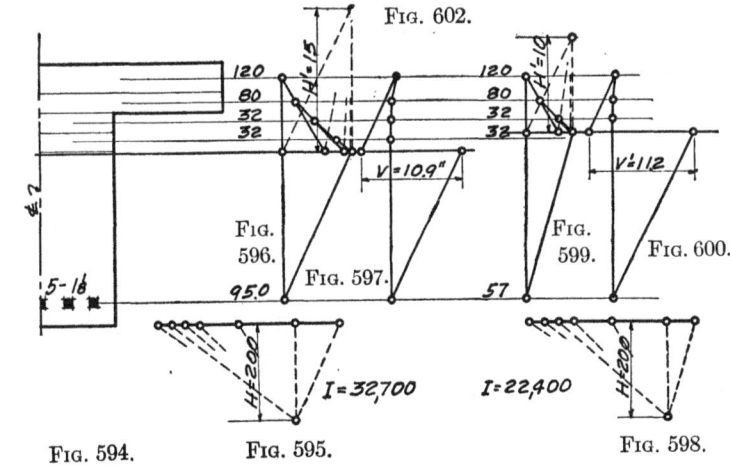

Fig. 594.   Fig. 595.   Fig. 598.

Consider a T-beam of 30 foot span and carrying a loading as shown in Fig. 591. The force polygon shown in Fig. 592 and the moment diagram, Fig. 593, are drawn as usual. The moment diagram is divided into slices, the areas of which are laid off along the load line of Fig. 601, and from the centroids of these slices the vertical lines are drawn. In order to compute the pole distances in Fig. 601, it is necessary that the moment of inertia of the T-beam be found. By the use of Figs. 594 to 597, the I of the central portion of the beam is found to be 32,700 inches[4], while by the use of Figs. 598 to 600 the I at the ends, after the two rods have been bent up, is found to be 22,400 inches[4]. Using these values of I, which are in terms of concrete, the pole distances H' and H'_1 are computed, using the formula,

$$\text{pole distance} = \frac{E \cdot I}{H \cdot n \cdot a},$$

in which n is any convenient number such as 2, 4 or 5, and E is the modulus of elasticity of concrete, which is about 2,000,000 when the ratio of E_s to E_c is 15.

From Fig. 601 the funicular polygon of Fig. 602, which represents the elastic curve, is drawn, and the vertical intercepts between this curve and the line m–m' give n times the deflection, the value of n being that used in the formula to compute H' and H'_1. The maximum deflection is found to be about .48 of an inch.

The student is referred to the Chapter on Beams, Chapter IV, for a more detailed explanation of the construction of the elastic curve. Reinforced concrete beams which are fixed at one or both ends, those having overhanging ends, and continuous reinforced concrete beams with unequal spans and a complicated loading, may be solved by the construction given in Chapter IV. It should be noted that a variation in the moment of inertia, caused by a change in the cross-section of the beam, or by a change in the amount or location of the reinforcement, produces no difficulty. It is necessary only to find the correct I for each of the various pole distances. If, in the judgment of the designer, a portion of the concrete on the tension side of the neutral axis should be considered effective in resisting tension, that part of the concrete should be used when finding the moment of inertia.

CHAPTER IX

DESIGN

The designer's work is only half complete when the loads and stresses are found, the choice of sections and the indication of details being a very important part. Previous chapters have considered loads and stresses rather than actual design, this chapter will, therefore, give special attention to the design of sections and details. The proper use of good tables and diagrams saves a great amount of time and labor. One of the objects of this chapter will, therefore, be to illustrate and explain the use of a few such labor saving devices.

131. Design of Beams.—In Chapter IV very little is said regarding design, attention being devoted to moments, shears, reactions, and deflections. The present article will deal with a few problems of design, will refer to various tables and diagrams, and will consider some points of interest not covered in Chapter IV. We will start with the general formula $\dfrac{M}{S} = \dfrac{I}{c}$, in which M is the bending moment in inch-pounds, I is the moment of inertia of the section, S is the unit stress, and c is the distance to the fiber which has the stress S.

In Fig. 603 the rectangle a–c–g–h shows a typical cross-section of a rectangular beam, x–x being the neutral axis about which bending occurs. From Mechanics we know that the stress increases directly as the distance from the neutral axis, this fact is illustrated by the diagram m–n–o–r–p. The line m–n represents the maximum compressive stress S_c to some convenient scale, and r–p the maximum tensile stress S_t to the same scale. Also it is evident that any intercept w or w_1, between the lines n–r and m–p, at distance z or z_1 from the neutral axis, gives the stress in the fibers at this distance from the neutral axis. In fact the direct stresses in a beam may be represented by two stress volumes, one positive on the compression side of the neutral axis and the other negative on the tension side. For a rectangular

281

beam with bending about an axis parallel to one of the sides, these two stress volumes are wedge-shaped, as shown in plan by *a–c–f–e* and *e–f–g–h*, and in elevation or side view by *m–n–o* and *o–p–r* (see Fig. 603).

The resultant of all the compressive stresses passes through the centroid of the upper stress volume, and the resultant of all the tensile stresses passes through the centroid of the lower stress volume. Since these volumes are wedges, their centroids will be at their third points, and the effective depth is found to be $\frac{2}{3}d$. The resisting moment of the beam may be found by multiplying R_c or R_t and the effective depth together, this product being the moment of the couple formed by R_c and R_t. The mag-

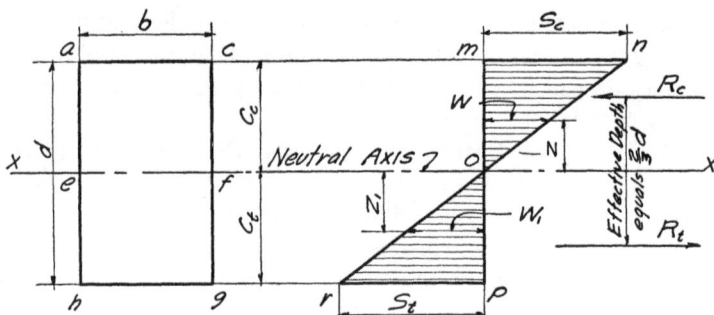

Fig. 603.

nitude of R_c is equal to area *a–c–f–e* times $\frac{1}{2}S_c$, and the magnitude of R_t is equal to area *e–f–g–h* times $\frac{1}{2}S_t$. The magnitude of R_c must be equal to that of R_t for all beams, but S_c will be equal to S_t only when c_c is equal to c_t.

The resisting moment of the beam is also given by the formula $M = \dfrac{I \cdot S}{c}$. When c is the distance to the outermost fiber and S is the allowable unit stress, the formula gives the maximum allowable resisting moment. In most handbooks there are tables giving the value of $\dfrac{I}{c}$ for a larger number of different beam sections, c being the distance to the outermost fiber. When the designer has determined the bending moment he may divide it by the allowable stress and obtain $\dfrac{M}{S}$. $M \div S$ is equal to the required $\dfrac{I}{c}$, and a beam may be picked out from the table which

has this required $\dfrac{I}{c}$. The value $\dfrac{I}{c}$ is often called the section

modulus of the beam.   In some handbooks the section modulus
is denoted by the letter S, and some other letter used for the unit
stress.

Fig. 604 shows the cross-section of a rectangular beam and
the shaded area indicates what is sometimes called the effective
area.   It is an area such that $w : g{-}h :: z_1 : z$, w being any
intercept parallel to the neutral axis and distant z_1 from it.   The
length $g{-}h$ is the width of the beam section where w is taken,
and z is the distance to the outermost fiber.   When c_c and c_t

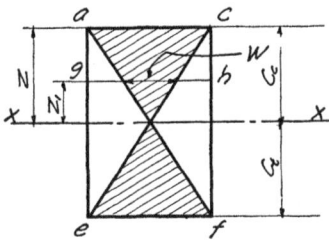 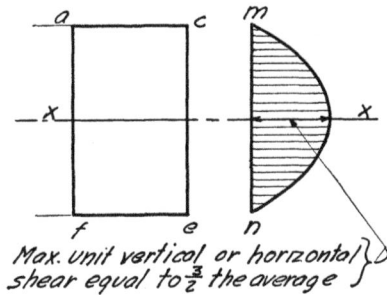

Max. unit vertical or horizontal
shear equal to $\frac{3}{2}$ the average

FIG. 604.   FIG. 606.

are not equal, z is equal to the larger of the two.   An important
thing about the effective area is that the centroid of the part
above the neutral axis is at the same distance from the neutral
axis as the resultant of all the compressive forces.   Also the centroid of the effective area on the tension side of the neutral axis
is the same distance from the neutral axis as the resultant of the
tensile stresses.   In other words, the effective depth of a beam
may be obtained by finding the vertical distance between the
centroid of the effective area above the neutral axis and the centroid of the effective area below.

In addition to the direct tensile and compressive stresses,
there are other stresses produced in a beam, such as vertical
shear, horizontal shear, diagonal tension, and diagonal compression.   One of the important things to keep in mind regarding
these four stresses is that their unit intensity at any given point

is the same. Consider dx, any small rectangular parallelopiped of the beam with each end a small square and with a length equal to the width of the beam. The vertical forces shown represent the vertical shear on the two vertical sides. When dx is very small the forces on these two vertical sides may be considered equal. In like manner the two horizontal forces which represent the horizontal shear on the two horizontal sides may be considered equal. There are two couples acting on dx, one consisting of the two horizontal shears and the other of the two vertical shears. Since dx is in equilibrium these two couples must produce the same moment, and since the vertical dimension of dx is equal to the horizontal dimension, H must be equal to V.

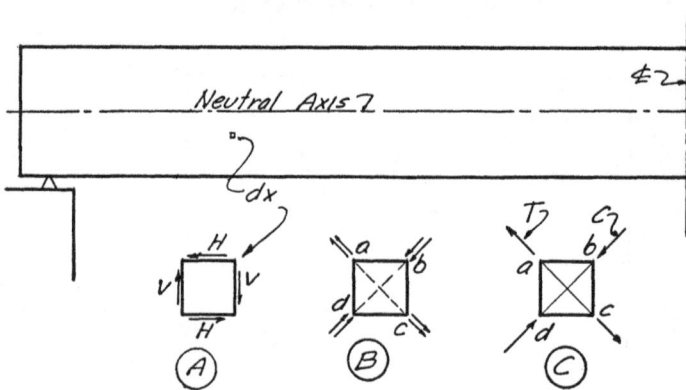

FIG. 605.

Also since the area of the side on which H acts is the same as the area of the side on which V acts, the unit horizontal shear must equal the unit vertical shear.

Now each one of the H and V forces shown in Fig. 605 (A) has a component normal to the diagonals a–c and b–d, as shown in Fig. 605 (B). Also each one of these components is equal to .707 of the original forces. These components may be combined as shown in Fig. 605 (C). each one of the four forces here indicated being equal to 1.414 of the original forces. Or we may say that the tension which must be resisted by the plane b–d is equal to T, which in magnitude is 1.414 times as large as H or V. Also the compression C which must be resisted by the plane a–c is 1.414 times as large as H or V. But the area of a–c equals the area of b–d equals 1.414 times the area of a–b or b–c. Therefore, at any point of the beam the unit diagonal compression equals the unit diagonal tension equals the unit vertical shear equals the unit horizontal shear.

Another important point in connection with shears and diagonal stresses is their distribution over the cross-section of the beam. For a rectangular beam the distribution may be indicated by a parabola, as shown in Fig. 606. This means that the unit horizontal shear and also the unit vertical shear increases from zero at the top to a maximum at the neutral axis and then decreases down to zero again at the bottom. The maximum unit shear for a rectangular beam is $\frac{3}{2}$ the average. The distribution of shear, of course, varies greatly with the shape of the cross-section.

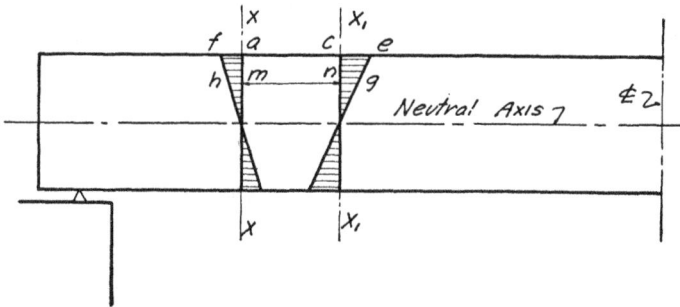

Fig. 607.

The unit horizontal shear may be easily found. Consider the beam shown in elevation by Fig. 607. Sections x–x and x_1–x_1 are any two vertical sections taken any convenient distance a–c apart. Let n–m be any horizontal plane between them and above the neutral axis. Now the resultant horizontal shear on the plane m–n is equal to the difference between the two stress volumes c–e–g–n and a–m–h–f. This value divided by the area (m–n times the width of the beam) gives the average unit horizontal shear over the plane m–n. Shears on planes below the neutral axis may be found in a similar way.

(*a*) *Wooden Beams.* Let it be required to design wooden floor joist having a 15-foot span, a 16-inch center to center spacing, and carrying a total load (live plus dead) of 105 pounds per square foot. The allowable fiber stress will be limited to 1600 pounds per square inch. The total load carried by each joist is $15 \times 1\frac{1}{3} \times 105 = 2100$ lbs.

Now the moment might be computed and then the required $\frac{I}{c}$, using the formula $\frac{M}{S} = \frac{I}{c}$, after which a joist could be chosen having an $\frac{I}{c}$, $\left(\frac{I}{c} \text{ equals } \frac{1}{6} b \cdot d^2\right)$ as large or slightly larger than required. When computing $\frac{1}{6} b \cdot d^2$ the student should remember that timbers are smaller than their nominal size, a 2×8 actually being $1\frac{5}{8} \times 7\frac{1}{2}$, etc.

A much easier method of designing is to use tables such as are given in the "Southern Pine Manual" published by the Southern Pine Association at New Orleans. Turning to these tables it is found that a 2×12 with a span of 15 feet is good for a uniform load of 2865 when stressed to 1800 pounds per square inch. By proportion it is found to be good for 2545 when stressed to 1600 pounds per square inch. This is more than is required, yet a 2×10 is not strong enough, therefore the 2×12 should be used. The tables also give information regarding deflection and horizontal shear.

Wood is weak in horizontal shear and the designer should remember that the horizontal shearing stresses in short heavily loaded beams may determine the safe load rather than the bending stresses. Deflection is often of importance especially in connection with long spans. Not only is excessive deflection unsightly, but when the beam or joist supports a plaster ceiling there is danger of cracks being produced in the plaster. A deflection of $\frac{1}{360}$ of the span is considered the maximum that should be allowed in most cases.

(b) *Steel Beams.* Bending moments and resisting moments are of great importance in designing steel beams, but there are several other things that should receive attention. The deflection should not be excessive, and the compression flange should have adequate lateral support or the allowable stress reduced. Also the web stresses should not be so high that there is danger of the web buckling.

Structural steel handbooks give the section modulus, $\frac{I}{c}$, for most rolled sections. Some of them also give tables of safe loads uniformly distributed. The advantage of using the table of safe uniform loads is that it is unnecessary to compute the moment. In many cases beams carry uniform loads, and when their loading is not uniform an equivalent uniform loading can often be quickly estimated. The designer should be careful not to use special beams because it is often difficult to obtain them without delay or extra expense.

The distribution of shear over the cross-section of an I-beam is illustrated by Fig. 608 (B), and the unit distribution is indicated by Fig. 608 (C). The diagram (B) differs from (C) because the width of the I-beam varies, the flanges being very wide and the web narrow. This tends to make both (B) and (C) approach rectangles. Most of the moment is carried by the flanges and most of the shear by the web.

The stress volumes for an I-beam are shown in Fig. 609.

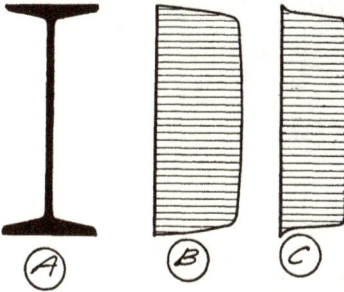
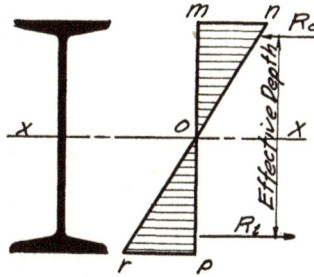

Fig. 608.   Fig. 609.

The triangle *m–n–o* is the side view of the upper stress volume, and the upper half of the I-beam section may be called its base. The centroids of these stress volumes are much nearer the top and bottom than was the case for rectangular beams, and the effective depth is, therefore, a much greater proportion of the depth of the beam.

The allowable maximum unit stress for steel beams is usually taken as 16,000 pounds per square inch, but this stress should be reduced when the compression flange is not supported laterally. A method of making this reduction is given in the Carnegie Steel Company's " Pocket Companion." A table for use in obtaining the deflection under uniform loads is given on the page just before the one referred to above. The student will find many other valuable tables in the " Pocket Companion."

(c) *Reinforced Concrete Beams.* The theory of reinforced

concrete and the development of formulas will not be given here, but the reader should be familiar with the principles of reinforced concrete design. The derivation of the following formula may

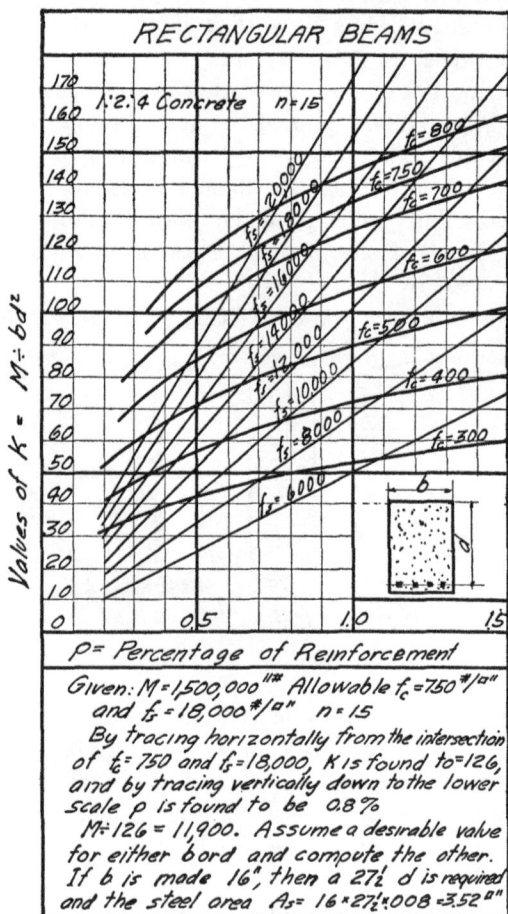

FIG. 610.

be found in almost any textbook on Reinforced Concrete: $M = K \cdot b \cdot d^2$. In this formula M is the moment in inch-pounds, b is the width of the beam in inches, d is the depth from the compression face to the center of the tension steel, and K is a value depending upon the percentage of steel and the allowable stresses. The letter R is used in place of K by some authorities. Values

Fig. 611 — SLABS + RECTANGULAR BEAMS

Resisting Moments, Steel Areas etc with fc=750#/☐ v=18000#/☐ n·15 Concrete 1:2:4

d″	M (×1000)	A (×1000)	d″	M (×1000)	A (×1000)
8	96.4	0.77	21	666.8	2.01
8¼	102.7	0.79	21¼	692.2	2.07
8½	108.9	0.81	22	729.6	2.11
8¾	115.4	0.84	22¼	763.2	2.16
9	122.1	0.86	23	782.0	2.21
9¼	129.0	0.89	23¼	833.8	2.26
9½	136.0	0.91	24	868.8	2.31
9¾	143.4	0.94	25	943.2	2.40
10	150.8	0.96	26	1020.0	2.50
10¼	158.5	0.99	27	1094.2	2.60
10½	166.3	1.01	28	1182.0	2.69
10¾	174.3	1.03	29	1272.0	2.79
11	182.4	1.06	30	1356.0	2.88
11¼	190.4	1.08	31	1452.0	2.98
11½	198.3	1.11	32	1548.0	3.08
11¾	208.8	1.13	33	1644.0	3.17
12	217.2	1.15	34	1740.0	3.27
12¼	235.2	1.20	35	1848.0	3.36
12½	254.4	1.25	36	1956.0	3.46
12¾	274.8	1.30	37	2068.0	3.56
13	245.2	1.36	38	2184.0	3.65
13¼	316.8	1.39	39	2272.0	3.75
13½	339.6	1.44	40	2412.0	3.84
13¾	362.6	1.49	41	2532.0	3.94
14	386.4	1.54	42	2664.0	6.04
14¼	410.4	1.59	43	2784.0	4.13
14½	0356	1.63	44	2916.0	4.23
14¾	462.0	1.68	45	3060.0	4.33
15	488.4	1.73	46	3192.0	4.42
15¼	6160	1.78	47	3339.0	4.52
15½	544.8	1.83	48	3468.0	4.61
15¾	573.6	1.87	49	3624.0	4.71
16	602.4	1.92	50	3768.0	4.81
16¼	633.6	1.97	51	3924.	4.90

FIG. 611.

Fig. 612 — SAFE LIVE LOADS IN LBS PER SQ.FT. FOR SOLID SLABS

fc=750#/☐ v=50#/☐ u=18,000#/☐ n·15 Concrete 1:2:4

Left descriptive columns (common to the sections):

Slab Thick. (in.)	Depth to cen. of Steel d	Area of Steel per ft. of slab	Main Steel (size & spacing)	Moment of Resistance (×1000 in. lbs)	Wt. of Slab (lbs)
3	2.4	.22	3/8″–6″	7.6	38
3½	2⅞	.26	3/8″–5″	11.4	44
4	3⅜	.31	1/2″–7″	15.9	50
4½	3⅞	.34	1/2″–6″	18.5	56
5	4⅜	.38	1/2″–5″	24.1	63
5½	4⅞	.43	5/8″–6¾″	30.6	69
6	5⅜	.48	5/8″–6″	37.6	75
6½	5⅞	.53	5/8″–5⅛″	45.5	81
7	6⅜	.55	3/4″–6″	49.9	88
7½	6⅞	.60	3/4″–5⅜″	59.0	94
—	—	.65	7/8″–6½″	68.7	100

Safe live loads by span (feet) — span headings: 4½ · 5 · 5½ · 6 · 6½ · 7 · 7½ · 8 · 8½ · 9 · 9½ · 10 · 11 · 12 · 13 · 14

For slabs with both ends supported but not continuous M = ⅛ WL

t	4½	5	5½	6	6½	7	7½	8	8½	9	9½	10	11	12	13	14
3	280	214	166	130	103	82	66	52	42	32	18					
3½	432	332	260	208	167	136	111	91	75	61	50	40	25			
4	614	475	378	300	245	201	166	139	116	97	81	68	56			
4½	718	554	437	350	286	236	196	163	137	115	98	81	56	32		
5	942	732	579	469	383	317	265	223	188	159	135	115	81	49		
5½	1206	939	746	608	497	414	347	293	240	213	183	157	127	73		
6	1442	1165	929	758	622	520	437	371	317	273	235	203	176	133	94	51
6½	1819	1422	1094	925	764	639	539	459	393	339	285	256	223	170	130	99

For slabs with one end continuous and one end supported M = 1/10 WL

t	4½	5	5½	6	6½	7	7½	8	8½	9	9½	10	11	12	13	14
3	364	278	216	173	139	113	92	75	60	50	33					
3½	551	430	336	272	220	181	150	125	104	87	74	62	51	39		
4	780	605	480	387	310	258	218	185	158	134	115	98	71	51		
4½	908	706	559	453	370	310	258	218	185	161	138	115	97	75	56	39
5	1192	937	742	607	445	416	348	296	213	186	143	108	82	56		61
5½	1527	1190	951	770	634	538	457	385	329	283	247	239	233	143	110	85
6	1790	1477	1100	963	795	671	565	495	415	389	343	328	312	246	201	158
6½			1530	1170	972	810	681	586	512	424	374	397	374	281	246	196

For slabs continuous at both ends M = 1/12 WL

t	4½	5	5½	6	6½	7	7½	8	8½	9	9½	10	11	12	13	14
3	439	334	267	214	174	142	118	95	80	67	54	46				
3½	670	519	411	333	273	226	189	160	134	113	96	82	70	53	44	
4	948	738	588	478	390	328	276	233	206	170	147	127	109	82	61	60
4½	1100	857	684	556	458	381	320	273	234	204	128	135	104	79		87
5	1445	1130	905	737	609	507	429	364	309	270	237	237	182	143	112	117
5½	1841	1440	1151	941	780	631	555	475	409	354	390	300	235	188	152	
6	1785	1425	1165	965	817	695	594	515	444	390	341	324	268	207	166	
6½			1557	1298	1072	929	794	692	607	527	465	410	324	207	254	206
7			1850	1530	1300	1106	966	851	750	633	658	658	658	628	254	206
7½			1810	1556	1300	1120	928	827	750	642	688	420	670	378	303	251

Values to the left and below lower Zig-Zag line give excessive diagonal tension stresses

Values to the left and below upper Zig-Zag line give excessive bond stresses

FIG. 612.

of K for different stresses and percentages of steel are given by the diagram in Fig. 610. Similar diagrams for other values of n may be found in various books on reinforced concrete.

The resisting moment of concrete beams of various depths, per foot of width, is given in the table of Fig. 611. The required steel area is also given. This table is based on a stress of 750 pounds per square inch in the concrete and 18,000 pounds per square inch in the steel. Similar tables for other stresses will be found in Thomas and Nichol's "Reinforced Concrete Design Tables."

The table in Fig. 612 gives the safe live loads in pounds per square foot, for slabs of various thicknesses and spans. The size and spacing of bars is also given. The student can easily make tables of his own for other stresses.

Many concrete beams and slabs are continuous over their supports, which means that there is tension in the top for some distance on either side of the support. These tensile stresses should be taken care of either by extra top bars or by bending a part of the main steel and carrying it past the support to approximately the quarter point. See the typical detail for continuous beams in Fig. 613.

A large proportion of the concrete beams used in buildings are T-beams, the T being formed by the slab. Usually there is a good deal more slab than is necessary for the T. As a result a large portion of the slab probably acts as the T but with a rather low stress in the concrete. The principal part of the problem then is to find the necessary steel area and see that there is enough concrete for diagonal tension.

When T-beams are continuous at the supports the designer should be sure that the concrete in the stem at the supports is not over-stressed in compression. If a continuous beam of uniform moment of inertia has equal spans, carries a uniform load and is fully fixed at the end supports, the moment at the supports is $\frac{1}{12} W L$ and at the center of each span $\frac{1}{24} W L$. In this expression W is the load on each span and L is the length of each span. When designing continuous reinforced concrete beams of equal spans it is customary to design for $\frac{1}{12} W L$ at the supports and also $\frac{1}{12} W L$ at the center of the spans in order to take care of unequal loading. Beams continuous over only one support are usually designed for a positive moment of $\frac{1}{10} W L$,

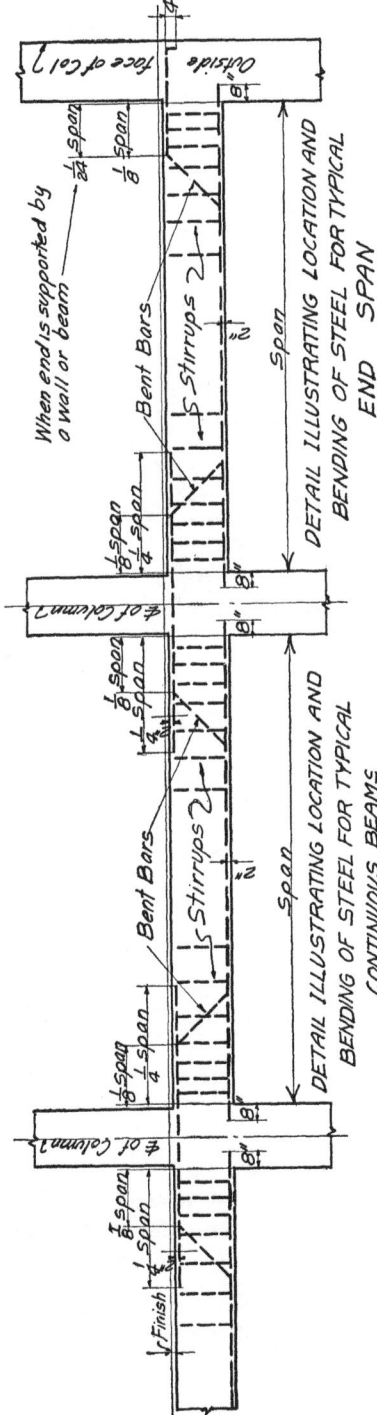

FIG. 613.

and for a negative moment of the same value above the support over which it is continuous.   The point is, that for continuous beams we usually design for at least as much moment at the sup-

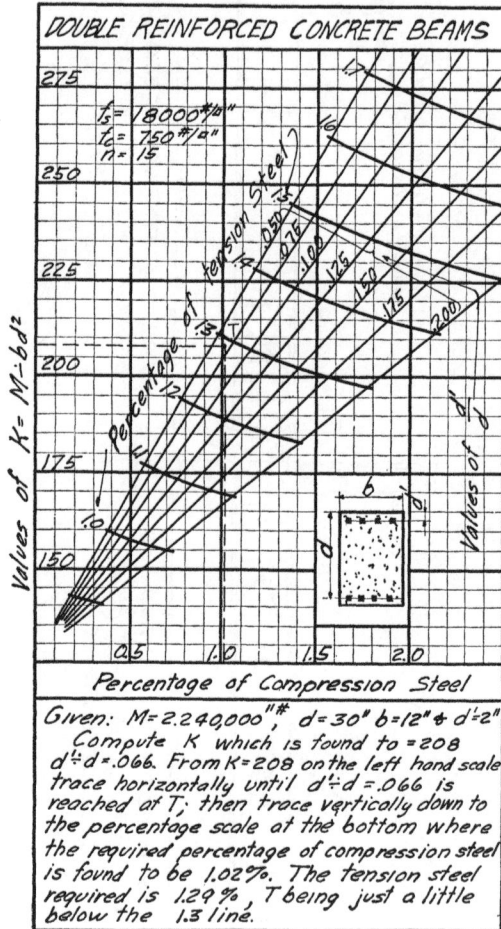

Fig. 614.*

port as we do at the center; in actual beams it is probable that more moment is often taken at the support than at the center. From all this it follows that the concrete in the lower part of the stem right next to the support is often the critical part of con-

---

*The writer is indebted to Harvey Hanna of Hanna, Zabriskie & Daron, Constr. Engrs., for ideas embodied in Figs. 614 and 623.

tinuous T-beams. Even when the allowable stress at this point
is increased 15 per cent and compression reinforcement used it
will seldom take a very wide T at the center to develop the same
moment.   Fig. 614 shows a convenient diagram for use in designing double reinforced concrete beams.   In connection with compression reinforcement the student should refer to Article 125.

The common formulas for T-beams do not take into account
the compression in the stem below the T and above the neutral
axis.   This error is, of course, on the safe side, but in some cases
it may be quite large.   The table in Fig. 615 shows how the values

T - BEAMS	
	Values of $K = M \div b d^2$ $f_c = 750 \#/\square''$
$t \div d$	$f_s = 18000\#/\square''$
.10	62
.12	71
.14	80
.16	88
.18	95
.20	101
.22	106
.24	111
.26	115
.28	118
.30	121
.32	123
.34	124
.36	125
.38	126

FIG. 615.

of K vary with different values of $\dfrac{t}{d}$ when $f_c = 750$ and $f_s = 18,000$.
pounds.

If one wishes to take account of the compression in the stem
below the flange it may be conveniently done in the following
way: First find the resisting moment of a rectangular beam
having the depth and width of the stem, subtract this moment
from the given moment and then compute the width of the
T-beam required to carry this remaining moment.   The sum
of these two widths will give the width required.

The following is a quick method of finding approximately
the necessary width of T required: Find b_1, the width of a rectangular beam which would carry the given moment, see Fig.

616. In the T-beam the areas A_1 and A_2 are taken away and the stress which they took in the rectangular beam must be taken by the areas B_1 and B_2. The average stress over areas B_1 and B_2 is much greater than that over areas A_1 and A_2, therefore, areas B_1 and B_2 will be correspondingly smaller. Knowing c_1 the experienced designer can estimate the dimension c with very small error.

Concrete is weak in tension and it is often necessary to use steel reinforcing for resisting a part of the diagonal tension stresses. Since the unit diagonal tension and the unit shear are numerically equal, it is often convenient to compute the vertical shear in order

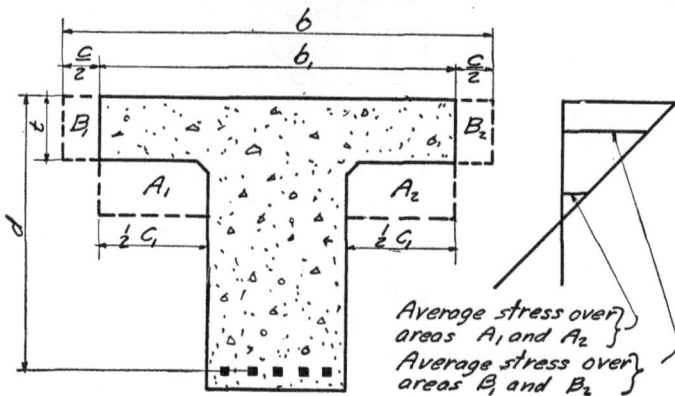

Average stress over areas A_1 and A_2
Average stress over areas B_1 and B_2

FIG. 616.

to obtain the values for the diagonal tension. In fact, most of the formulas used for designing web reinforcing deal with vertical shear and often nothing is said about diagonal tension. However, the student should keep in mind it is for diagonal tension that he is reinforcing and he deals with the vertical shear because it has the same numerical unit value and is more convenient to use.

Figs. 618 and 619 show enlarged portions of the beam given in Fig. 617. Let z–z be any zigzag section as shown. The diagonal tension tends to pull the lower portion down and towards the right, as indicated by the arrow below. When the diagonal tension becomes greater than about 40 pounds per square inch it is thought necessary to provide steel reinforcing to take a part of it. Some authorities assume that this reinforcing takes $\frac{2}{3}$ of

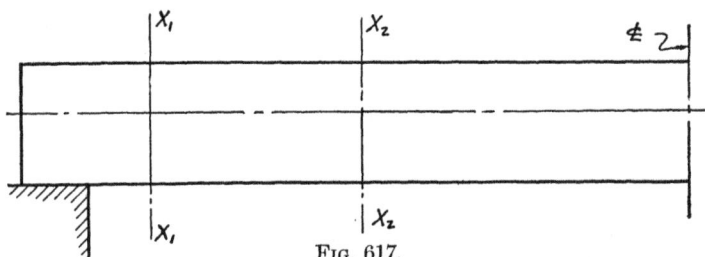

FIG. 617.

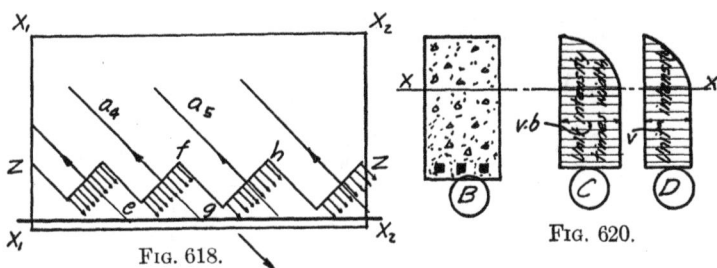

FIG. 618.

FIG. 620.

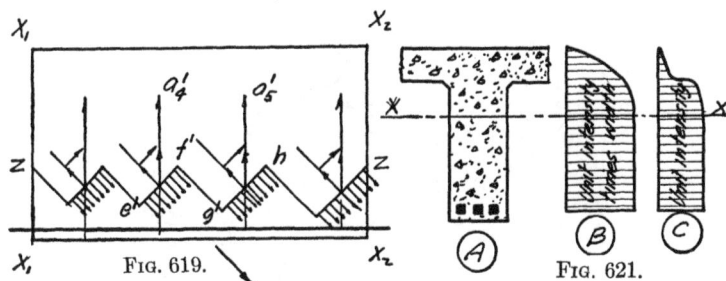

FIG. 619.

FIG. 621.

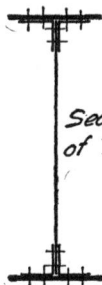

Section x-x
of FIG. 624.

FIG. 625.

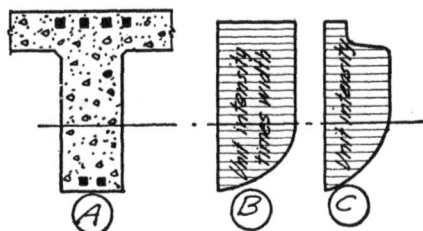

FIG. 622.

the diagonal tension, while others assume that it takes only the excess above 40 pounds. Fig. 618 illustrates diagonal stirrups while Fig. 619 shows vertical stirrups. Stirrup a_4 is supposed to resist that portion of the diagonal tension along e–f that is not taken

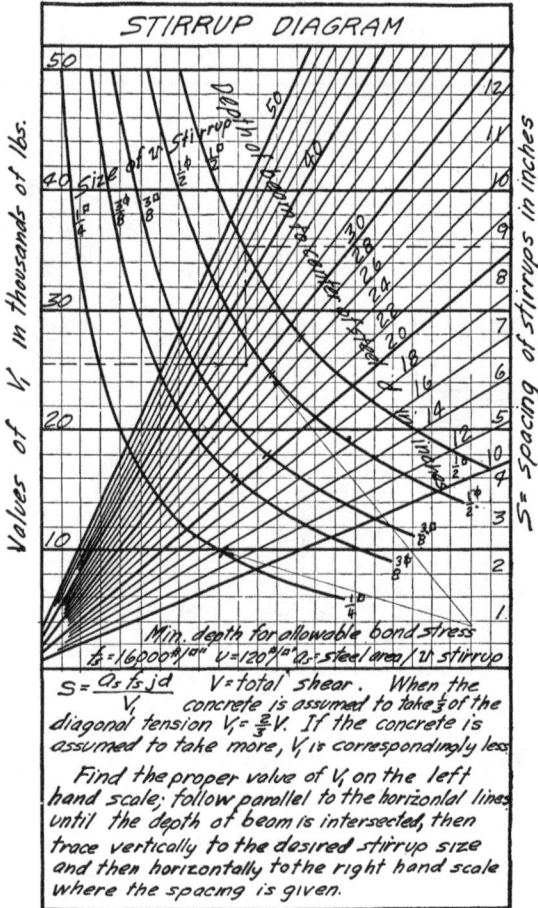

Fig. 623.

by the concrete. In Fig. 619 stirrup a'_4 is supposed to resist that portion of the diagonal tension along e'–f' which is not taken by the concrete. But, since the direction of this stirrup makes an angle of 45° with the lines of stress, the stress in a'_4 must be such that its component parallel to the lines of stress is equal to the force to be resisted. This means that the load on vertical stirrups is 1.414 times as great as the load on stirrups inclined at 45°, but the inclined stirrups are 1.414 times as long as the vertical. Therefore, the same percentage of web reinforcing is required for diagonal as for vertical stirrups. The diagonal stirrups, being parallel to the diagonal tension stresses, take stress quicker, that is, for a given web deformation they will have a higher stress. However, they are difficult to connect to the main tension steel, so vertical stirrups and bent bars are usually used in place of diagonal stirrups. The diagram shown in Fig. 623 will be found convenient for determining the spacing and size of vertical stirrups required under various conditions.

It is interesting to note how v, the unit intensity of the shearing stresses, varies over the cross-section of the beam. The variation of the unit intensity for a rectangular beam is shown in Fig. 620 Ⓓ. Fig. 620 Ⓒ shows the variation of $v \cdot b$. The horizontal dimensions of Ⓒ measured to scale are equal to the horizontal dimensions of Ⓓ times the width of the beam. Figs. 621 Ⓑ and Ⓒ are for a T-beam such as is shown in Fig. 621 Ⓐ and correspond to Figs. 620 Ⓒ and 620 Ⓓ respectively. For the depth of the T the intensity of the unit shear is greatly reduced because of the greater width. This accounts for the small lateral dimension in Fig. 621 Ⓒ just opposite the T. Figs. 622 Ⓐ to Ⓒ illustrate the distribution of shear over a section near the support of a continuous T-beam, the bottom having compression. Note the reduced lateral dimension of Fig. 622 Ⓒ opposite the T.

The areas of the diagrams given in Figs. 620 Ⓒ, 621 Ⓑ and 622 Ⓑ measured to scale represent in each case the total shear over the corresponding beam section.

132. Design of Plate Girders.—When loads and moments are very large it is often necessary to use a beam built up out of plates and angles, which is called a plate girder. See Figs. 624 and 625. The resistance of such a girder to bending is perhaps most accurately measured by computing its section modulus. This can be very easily done by the use of tables such as are given in Ketchum's "Structural Engineer's Handbook." When computing the value of $\dfrac{I}{c}$ deduction should be made for rivet holes.

There is a large table in the Carnegie Steel Co.'s " Pocket Companion " which gives the section modulus for a number of

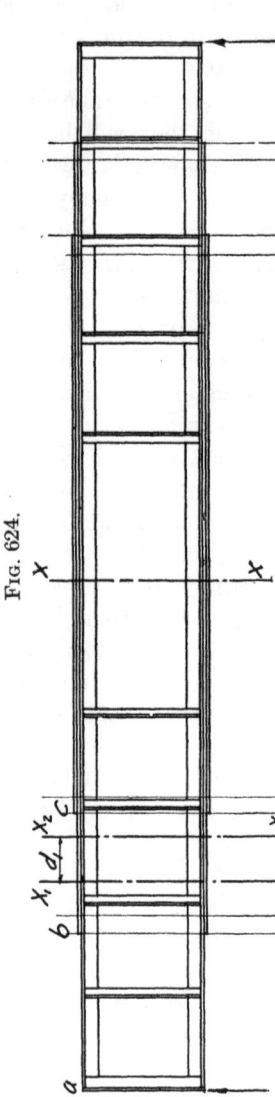

Fig. 624.

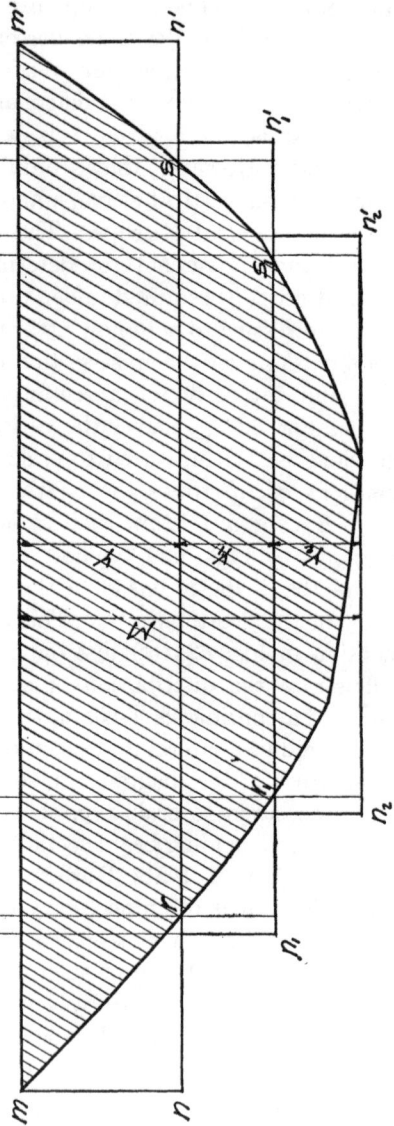

Fig. 626.

small plate girders. However, when using these tables, the designer should keep in mind that the deduction for rivet holes has not yet been made.

There are, of course, approximate methods which are quicker than computing the section modulus. One of these approximate methods is to assume that the effective depth is 2 inches less than the distance back to back of angles. If the moment is divided by the effective depth the force carried by each flange is obtained, and this value divided by the allowable flange stress gives the approximate net area required for each flange. When figuring the flange area for this method, $\frac{1}{8}$ of the area of the web is sometimes considered as flange area for resisting moment.

After the cross-section of the girder has been determined, the next step is to find the necessary length for the cover plates. This may be done conveniently by the use of the moment diagram. See Fig. 626. In connection with moment and shear diagrams the student will do well to refer to Chapter IV. In Fig. 626 the line n–n' is drawn parallel to m–m' and Y distance from it, the distance Y representing the moment, which the web and angles can take, to the scale at which the moment diagram was drawn. The distance Y_1 represents the moment that the first pair of cover plates is able to take, and Y_2 represents the moment that the second pair of cover plates can take. If there were more than two pairs of cover plates there would be a Y_3 and possibly a Y_4, the sum of the Y dimensions in any case being equal to M. The first pair of covers might be cut off so that the end rivets are at r and s, the points where n–n' intersect the moment curve. The end rivets for the second pair might be at r_1 and s_1. However, it is customary to run the cover plates several inches or a foot past these points in order to get a few rivets in the ends of the plates before the section, where they are really needed, is reached.

The webs of plate girders are usually thin and deep, and it is often necessary to stiffen them by means of vertical angles, called stiffeners, in order to prevent them from buckling. When stiffeners are necessary they should be spaced so that the distance between them is not greater than about 60 times the thickness of the web. When the distance between flange angles is less than 60 times the thickness of the web, the stiffeners are often omitted except where concentrated loads are applied.

One of the important things in connection with plate girders, and one which is often not given proper attention, is the spacing of rivets. Since the web and flange angles are working to capacity at b, there should be at least enough rivets connecting the flange angles to the web between a and b to develop the full strength of the angles. Also there must be at least enough rivets between b and c connecting the cover plate to the angles to develop the full strength of the cover plate.

There are many more or less approximate methods and rules for determining rivet spacing, but the results from any of them may be checked up in the following way. Take any two sections x_1–x_1 and x_2–x_2 a distance d_1 apart. The intensity of stress over any part of either section can easily be found from the moment diagram and the properties of the section. The designer should, however, keep in mind that there can be no more stress in the cover plates at x_1–x_1, than can be put in by the rivets between b and x_1–x_1. The total load or stress carried by the top cover plate or plates at section x_2–x_2, less the load they carry at section x_1–x_1, gives the horizontal shearing force which must be resisted by the rivets connecting the top covers to the angles along the length d_1. The total stress carried by the top cover plates and angles at section x_2–x_2, less the amount carried at section x_1–x_1, gives the horizontal shearing force which must be resisted by the rivets connecting the angles to the web. This method may be extended to take care of the most complicated cases.

133. Steel Trusses; Design and Details.—In Chapter V especial attention was given to the construction of stress diagrams for various trusses with more or less complicated loadings. In other words, Chapter V was devoted largely to the determination of stresses by the use of stress diagrams. The present article will indicate how a simple truss may be designed after the stresses have been determined, and show by means of a detail drawing the way the results may be presented.

Consider the truss shown by center lines in Fig. 627. After carefully computing the loads and drawing-stress diagrams the maximum stresses are found to be those shown in the table of Fig. 628. In case it seems desirable the stresses may be marked on the diagram of the truss and the table omitted.

Many trusses are designed from stresses produced by vertical loads, the wind load being neglected. For low flat trusses the stresses produced by the wind are small and no large error is produced by neglecting them, in fact, all of the maximum stresses may be produced by the dead and maximum snow loads. However, when the truss is high or is supported by steel columns and knee-braced to them, it is important to consider the wind stresses.

A vertical live load of 30 or 40 pounds per square foot is sometimes used. This live load is supposed to produce stresses at least as great as those which would be produced by snow, wind, and special loads.

Referring to Fig. 627 the student will note that all of the upper chord members have compression, all of the lower chord members have tension, and about every other web member has tension. Steel trusses of ordinary proportions are usually built up out of angles with gusset plates at the panel points. It is desirable to use two angles for each chord member and for each important web member, one on each side of the gusset.

When designing tension members the net section should be used, that is, the rivet holes should be deducted because all of the stress must go around the first rivet or rivets and be applied to their far side or to other rivets. The tension which any given member must carry may be divided by the allowable working stress (usually 16,000 pounds per square inch), and the required net area obtained. Then a pair of angles may be picked out which will have the required area after deduction has been made for rivet holes. When only one leg is connected, the connection is, of course, eccentric, and many authorities contend that the average stress should be reduced, some going so far as to recommend figuring only the connected leg. The eccentricity is most severe when the member consists of one small angle and the connection is short. When the member consists of two angles, one on each side of the gusset, and has a connection at least 5 or 6 rivets long, it is probable that most of the eccentricity is taken up in the connection even when clip angles are not used. The student should keep in mind that deduction is made for a hole $\frac{1}{8}$-inch larger than the nominal size of the rivet. The hole is punched $\frac{1}{16}$-inch larger than the normal size of the rivet and it is assumed that a certain amount of the metal next to the hole is injured. The design of tension members is simplified by use of a table given in The Carnegie Steel Co.'s " Pocket Companion."

This table gives the tension value in thousands of pounds for various angles after one or two rivet holes have been deducted.

The design of compression members is a little more com-. plicated, since the allowable unit stress varies with the ratio $\frac{L}{r}$. A common formula for the allowable stress S is

$$S = 16,000 - 70\frac{L}{r};$$

in which r is the least radius of gyration and L is the unsupported length in inches. Since r varies with different sections the allowable stress for any given member cannot be determined until the member has been designed. This makes a trial method necessary unless tables or diagrams are used. Ketchum, in his "Structural Engineers' Handbook," gives tables for use in designing angle compression members or struts. A diagram for use in designing such members will be found in Fig. 645 in connection with Article 137. When designing compression members no deduction is made for rivet holes because it is assumed that well driven rivets can transmit compression from one side of the hole to the other.

The members of the truss are now designed using the diagram of Fig. 645, and the tables for tension values of angles given in the Carnegie Handbook. The sizes for the various members are marked on the truss diagram in Fig. 627. The truss is too long to ship conveniently in one piece, a field splice will therefore be made at the center. Members B–3, C–5, and D–7 are made of one pair of continuous angles, because it is cheaper to use a little additional metal rather than make a splice at each panel point. One continuous pair of angles is used for members 1–Q, 2–P, 4–O, and 6–N for the same reason.

A detail drawing of the truss is given in Fig. 629. First the center lines of the truss are drawn. The angles for the various members are usually placed so that their gauge lines coincide with these center lines; when the angle leg next to the gusset has two gauge lines the one next to the outstanding leg should be placed over the center line. It is desirable to use two unequal leg angles for a compression member, and place them with the long legs together, one on either side of the gussets. This makes the radius of gyration about the different axes more nearly equal.

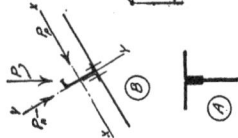

FIG. 627.

The stresses noted above are in thousands of pounds.
One angle is often considered satisfactory when the stress is
small, like that in members 1-2, 6-7, 7-8, 8-9, and 13-14.
2-L₂ 4"×3½ might be used for members 3-4, 5-6, 9-10 and
14-12 if they could be obtained.

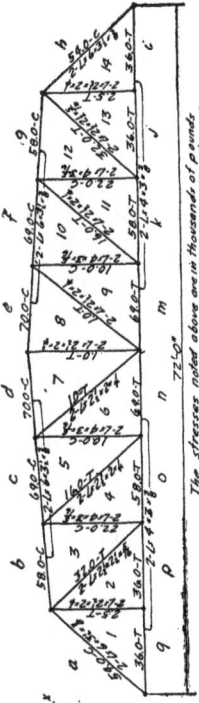

FIG. 630.

FIG. 628.

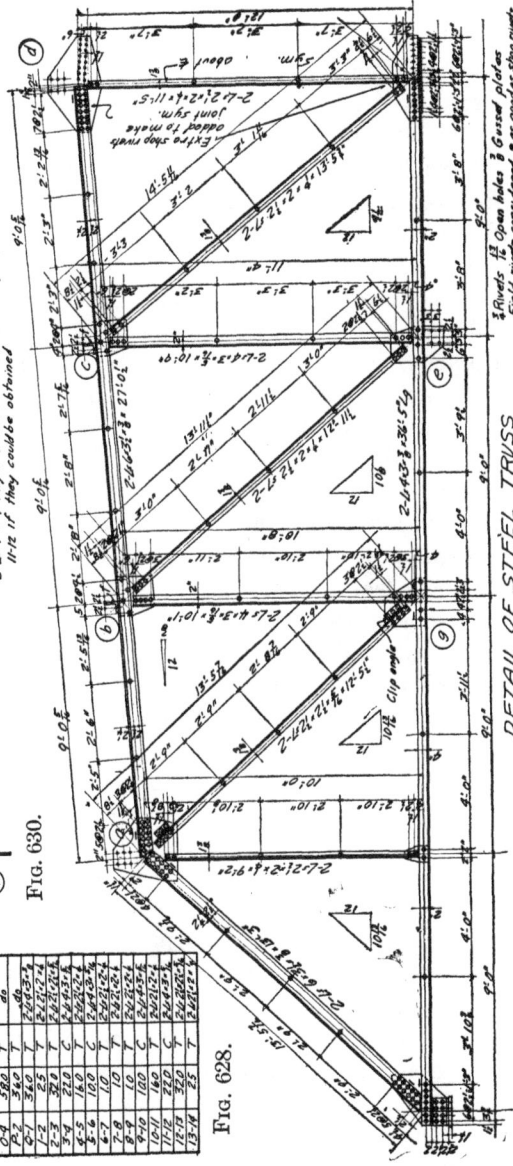

DETAIL OF STEEL TRUSS

FIG. 629.

In special cases when a compression member has lateral support
in one direction at closer intervals than in the other, it may be
desirable to place the short legs of the angles together. When
the compression member is very long, so long that two 6 × 6
angles do not give a large enough radius of gyration, two angles
and a plate as illustrated in Fig. 630 Ⓐ may be used to advantage.

It would be very nice if the members of the truss could be
placed so that the center of gravity of each section coincided
with the center line for that member. Not only is this desirable,
but it is also important that the connection of the member to the
gusset be centered around this line. The truss center lines
should, of course, always intersect at the panel points. It is
not practical to realize all three of these desirable features in ordinary trusses. The center lines can usually be made concurrent
at the panel points, but difficulty is encountered in connection
with the other two requirements. If the gauge line for small
angles were moved over to the centroid it would be so near the
outstanding leg that the rivets could not be driven. However,
it is often possible to move the standard gauge line for 4- and 3-
inch angle legs a short distance towards the centroid. When
there are good deep gussets the designer does not need to worry
much about small eccentricities at the panel points. The moment
produced will be taken up largely by the gusset, since the moment
from some members at any one joint will often be of different
sign than that produced by others.

The determination of the proper number of rivets to connect
each member to the gussets at its ends is an important matter.
There are three things to investigate, the bearing on the gusset,
shear between the gusset and the member, and bearing on the
member. The smaller of these three values governs, that is,
gives the allowable load per rivet. When the member consists
of two angles, one on either side of the gusset, the rivets are in
double shear. Shear and bearing values for rivets of different
sizes with various unit stresses are given in the handbooks. The
standard size rivets for ordinary structural work are $\frac{3}{4}$-inch and
$\frac{7}{8}$-inch. Gusset plates vary in thickness from $\frac{1}{4}$ to $\frac{5}{8}$, with $\frac{3}{8}$ being
more or less of a standard for average trusses. It is desirable
to use the same sized rivets and the same thickness gussets throughout any one truss.

Clip angles similar to those shown at the ends of member

2–3 are sometimes used, and certain authorities think they are of great value, while others think their value is small. Tests have not been complete enough to prove which contention is right. If the clips are considered a part of the member, then the connection of the member to the gusset when clips are used may be more eccentric than when clips are not used. But if the clip is considered a part of the gusset its addition decreases the eccentricity of the connection. It would seem just about as reasonable to consider the clip a part of the member as to consider it a part of the gusset, therefore its value is uncertain. One advantage of clips is that they shorten the connection, and this is of special value when the stress is large and a great many rivets are needed in the connection. When clips are used there should be enough rivets connecting the member and clip to the gusset to take out the entire load, also the connection of the clip to the member should be as strong as its connection to the gusset. In case the connection of the clips to the member is weaker than their connection to the gusset, the connection to the gusset should not be figured as taking any more load than can be put into the clips by their connection to the member.

At panel points ⓐ and ⓓ the gusset plates are shown extending above the upper chord to receive the purlin connection. This makes a very strong connection. At panel points ⓑ and ⓒ a much simpler purlin connection is shown, the channel purlin rests directly upon the upper chord and is riveted to it and the clip angles f. Panel points ⓔ and ⓖ show methods of connecting ceiling beams. When the upper chord makes a large angle with the horizontal, the parallel component of the loading P_p becomes rather large, see Fig. 630 ⓑ. This parallel component produces bending about the axis $Y–Y$, an axis about which I-beams and especially channels have little strength. The purlins are, therefore, usually supported laterally by tie rods spaced 4 to 8 feet apart. These tie rods carry an additional load up to the purlin at the ridge, this additional load should be taken into account when that purlin is designed.

134. Wooden Trusses, Design and Details.—Wooden trusses present many interesting problems especially in connection with the details. It is very difficult to connect wooden tension and compression members in such a way as to develop their full strength. It is also difficult to connect the web members to the chords. The chords of wooden trusses are usually made of wood, also the web members which have compression, but steel or wrought iron rods are often used for the tension web members.

Consider the truss indicated by the line diagram of Fig. 631. This truss is similar to the one shown in Fig. 627. The stresses are marked on the diagram. When designing rods for tension members, the student should remember that the weakest section is at the root of the threads, unless the rods are upset. If the rods are large a saving may be effected by enlarging the ends before cutting the threads so that the area at the root of the threads will be at least as great as the area of the rod proper. This is called upsetting; rods smaller than one inch in diameter are seldom upset, and when the length of the larger rods is short the cost of upsetting may more than balance the saving.

The lower chord, usually a wooden tension member, should be designed so that the unit stress on the net section does not exceed the allowable tension for the wood. In making such computations the actual size and not the nominal size of the timbers should be considered. In some cases as much as a third of the member may be cut out in order to make the connection at joints or splices, thus leaving the net area only about two-thirds of the gross. The designer should do better than this if possible.

The compression members are made of wood and are designed as struts or columns, using some column formula such as

$$S = C\left[1 - \frac{L}{80\,D}\right],$$

in which L is the unsupported length in inches, D is the least lateral dimension in inches, C is the allowable compression on short blocks, and S is the allowable average unit stress for the column. As was the case in connection with steel columns and struts, the allowable stress depends upon the section chosen. A trial solution is therefore necessary, unless some kind of table or diagram is used. A convenient diagram is shown in Fig. 643 and its use explained in connection with Article 135. Note that this diagram is for the actual sizes rather than the nominal size.

After having determined the size of the different members the detail shown in Fig. 632 is drawn, the members being centered on the center lines. The details will work out much better if the dimension of the upper chord normal to the plane of the truss is kept constant, and also the same as that of the lower chord. The appearance of the truss will be neater if the dimension of the wooden web members normal to the plane of the truss is kept the same as that of the chords. The connection of the upper and lower chord at the end of the truss is pe. haps the most difficult detail. For the truss in Fig. 632, two steel plates and bolts are used for this connection. The bolts are in a sense little beams supported at their ends by the plates, and loaded between the plates with a distributed load from the member. These bolts should, therefore, be investigated for bending as well as for bearing on the plates. The shearing stresses in the bolts will seldom be high, but the bearing between the wood and the bolts should be investigated because it may determine the load per bolt and therefore the number of bolts. The splice at ⓒ is also made with plates and bolts. Plates have been used at joint ⓓ, but their principal function is simply to hold the members in position, most of the load being transmitted by direct bearing from the end of one member to the end of the other.

The rods are extended through the chord members and have washers of some kind at their ends in order to spread the load over a larger area. One of the characteristics of wood which gives the detailer a good deal of difficulty is its weakness in taking compression at right angles to the grain. Given the allowable unit pressure parallel to the grain and also the allowable at right angles to the grain, the allowable pressure on a surface making any angle with the grain may be found as illustrated in Fig. 633. Let it be required to find the allowable pressure on the surface $A-C$ which makes an angle α with the grain. Draw the line $C-B$ parallel to the grain and $A-B$ at right angles to the grain. The force F_p represents the resultant of the allowable pressure for $A-B$, and is numerically equal to $A-B$ times the thickness of the timber times the allowable pressure parallel to the grain. Similarly F_n is the resultant of the allowable pressure for $B-C$. The force F'_n has the same magnitude and action line as F_n but opposite sense, also F'_p has the same magnitude and action line as F_p but an opposite sense. The resultant of F'_n and F'_p is F_r, and F is the component of F_r normal to the surface $A-C$. We will therefore consider F divided by (the length $A-C$ times the width of the timber), the allowable unit pressure on a surface

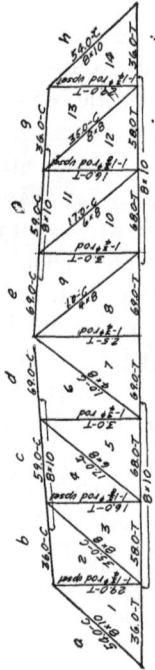

Fig. 631.

The stresses given above are to the nearest thousand. The allowable stress for rods will be taken as 12,000 #/□", and the allowable stress for short blocks as 1300 #/□".

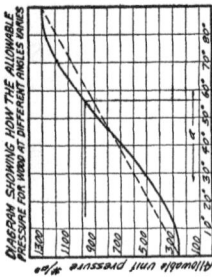

Fig. 634.

DIAGRAM SHOWING HOW THE ALLOWABLE PRESSURE FOR WIND AT DIFFERENT ANGLES VARIES

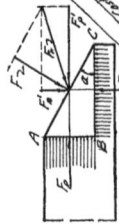

Fig. 633.

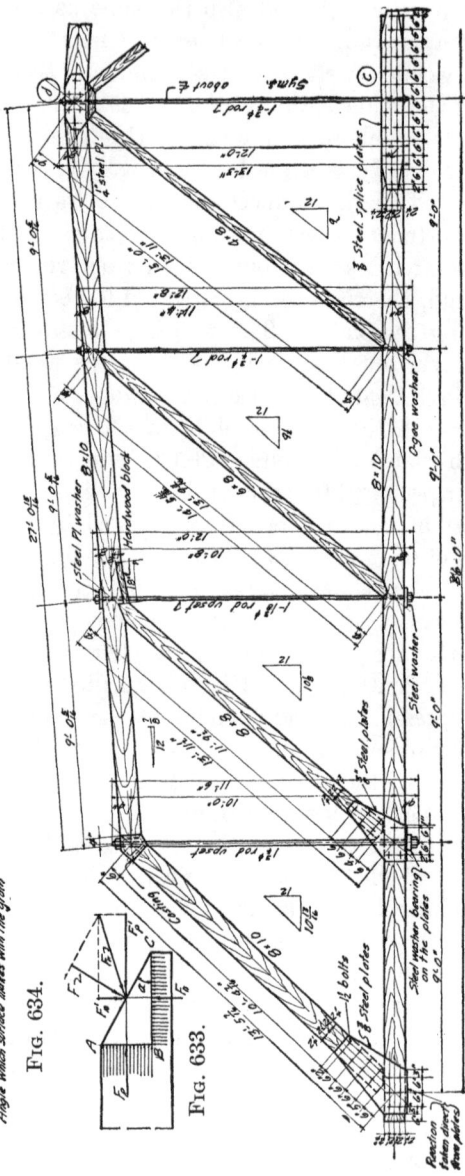

DETAIL OF WOODEN TRUSS

Fig. 632.

making the angle α with the grain. In Fig. 634 a simple diagram is given which shows the variation, in allowable normal pressure as the angle which the surface makes with the grain varies from zero to 90 degrees. This diagram was drawn for an allowable pressure of 1300 pounds per square inch parallel to the grain and 260 pounds per square inch at right angles to the grain. Other diagrams may be drawn for other allowable pressures.

The full line curve shows the variation obtained from the method illustrated in Fig. 633. The straight dotted line deviates from the curve only a small amount and is preferred by some designers.

The washers used at the ends of the rods may be of either cast-iron or steel plates. An interesting diagram for use in design-

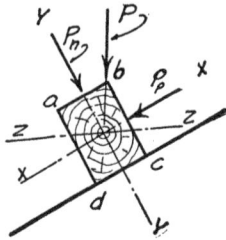

FIG. 635.

ing cast-iron o-gee washers is given in Fig. 644 and explained under Article 136.

The design of the purlins is more complicated than the design of ordinary beams because the resultant bending may be about some axis z–z making an angle with the sides. The effect of this is most pronounced when the upper chord is steep and the purlin is set as shown in Fig. 635. In such cases the resultant load P may be broken into the two components P_n and P_p, P_p producing moment about axis Y–Y, and P_n producing moment about the x–x axis. The effect of P_p is to produce compression along the side b–c and tension along the side a–d. At the same time that P_p is producing these stresses, P_n produces tension along d–c and compression along a–b. The maximum stress produced by P_p plus the maximum stress produced by P_n must therefore be added together in order to obtain the stress at b and at d. A member carrying a loading similar to that indicated in Fig. 635

may be conveniently designed by the use of the diagram of Fig. 636. Compute the moment about the x–x axis and also about

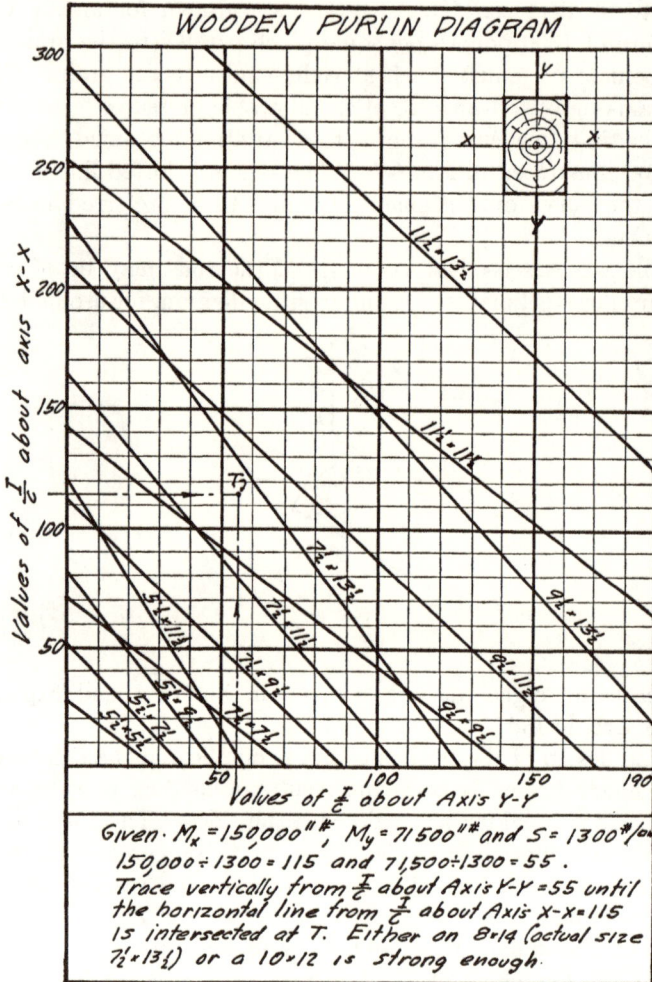

FIG. 636.

the Y–Y axis, and then divide these moments by the allowable working stress for the material, thus obtaining what may be called the required $\dfrac{I}{c}$ about each axis. These values are used in the diagram and any section above the intersection T is strong enough. A similar diagram for steel I-beams and channels is given in Fig. 637.

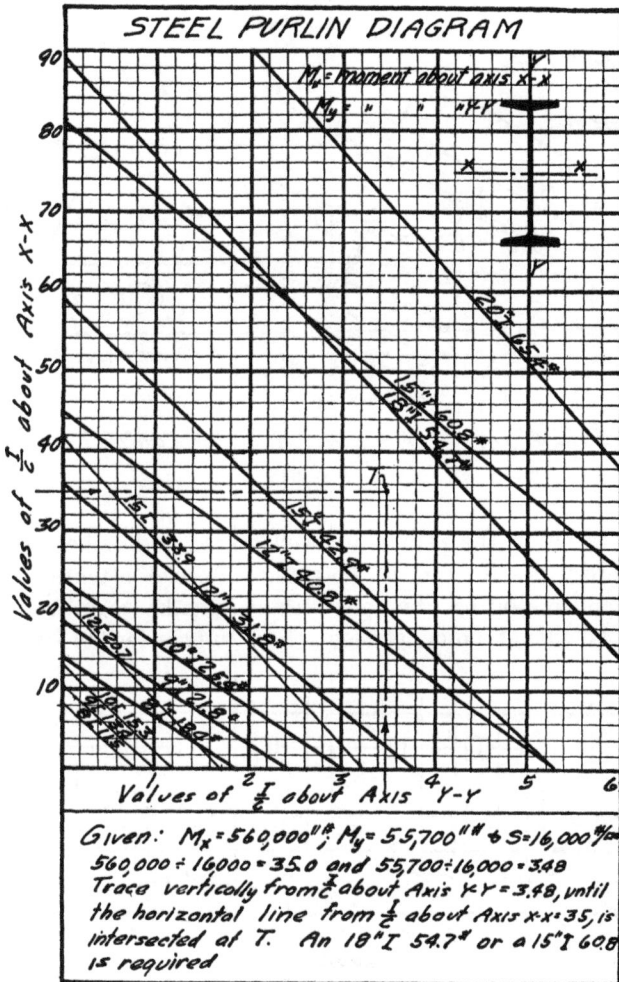

FIG. 637.

Figs. 638 to 642 show a number of wooden truss details. The difficulty with a number of them is that they are rather hard to make. This is especially true of Figs. 638 and 639.

Small trusses are sometimes built up out of 2-inch plank. This does not give as rigid a truss as one in which solid timbers

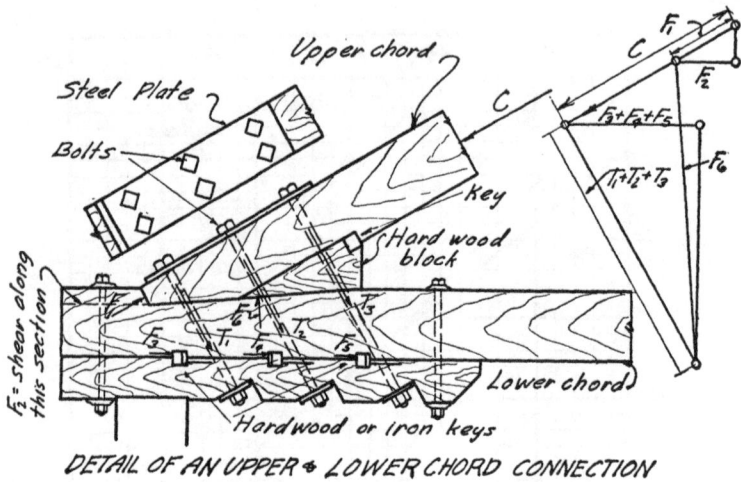

DETAIL OF AN UPPER & LOWER CHORD CONNECTION

Fig. 638.

are used, but if the planks are well bolted and spiked together, satisfactory results can be obtained. Special attention should be given to bolting the lower chord planks together. There should be enough bolts between splices to transfer the stress from

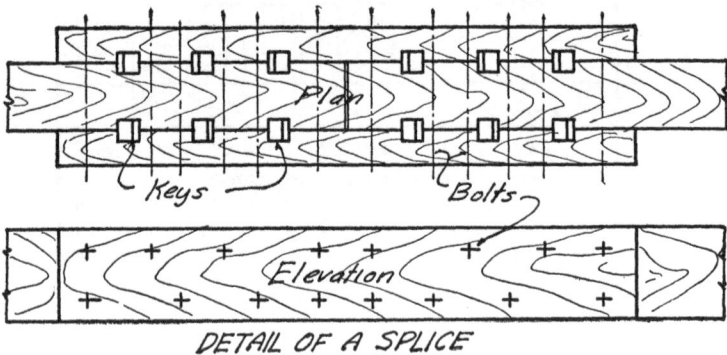

DETAIL OF A SPLICE

Fig. 639.

the plank which is spliced to those which run past the splice. The two inch planks are really only about $1\frac{5}{8}$ inches thick so that

6 — 2 × 12 planks are only equivalent to a 9¾ × 11½ solid timber, yet the charge would be made for 12 and not 10 board per ft. The point is that the designer may think he is saving by using plank at a lower price per board foot, when in reality he may

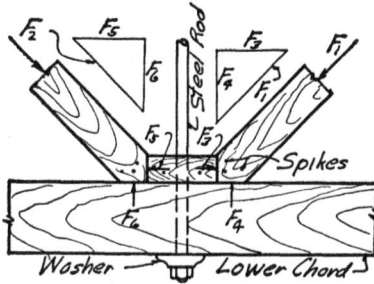

FIG. 640.

get so much less wood per board foot that the actual cost per unit cross-section may be almost as great as for solid timbers.

135. Diagram for Designing Wooden Compression Members.— The diagram given in Fig. 643 will be found convenient for design-

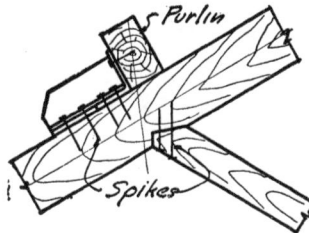

FIG. 641.

ing wooden struts and columns. This diagram has been drawn for the formula

$$S = C\left[1 - \frac{L}{80\,D}\right],$$

but the designer can easily draw similar diagrams for other straight-line formulas. The inclined lines on the Ⓐ side are the load lines, while the inclined lines on the Ⓑ side are for the various sections. The actual size is given in place of the nominal size but the area is for the nominal size because that is the area that must be paid for. The line for a 10 × 12 timber is marked

$9\frac{1}{2} \times 11\frac{1}{2}$ because that is the actual size of the timber, but the area used in computing the number of board feet is not $9\frac{1}{2}$ times $11\frac{1}{2}$ but 10 times 12 or 120 square inches or 10 board feet per foot.

Let it be required to design a compression member which must support a load of 85,000 pounds and has an unsupported length of 16 feet. The allowable compression for short block, C, varies with the kind of wood, and is specified in building laws and standard specifications. A stress of 1300 pounds per square inch will be used for C in this problem. The C line marked

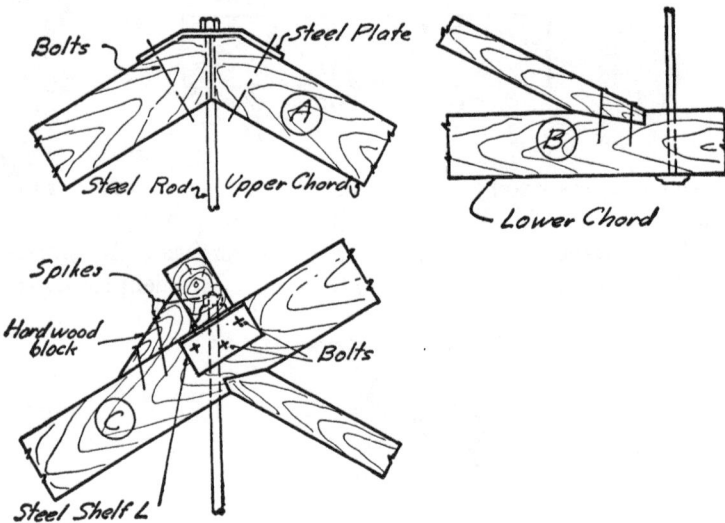

Fig. 642.

1300 is located in the Ⓐ half of the diagram, and its intersection T_1 with the 85,000 pound load line is found. From T_1 move to the right parallel to the guide lines until the 16 foot length line is intersected at T_2. There are two section lines a short distance above this intersection, one marked $9\frac{1}{2} \times 9\frac{1}{2}$ and the other $7\frac{1}{2} \times 13\frac{1}{2}$, and their nominal areas are 100 and 112 respectively. The 10×10 having actual dimensions of $9\frac{1}{2} \times 9\frac{1}{2}$ would therefore be chosen as the more economical section.

136. Diagram for Designing Cast-iron O-gee Washers.— An interesting diagram for use in designing cast-iron o-gee washers is shown in Fig. 644. Cast iron is weak in tension but strong in compression, the compression side of the washer which

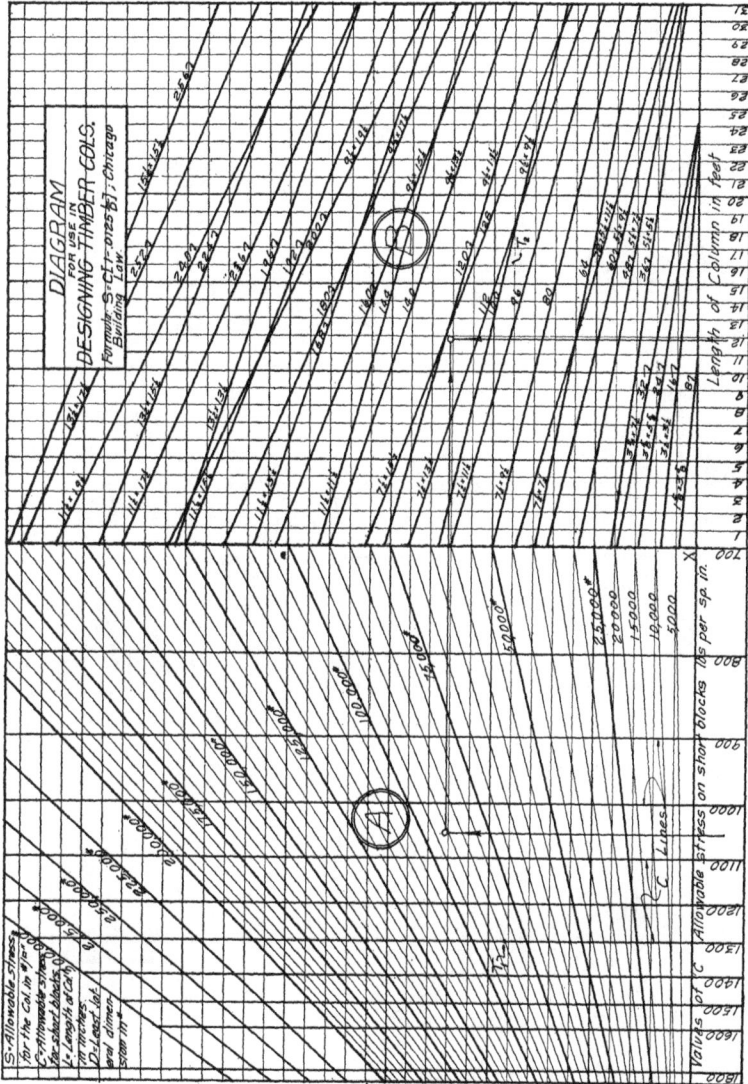

FIG. 643.

is next to the nut can, therefore, be made of small diameter as shown in the figure without any large decrease in the strength of the washer. In fact the strength per pound of metal is increased.

Problem. The load which the rod must carry is 19,000 pounds.

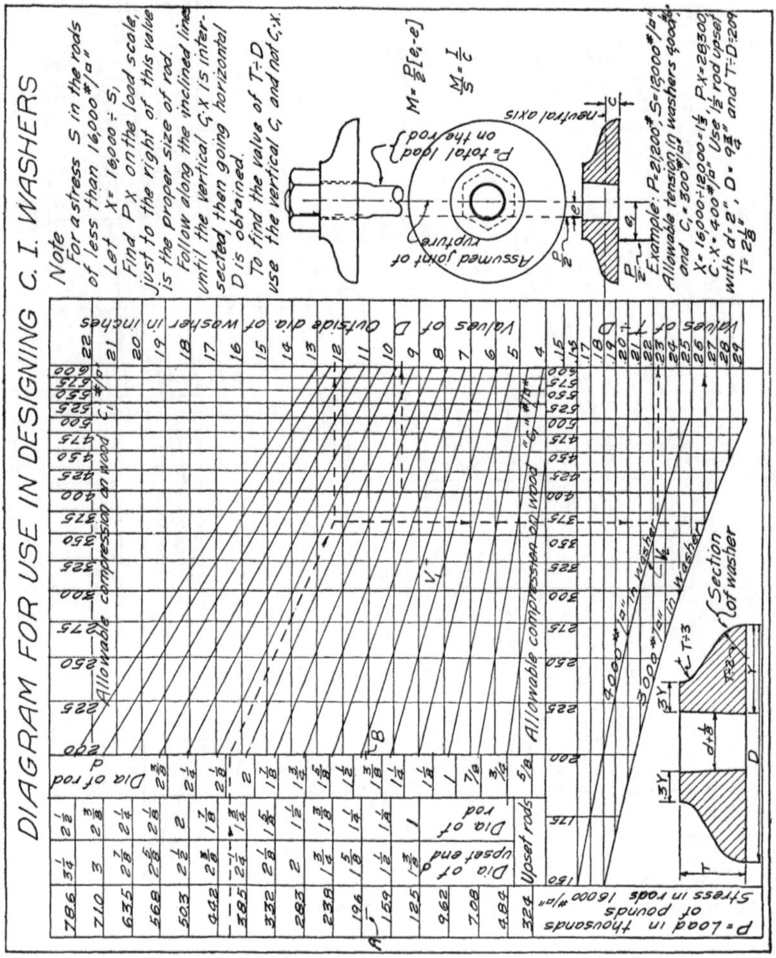

Fig. 644.

Let the allowable stress in the rod be 16,000 pounds per square inch, and the allowable compression on the wood 325 pounds per square inch. Along the load scale at the left edge of the sheet 19,000 is found to be at A. Moving over to the right it is found that a $1\frac{1}{4}$ rod upset to $1\frac{5}{8}$ or a $1\frac{1}{2}$ rod not upset will take the load.

The horizontal line from A intersects the 200-pound C_1 line at B. From B move parallel to the inclined lines until the 325-pound C_1 line is intersected at V_1. From V_1 go horizontally to the D scale where the required diameter is found to be $8\frac{3}{4}$ inches. In order to find the necessary thickness for a tension of 4000 pounds per square inch in the cast iron follow down the 325 C_1 line until the 4000-pound line is intersected at V_2. Then moving horizontally to the $T \div D$ scale, the value .215 is found for $T \div D$. The diameter D is known and .215 times $8\frac{3}{4} = 1\frac{7}{8}$, the required thickness.

When the allowable stress for the rod is less than 16,000 pounds per square inch, the washer may be designed as already explained and the size of the rod determined by a separate computation, or the method illustrated in the figure may be used.

137. Diagram for Designing Steel-angle Struts.—A simple diagram for use in designing steel-angle compression members is given in Fig. 645. The inclined lines are for the various sections, one line for each section. The short dash cutting each line indicates the length at which $\dfrac{L}{r}$ for that particular section is 120. The weight of the section per foot is also given on each line. This weight assists the designer in picking economical sections quickly. An illustrative problem is given in the figure. Similar diagrams may be drawn for plate and angle columns or for Bethlehem H columns if desired.

The diagrams given in Figs. 653 and 654 and explained in Article 140 may be used for direct load only. This is simply a case in which M, and therefore P', is zero.

138. Combined Stresses.—Many structural members resist bending and direct load at the same time. Such problems have been considered in connection with masonry, and also in the chapter on reinforced concrete. The reader will recall the formulas $S_M = \dfrac{P}{A} + \dfrac{M \cdot c}{I}$ and $S_m = \dfrac{P}{A} - \dfrac{M \cdot c}{I}$ as giving the maximum and minimum unit stresses respectively. When the material cannot take tension these formulas should not be used if $\dfrac{Mc}{I}$ is larger than $\dfrac{P}{A}$.

It may be well to call attention to the fact that the above

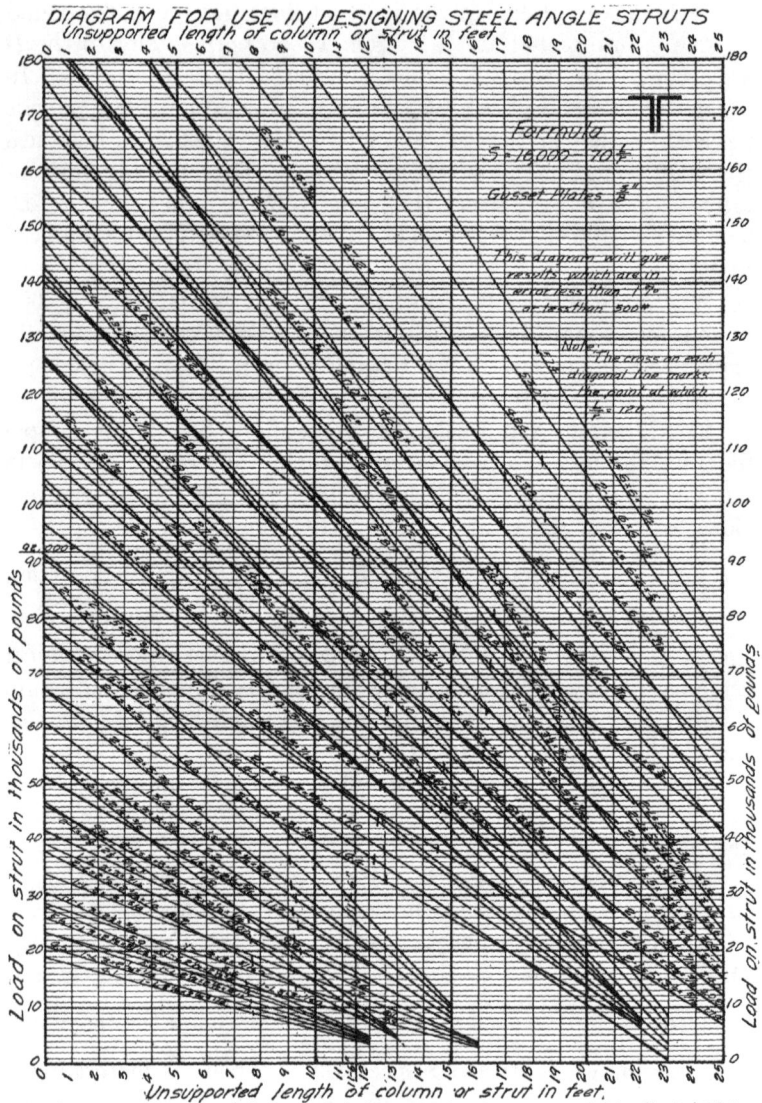

DIAGRAM FOR USE IN DESIGNING STEEL ANGLE STRUTS

FIG. 645.

formulas do not take account of the increased eccentricity, and therefore increased moment due to deflection. However, the error produced by this approximation is so small for ordinary columns that it is not worth considering. When the member is a column and the moment is produced by the eccentricity of the load P, the moment may be computed and used in the above formulas, or $P \cdot e$ may be substituted for M, e being the eccentricity of P.

There is often some question as to what the allowable unit stress should be for members with combined stresses. The allowable stress given by the column formula is usually used but this seems rather conservative when most of the stress is produced by bending.

139. Trussed Beams.—In some cases the designer meets with difficulty in getting wooden beams which will carry the load, when he does not wish to use either steel or concrete. In such cases a trussed beam may be desirable. There are many different forms of trussed beams consisting of rods, castings and timber. However, the detail shown in Fig. 646 will serve as an illustration. The center lines are shown in Fig. 647. The beam consists of a timber member on top, two short cast-iron struts, and tension rods at the bottom. The timber member acts somewhat as a three-span continuous beam, supported at the ends by R_1 and R_2 and at two intermediate points by the struts S_1 and S_2. The compression in the struts S_1 and S_2 is, of course, produced by the reactions at these points from the continuous wooden beam. Knowing the stress in the strut S_1 the stress in the rods on each side of the strut can be found by drawing a force triangle. The vertical component at the left end of the rods goes into the reaction R_1. This vertical component added to the reaction from the continuous timber beam at this point gives the magnitude of R_1. The magnitude of R_2 may be found in a similar way. The horizontal component from the rods goes into the timber member and produces compression for its entire length. Therefore this timber member must resist both bending and direct compression.

Example. Let it be required to design a trussed beam having a 30-foot span and an effective depth from the center of the timber to the center of the rod of 18 inches. See Fig. 647. A uniform load of 1200 pounds per foot will be considered and the struts placed

at the third points. A shear diagram for the continuous timber
beam is shown in Fig. 648, assuming that its supports are all at
the same level. *W* is the load for one span of the continuous

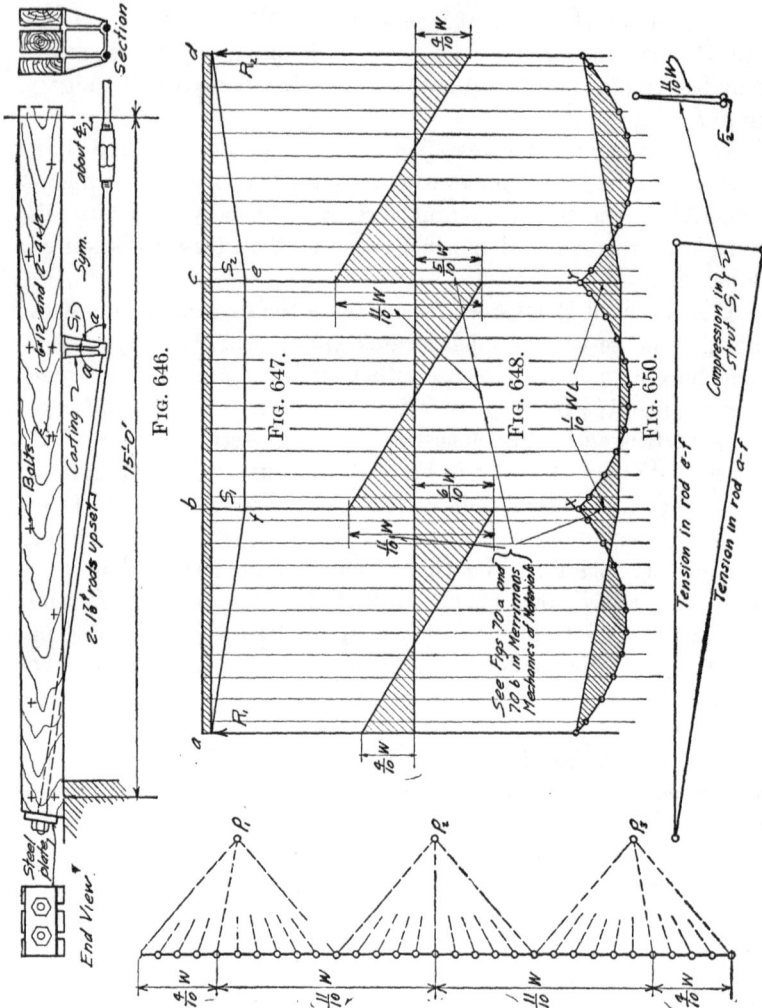

beam or one-third the total load for the trussed beam. The moment
diagram of Fig. 650 is easily drawn from the force polygon of Fig.
649 and points *X* and *Y* located from a table given by Fig. 70*a*
in Merriman's " Mechanics of Materials."

The reaction at b is found to be $\frac{11}{10} W$ = 13,200 pounds. Since the strut S_1 has a slight inclination, its stress is found by the force triangle of Fig. 651 to be slightly greater than 13,200 pounds. The vertical component of this stress balances the $\frac{11}{10} W$, and the horizontal component produces a small amount of compression between b and c. Knowing the stress in the strut, the stress in the rod from e to a may be found by drawing the force triangle of Fig. 652. The vertical component of the stress in f–a is found to be just 13,200 pounds, which added to the 4800 from the continuous beam at a, gives the reaction R_1 equal to 18,000 pounds, which is just half the total load carried by the trussed beam. The tension in a–f produces a compression in a–b equal to its horizontal component. The compression in a–b plus the force F_2 of Fig. 651 gives the compression from b to c. Since the trussed beam is symmetrical the stresses for the right end will be the same as those just found for the left end. The maximum compression for the timber is found to be about 88,000 pounds and the maximum moment 144,000 inch-pounds.

There may be some question regarding the allowable stress which should be used. If there is no lateral support the 30 feet shall be considered the unsupported length. But such beams usually have continuous lateral support from floor or roof construction. In such cases the unsupported length would be only 10 feet. If C, the allowable compression on short blocks, is taken as 1300 pounds per square inch and the depth of the timber as $11\frac{1}{2}$ inches, the allowable stress S is found from the column formula to be 1130 pounds per square inch.

Two 4 × 12's and one 6 × 12 as shown in Fig. 646 make a good design, and their stress amounts to

$$\frac{P}{A} + \frac{Mc}{I} = \frac{88,000}{149} + \frac{144,000}{287} = 590 + 501$$

$$= 1091 \text{ pounds per square inch.}$$

The actual size of the timbers was used in computing A and $\frac{I}{c}$. With an allowable stress of 16,000 pounds per square inch in the steel rods, two $1\frac{7}{8}$ rods upset at the ends are found to be satisfactory.

Special attention should be given to the design of the plate or large cast-iron washer through which the rods deliver their load to the ends of the timbers. The shear and moment diagrams shown in Figs. 648 and 650 are for the case when the four supports of the continuous timber beam are at the same elevation. If the rods are tightened so that points b and c are high, the negative moments at these places will be increased, the positive moments will be decreased, and the stress in the struts and rods will increase. The opposite will be true if b and c are low.

140. Diagram for Designing Eccentrically Loaded Steel Columns.—The diagrams given in Figs. 653 and 654 will be found convenient when designing steel columns which carry bending as well as direct stress. The diagrams give lines for two groups of plate and angle columns, one group having 12-inch webs and the other 14-inch webs. The designer can easily make diagrams for as many additional sections as he wishes. For every given moment applied to any one section there is a direct load which, when applied at the centroid of the section, will produce the same maximum unit stress. Let this direct load be called P', then $\dfrac{P'}{A} = \dfrac{M \cdot c}{I}$ and solving the equation for P', we find that $P' = \dfrac{M \cdot A \cdot c}{I} = M \div \dfrac{I}{A \cdot c}$. In other words, the moment divided by $\dfrac{I}{A \cdot c}$ gives a direct load which will produce the same maximum stress. The values $\dfrac{I_x}{A \cdot c_x} = X$ and $\dfrac{I_y}{A \cdot c_y} = Y$ have been computed for each section and recorded in the diagrams on the corresponding line. Now if the moment is divided by X or Y, as the case may require, and the equivalent direct load P' is added to the direct load P, the resultant direct load applied at the centroid will produce the same maximum unit stress as the given direct load and moment. In the diagrams, values of $P + P'$ are given along the vertical scale, and unsupported lengths along the horizontal scale.

Several different types of problems may be solved by the use of Figs. 653 and 654, three of which will be illustrated below.

Suppose the column has been designed and consists of the following material: a $14 \times \frac{5}{8}$ web, 4 angles $6 \times 6 \times \frac{5}{8}$, and cover plates on each side $16 \times 2\frac{1}{4}$. Let the unsupported length be 14 feet and the direct load 1,000,000 pounds. How much moment can this column take about the x–x axis without being overstressed? Follow up along the 14 foot line in Fig. 653 until the inclined line

DIAGRAM FOR USE IN DESIGNING COLUMNS HAVING ECCENTRIC LOAD

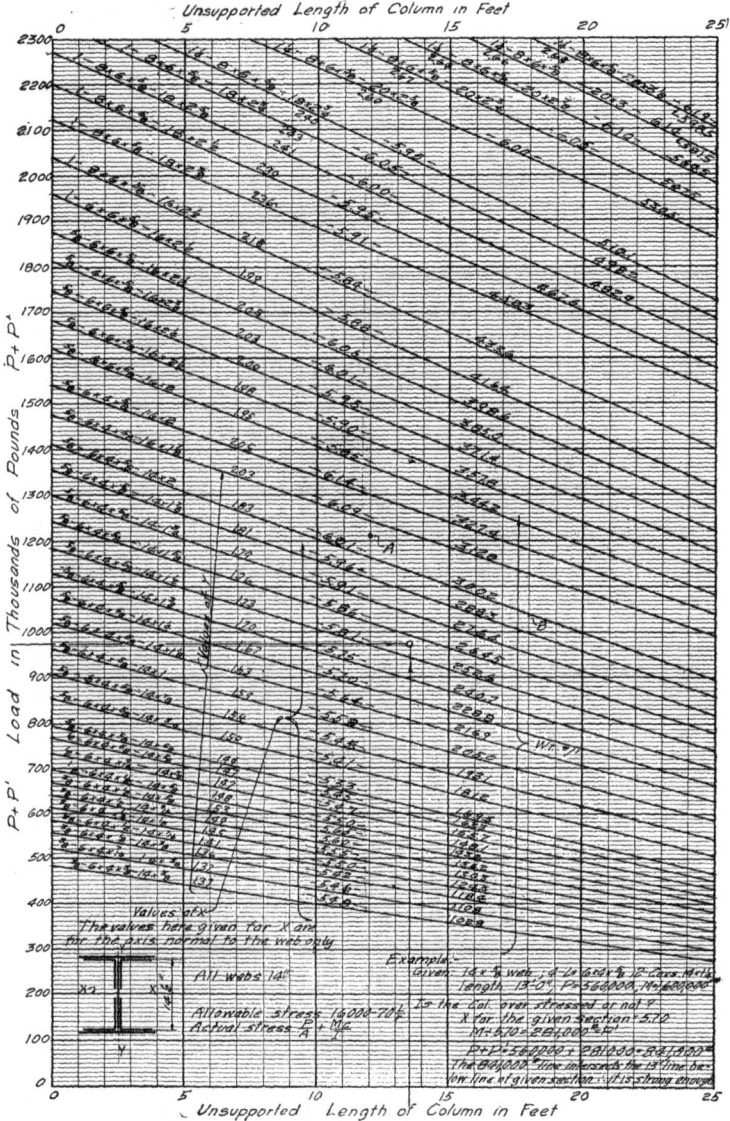

Fig. 653.

for the given section is intersected. From this intersection move parallel to the horizontal lines to the left-hand scale where the allowable $P + P'$ is found to be 1,426,000 pounds. Subtracting P from this value, the value of P' is found to be 426,000 pounds. The value of X marked on the inclined line for the given section is 5.95. $5.95 \times 426,000 = 2,534,700$ inch-pounds, which is the allowable moment.

Given the column and the loading to find whether or not the column is over-stressed. Column section; one $14 \times \frac{5}{8}$ web, four L's $6 \times 4 \times \frac{5}{8}$ and cover plates on each side $16 \times 1\frac{7}{8}$. The unsupported length is 12 feet, $P = 800,000$ pounds and $M = 2,500,000$ inch-pounds about the x–x axis. The value of X for the given section is 6.09. $P' = 2,500,000 \div 6.09 = 410,000$ approximately. $410,000 + 800,000 = P + P' = 1,210,000$ pounds. The intersection of the 12-foot line and the horizontal load line 1,210,000 pounds is at A, a short distance below the inclined line for the given section. The column is, therefore, not overstressed, and yet the stress is almost up to the allowable.

If the moment was produced by wind and the designer wished to increase the working stresses, say 25 per cent, the value obtained for $P + P'$ should be divided by 1.25, and the solution continued with this new value.

Let P, M, and the unsupported length be given and the design of the column required. Let $M = 2,000,000$ inch-pounds, $P = 700,000$ pounds and the unsupported length be 18 feet.

The values of X in Fig. 653 vary from about five and a half to a little over six. M divided by a little less than six would give P' about 350,000 and $P + P'$ about 1,050,000. Follow along 1,050,000 load line until the 18-foot line is intersected. This intersection is just above the line for a certain section having X equal to 6.01. This section which consists of a $14 \times \frac{5}{8}$ web, four L's $6 \times 4 \times \frac{5}{8}$ and covers on each side 14×2, is, therefore, assumed as a trial section. $M \div 6.01 = P' = 333,000$ pounds approximate and $P + P' = 1,033,000$ pounds. The intersections of this load line and the 18-foot line is at B, the trial section is, therefore, satisfactory.

If the moment were about the Y–Y axis the values of Y would be used in place of the values of X.

141. Diagram for Designing Eccentrically Loaded Reinforced Concrete Columns.—Reinforced concrete columns often have

DIAGRAM FOR USE IN DESIGNING COLUMNS HAVING ECCENTRIC LOADS

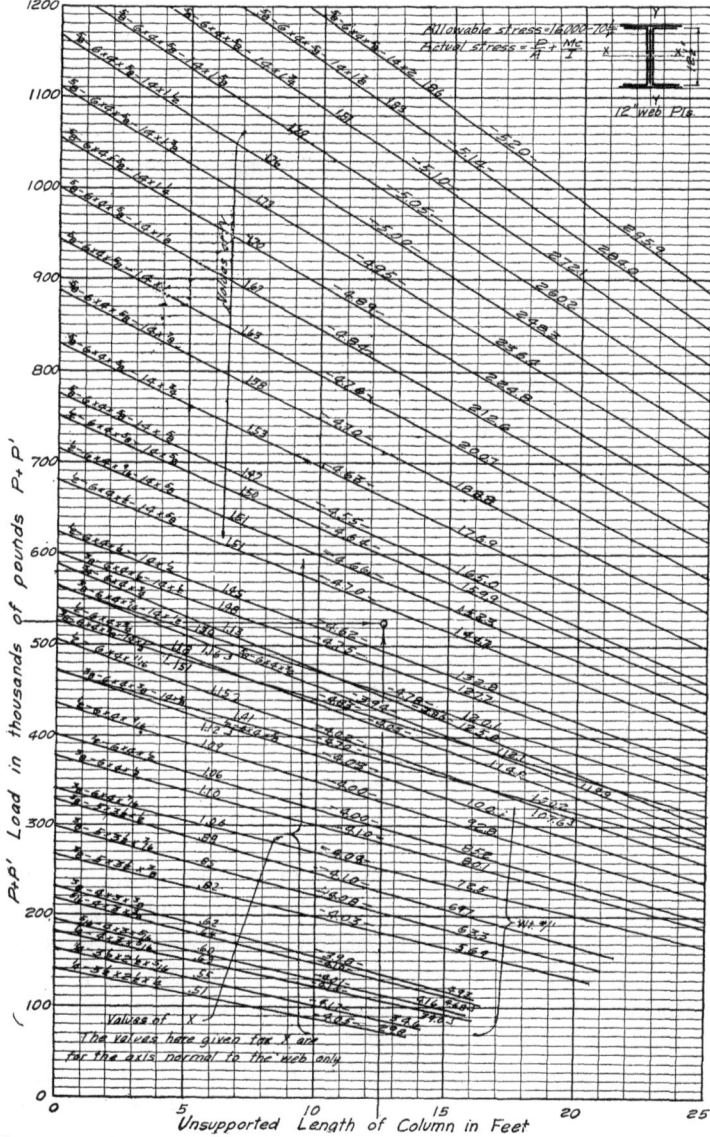

FIG. 654.

eccentric loading, or in other words, resist both direct load and moment. Some Building Laws require that wall columns supporting flat slabs be designed to resist certain moments as well as the direct load. An increased working stress in the concrete, above that allowed for columns on which no moment is considered, is often permitted.

In Fig. 655, the load scale is laid off along the Y–Y axis and the moment scale along the X–X axis. Let any column section be assumed. Find the allowable direct load P which this assumed column is good for and locate A along the load scale. Now when

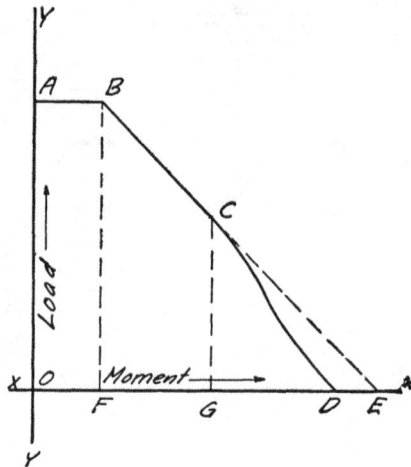

Fig. 655.

moment is added and the direct load kept the same, the point representing the section will move from A horizontally towards B. The point B is directly above F, the moment which produces a stress equal to the allowable increase in the column stress for the case when moment is considered. If more moment is added P must be decreased in order to keep the column from being overstressed. Therefore, as the moment increases beyond F the point representing the assumed section moves along the inclined line B–E towards C. The point C is directly above G, the moment which is just large enough to reduce the stress along one side of the column down to zero. Now if all the material composing the column were able to take tension as well as compression, the line would continue from C on down to E, B–E being a straight line. But since the concrete in a reinforced concrete column is assumed to be unable to take tension, the line from C on down to D is a curve of the shape indicated in the figure. Note that the deviation from the dotted line is small for some little distance below C.

The value for A may be found by multiplying f_c, the allowable unit concrete stress for columns, by the area of concrete plus n times the area of the steel. It is often more convenient to find the area of the concrete plus one times the area of the steel than to find just the area of the concrete. When this is done there is only n–1 times the area of the steel left to be added in.

Standard values of n are 15, 12, and 10 for $1 : 2 : 4$, $1 : 1\frac{1}{2} : 3$, and $1 : 1 : 2$ concrete, respectively.

Suppose an increase of 20 per cent is allowed in the working stress when moment is considered, the direct load never to be greater than O–A. Point F would be located by finding the moment which would produce a stress just equal to 20 per cent of f_c. B is directly above F on the horizontal line from A. The line A–B may now be drawn and, since B–C is a straight line, it may be drawn as soon as C is located. As long as there is compression over the entire section the maximum unit stress is given by the formula $\dfrac{P}{A} + \dfrac{M \cdot c}{I}$, and the minimum unit stress is given by the formula $\dfrac{P}{A} - \dfrac{M \cdot c}{I}$. At point C, $\dfrac{P}{A}$ is equal to $\dfrac{M \cdot c}{I}$ and each is equal to $\dfrac{f_c + 20 \text{ per cent } f_c}{2}$. Knowing this, the M and P co-ordinates of C can be easily computed. Below C the formula $\dfrac{P}{A} + \dfrac{M \cdot c}{I}$ would be in error an amount indicated by the deviation of the curve from the dotted line. Points on this curve C–D may be located by the method given in Chapter VIII for eccentrically loaded reinforced concrete columns, and a smooth curve passed through them.

Wall columns supporting flat slabs often have their width determined by the steel sash, and in other columns either the width or depth can be assumed and the other solved for. The heavy lines drawn in Fig. 656 are, therefore, for columns 1 inch wide, for different depths as shown, and for 1, 2, and 3 per cent of steel. There are three lines for each depth one for 1 per cent steel, another for 2 per cent, and the third for 3 per cent steel.

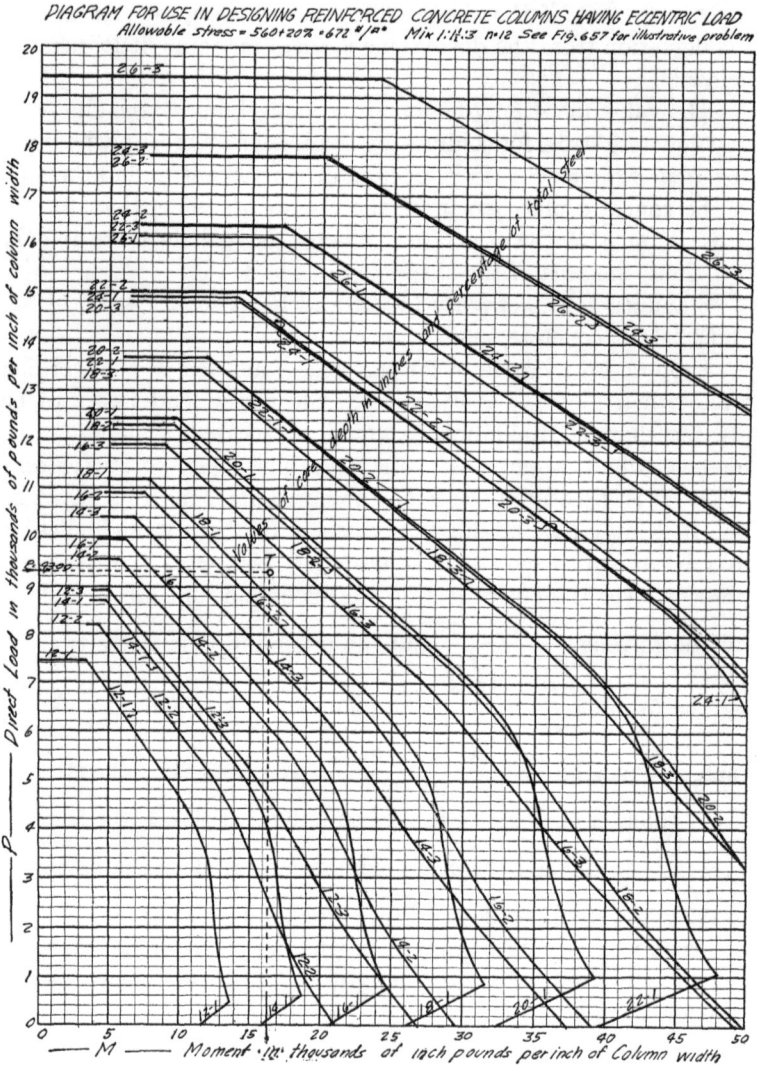

FIG. 656.

The designer is expected to interpolate in order to get 1.3 per cent, 2.7 per cent, 3.2 per cent, etc.

When designing, the direct load and the moment are each divided by the assumed width of column, thus obtaining the direct load and moment per inch of width. These values are located on the M and P scales in Fig. 656, and the intersection of the horizontal P line with the vertical M line gives the desired point.

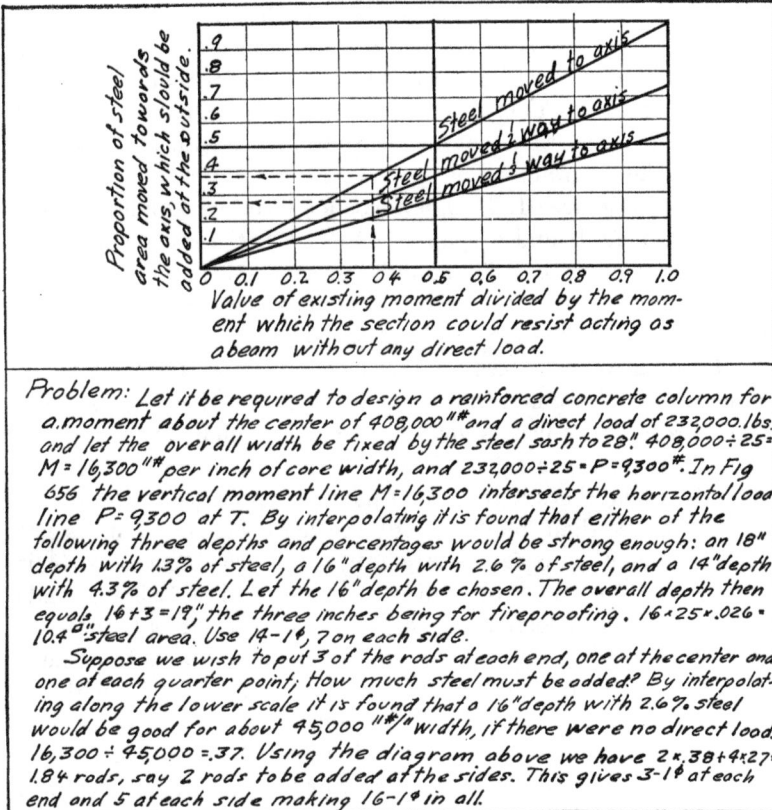

Problem: Let it be required to design a reinforced concrete column for a moment about the center of 408,000 "# and a direct load of 232,000 lbs. and let the overall width be fixed by the steel sash to 28." 408,000 ÷ 25 = M = 16,300 "# per inch of core width, and 232,000 ÷ 25 = P = 9300 #. In Fig 656 the vertical moment line M = 16,300 intersects the horizontal load line P = 9300 at T. By interpolating it is found that either of the following three depths and percentages would be strong enough: an 18" depth with 1.3% of steel, a 16" depth with 2.6% of steel, and a 14" depth with 4.3% of steel. Let the 16" depth be chosen. The overall depth then equals 16 + 3 = 19," the three inches being for fireproofing. 16 × 25 × .026 = 10.4° steel area. Use 14-1#, 7 on each side.
   Suppose we wish to put 3 of the rods at each end, one at the center and one at each quarter point; How much steel must be added? By interpolating along the lower scale it is found that a 16" depth with 2.6% steel would be good for about 45,000 "#/" width, if there were no direct load. 16,300 ÷ 45,000 = .37. Using the diagram above we have 2 × .38 + 4 × .27 = 1.84 rods, say 2 rods to be added at the sides. This gives 3-1# at each end and 5 at each side making 16-1# in all.

FIG. 657.

Any section line above and to the right of this point gives a depth and percentage of sufficient strength. If the designer so desires he may pick out some depth, and by interpolation determine just the right percentage of steel.

The diagram in Fig. 656 was drawn for 1 : 1½ : 3 concrete, assuming a working stress of 560 pounds per square inch for columns, with a 20 per cent increase for combined stresses. This means that the allowable stress may increase as moment is added until it reaches $560 \times 1.20 = 672$, at which point the lines in the diagram change direction and begin to slope. The depths given in the diagram are core depths only, and do not include fireproofing. All of the steel was equally divided between the two sides parallel to the axis about which bending occurs. Fig. 657 indicates how much steel should be added when it seems desirable to put a part of the steel on the other two sides.

If the designer wishes to use higher stresses than those used in drawing the diagram, he may reduce his load and moment in the same proportion that he wishes to increase the stresses, and then use the diagram.

142. Problems.—1. Design a wooden beam for a uniform load of 15,000 pounds and a span of 19'–0''. Let the allowable unit stress for bending be 1600 pounds per square inch and the allowable stress for horizontal shear 175 pounds per square inch. The deflection will be limited to $\frac{1}{360}$ of the span.

Turning to the tables for safe loads given in the " Southern Pine Manual " we find that a timber 10 inches wide and 14 inches deep spanning 19 feet is good for 18,225 pounds when stressed to 1800 pounds per square inch. This value reduces to 16,200 pounds when the stress is 1600 pounds. A capacity of 16,200 is 1200 larger than required, but the next smaller beam is not strong enough. As far as strength is concerned the 10×14 is satisfactory, but the table indicates that a load of 12,949 pounds will produce a deflection $\frac{1}{360}$ of the span. Therefore, the next size beam, a 12×14, must be used if the deflection is to be kept within the given limit. In many cases the 10×14 with its slightly larger deflection would be considered satisfactory. The horizontal shearing stresses are much below the allowable.

An 8×16 would also be a satisfactory design, in fact it is a more economical beam because of its smaller cross-section. If the 16-inch depth could be obtained without excessive delay or expense it should be used.

In the case the designer were limited to a 12-inch depth, two 8×12 would satisfy the requirements for strength and two 10×12 would keep the deflection less than $\frac{1}{360}$ of the span.

2. Design a steel beam for a uniform load of 65,000 pounds and a span of 23 feet. It will be assumed that the compression

flange has adequate lateral support and a stress of 16,000 pounds per square inch will, therefore, be used.

The safe load tables in the Carnegie Steel Co.'s " Pocket Companion " show that either a 20-inch I 80-pound or a 24-inch I 80-pound is strong enough. If there is no objection to the extra 4 inches of depth the 24-inch I 80-pound should be chosen because it would not cost any more, it has extra strength, its deflection would be less, and it is more of a standard section. The 24-inch I 74-pounds might be used if it could be obtained without difficulty.

3. A certain steel beam will be required to resist a bending moment of 6,200,000 inch-pounds. What section can be used? There are no rolled sections listed in the " Pocket Companion " which are able to resist such a bending moment. Therefore, either a plate girder or a Bethlehem Girder Beam should be used unless the designer prefers two smaller beams. 6,200,000 ÷ 16,000 = 387 equals the required section modulus. Use a 26 B.G. 150.

4. Design a continuous reinforced concrete slab for a clear span of 12 feet and a live load of 175 pounds per square foot. Let the concrete be 1 : 2 : 4 and the allowable stresses 750 pounds per square inch for the concrete and 18,000 pounds per square inch for the steel.

Using the table given in Fig. 612, a 6-inch slab is found to be satisfactory. The depth to the tension steel is 5 inches, the main reinforcement is $\frac{1}{2}$ inch square bars 6 inches on centers and the temperature reinforcement is $\frac{3}{8}$-inch round rods 16 inches on centers.

5. Design a simply supported slab having a span of 12 feet and a live load of 125 pounds per square foot. Use the stresses given in Problem 4.

By the use of the table in Fig. 612 a $6\frac{1}{2}$-inch slab with $\frac{1}{2}$-inch square bars $5\frac{1}{2}$ inches on centers is found to be satisfactory. Temperature reinforcing $\frac{3}{8}$-inch round rods 16 inches on centers.

6. Design the cross-section of a rectangular reinforced concrete beam having a bending moment of 1,700,000 inch-pounds. Let the allowable $f_c = 750$, the allowable $f_s = 18,000$, and the concrete mix 1 : 2 : 4.

Assume some desirable width, say 15 inches, and use the table in Fig. 611. This is for a 12-inch width. $1,700,000 \times \frac{12}{15} = 1,360,000$ inch-pounds. This moment requires a depth down to the

steel of 30 inches. The beam may, therefore, be made 15 inches
wide and 32 inches over all depth. The steel area required is
$2.88 \times \frac{15}{12} = 3.6$ square inches. Use two 1-inch round rods and
two $1\frac{1}{8}$-inch round rods.

7. In Problem 6 suppose the diagonal tension to be taken
by the stirrups is represented by a total vertical shear of 25,000
pounds. What size and spacing of stirrups may be used?

In Fig. 623 find 25,000 on the V_1 scale, then follow the horizontal lines until the 30-inch depth line is intersected. From
this intersection follow the vertical lines until the stirrup size is
intersected and then move horizontally until the spacing scale
is reached. If $\frac{3}{8}$ square stirrups are used the end spacing is about
$4\frac{1}{2}$ inches, when $\frac{1}{2}$ round stirrups are used the end spacing is $6\frac{1}{2}$
inches.

8. Determine the rivet spacing from a to c for the girder
shown in Fig. 624. Let the web be $\frac{1}{2}$ inch thick, the flange angles
$6 \times 6 \times \frac{5}{8}$ and the first cover $16 \times \frac{3}{4}$. Use $\frac{7}{8}$ rivets. The distance from a to b is 5 feet and from b to c, 4 feet.

The moment diagram indicates that the flange angles carry
an average stress of about 15,000 pounds per square inch at b
and, of course, they have no stress at their end a. The net area
of two $6 \times 6 \times \frac{5}{8}$ angles times 15,000 equals 194,000 pounds.
This load must be taken by the rivets connecting the angles to
the web between a and b. These rivets are in double shear,
therefore, bearing on the $\frac{1}{2}$-inch web will govern. At 20,000
pounds per square inch for bearing, each $\frac{7}{8}$ inch rivet is good for
8750 pounds and 22 are required. The shear is almost constant
from a to b because the moment curve from m to r is almost a
straight line, therefore, a $2\frac{1}{2}$ inch spacing giving 24 rivets will
be satisfactory. Use two gauge lines with 12 rivets in each line.

At c the moment diagram indicates the average stress in the
cover plate to be about 15,500 pounds per square inch and in the
angles about 15,000 pounds per square inch. The net cross-
section of the cover times $15,500 = 163,000$. This load must
be transmitted to the angles below by the rivets connecting
the cover to the angles. These rivets are in single shear, and
at a shearing stress of 10,000 per square inch each rivet is good
for 6010; 28 are required. Use four gauge lines with a quarter
of the rivets in each. The tendency is for the stress to come into
the cover very rapidly near its end until its stress is approximately

the same as the stress in the angles directly below. This means that it is desirable to have a very close spacing near the end of the cover plate. Use a 4-inch spacing in each gauge line for, say the first 4 spaces at the end, and then 8 inches for the rest. This will give a few extra rivets but it is better not to have the spacing too great.

The stress in the angles at *c* is just about the same as at *b*. This means that the horizontal shear transmitted to the angles by the cover plate has been passed on to the web by the rivets connecting the angles to the web. Each of these rivets is good for 8750 pounds. 163,000 ÷ 8750 gives 19 required. 48 ÷ 19 = 2.52. Continue the $2\frac{1}{2}$ spacing using two gauge lines with the rivets staggered and a 5 inch spacing along each gauge line.

9. A compression member in a steel truss has an unsupported length of 12 feet and carries a load of 64,000 pounds. Design the member.

In Fig. 645 find the intersection of the 64,000 horizontal load line with the 12-foot vertical length line. Two angles $5 \times 3\frac{1}{2} \times \frac{7}{16} \times 24$ pounds per foot or 2 angles $6 \times 4 \times \frac{3}{8} \times 24.6$ pounds per foot will be satisfactory.

10. How many $\frac{3}{4}$-inch rivets will be required to connect the member of problem 9 to a $\frac{3}{8}$-inch gusset. For rivet stresses use 10,000 pounds per square inch in shear and 20,000 pounds in bearing.

Since the rivets are in double shear and the gusset is only $\frac{3}{8}$ of an inch thick, bearing in the gusset will govern. The bearing value for a $\frac{3}{4}$-inch rivet on a $\frac{3}{8}$-inch gusset is 5630 pounds.

$$64,000 \div 5630 = 11.4,$$

say 12 rivets required.

11. In Problem 10 suppose clip angles are used. Let 7 of the rivets connect the angles of the member direct to the gusset and the remaining 5 connect the clip angles to the gusset. How many rivets should be used to connect the clips to the angles of the member? $64,000 \div 12 = 5333$ pounds = load per rivet, $5333 \times 5 \div 2 = 13,333$ pounds = load on each clip. The rivets connecting the clips to the member are in single shear and the value of a $\frac{3}{4}$-inch rivet in single shear is 4420 pounds. $13,400 \div 4420 =$ approximately 3, but 4 will make a better detail, therefore use 4.

12. Design an angle tension member to carry a load of 78,000

pounds. Allowable stress 16,000 pounds per square inch, $\frac{7}{8}$-inch rivets.

The table in the Carnegie Steel Co's. " Pocket Companion " giving allowable tension values for angles is used and 2 angles $4 \times 4 \times \frac{3}{8}$ are found to be satisfactory.

13. Fig. 658 shows a splice at which the size of the angles changes. Determine the number of rivets required to connect the various parts together.

All the load in the 3×3 angles must be taken out, part going into the gusset and part into the bottom splice plate. The entire load in the 4×3 angles must also be taken into the gusset and splice plate. It is evident that no more load can be put into one end of the splice plate than is taken out at the other.

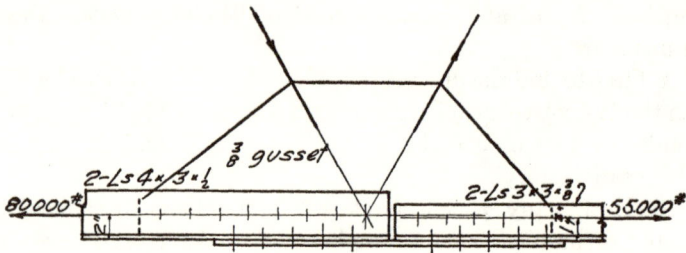

Fig. 658.

The difference between the stress in the two members goes into the gusset and is taken out by the diagonals above.

Let it be assumed that rivets connecting the splice plate to the members are field driven, also the rivets connecting the 3×3 angles to the gusset. Field rivets will be considered 80 per cent as good as shop rivets.

$\frac{3}{4}$ Shop rivets bearing on $\frac{3}{8}$-inch plate. . . .5630 pounds
$\frac{3}{4}$ Field rivets bearing on $\frac{3}{8}$-inch plate. . . .4500 pounds
$\frac{3}{4}$ Field rivets, single shear. .   3540 pounds

Use 5 rivets on each side, ten in all, to connect the splice plate to the 3×3 angles and 5 rivets connecting these angles to the gusset. The 10 rivets connecting to the splice plate are good for $3540 \times 10 = 35,400$ and the five connecting to the gusset are good for $4500 \times 5 = 22,500$ pounds. $22,500$ pounds $+ 35,400 = 57,900$, which is slightly greater than the stress in the member.

Now consider the 4×3 angles. The splice plate with the ten field rivets at each end, as shown, can take out 35,400 pounds. $80,000 - 35,400 = 44,600$; $44,600 \div 5630 = 8$ rivets required for the connection of the angles to the gusset.

14. Design a wooden strut for a load of 65,000 pounds and an unsupported length of 18 feet. Let $C = 1500$ pounds per square inch. In Fig. 643 find the 1500-pound C line, follow along it until the 65,000-pound load line is intersected, then follow parallel to the transverse lines until the 18-foot length line is intersected. The nearest section above is an 8×10 actual size $7\frac{1}{2} \times 9\frac{1}{2}$, it will be satisfactory.

15. A web tension member in a wooden truss has a load of 23,000 pounds. How large a steel rod is required?

The stress, $23,000 \div 16,000 = 1.44$ square inches of cross-section required. Use either a $1\frac{3}{8}$ round rod upset or a $1\frac{5}{8}$ round rod not upset. If the rod is long it may be cheaper to upset.

16. Design a cast-iron o-gee washer for a rod which has a load of 20,000 pounds. Let C_1 be 450 pounds per square inch and use 4000 pounds per square inch for the allowable tension in cast iron. Locate 20,000 on the left-hand load scale of Fig. 644, move horizontally to the 200 pound C_1 line, then follow the inclined lines until the 450 C_1 line is intersected, then move horizontally to the D scale where the required diameter is found to be $7\frac{3}{4}$ inches. In order to determine the thickness follow down the 450 C_1 line until the 4000-pound curve is intersected, then moving horizontally to the right-hand scale, $T \div D$ is found to be .24. Therefore $7\frac{3}{4} \times .24 = 1\frac{7}{8}'' = T$.

17. Design a detail similar to the one shown in Fig. 632 for the connection of the upper and lower chord members. Let the stress in the upper chord be 74,000 pounds compression, the stress in the lower chord 64,100 pounds tension, and the reaction 37,000 pounds. The timbers will be considered $9\frac{1}{2}$ inches wide which will give the bolts a span of about 10 inches.

The bolts act very much as small beams supported at their ends. When they are stressed in bending they will deflect slightly, this will tend to reduce the bearing between the bolts and the wood at their center and increase it near the ends of the bolts. This un-uniform distribution of bearing on the bolts will relieve the bending stresses somewhat. Use an allowable stress of 20,000 pounds per square inch in the bolts and consider the bearing uniformly distributed, keeping in mind that the actual stress will be somewhat less because the bearing is higher at the ends of the bolts than at the center. Let the allowable average bearing, parallel to the grain, between the bolts and the timber, be 1000 pounds per square inch.

A $1\frac{1}{2}$-inch bolt would be good for 14,250 pounds in bearing. Turning to the table for bending moments in pins given in the Carnegie Steel Co.'s " Pocket Companion," the allowable moment for a $1\frac{1}{2}$-inch bolt with a stress of 20,000 pounds per square inch is found to be 6600 inch-pounds. Solving the equation $M = \frac{1}{8} WL$, the allowable total load W is found to be only 5550 pounds

For a $1\frac{3}{4}$-inch bolt the allowable bearing is $1\frac{3}{4} \times 9\frac{1}{2} \times 1000 = 16,600$ pounds; the allowable M is 10,500 inch-pounds, and the allowable W is found to be 8840. For a 2-inch bolt the allowable bearing is 19,000 pounds and W is 13,200 pounds.

There must be enough bolts connecting the upper chord to the plates to take out its entire load. If 2-inch bolts are used, $74,000 \div 13,200 = 5.6$, six will be required. If $1\frac{3}{4}$-inch bolts are used, $74,000 \div 8840 = 8.38$, nine will be required.

If the reaction is taken directly into the plates and does not go through the lower chord, the lower chord will need just enough bolts connecting it to the plates to take out its stress of 64,100. Five 2-inch bolts or eight $1\frac{3}{4}$-inch bolts would be required.

The shear in the bolts is seldom high and the thickness of the plates is more a matter of judgment than anything else, the bearing between the bolts and the plates seldom being excessive. Plates $\frac{1}{2}$-inch thick would seem desirable for timbers and stresses as large as those in the present problem.

The minimum allowable spacing of the bolts in the same line parallel to the grain is determined by the allowable shearing stress for the wood parallel to the grain. If this stress is 175 pounds per square inch, $13,200 \div 175 = 75.5$ square inches required. In order to have the blocks between bolts sheared out, two planes $9\frac{1}{2}$ inches wide would have to shear, $75.5 \div (9\frac{1}{2} \times 2) = 3.97$, say 4 inches. The shear planes are, of course, a little longer than the distance between edges of bolts, but it would seem desirable to space the 2-inch bolts not less than 4 inches apart, or in other words, 6 inches center to center.

18. A steel column suports a concentric load of 1,450,000

pounds and has an unsupported length of 18 feet.   Design the column.

This is a special case of an eccentrically loaded column, the eccentricity being zero, therefore, use the diagram given in Fig. 653.

Follow along the 1,450,000-pound load line in Fig. 653 until the vertical length line 18 feet is intersected.   This intersection indicates that a section consisting of a $14 \times \frac{5}{8}$ web, 4 — angles $6 \times 6 \times \frac{5}{8}$ and $16 \times 2\frac{1}{2}$ covers on each side, will be satisfactory since it has about $99\frac{1}{2}$ per cent of the required strength.

19. A certain steel column must support a concentric load of 1,150,000 pounds at the same time it resists a bending moment of 1,300,000 inch-pounds about the Y–Y axis.   The unsupported length is 15 feet.   Let the allowable stress be $16,000 - 70\dfrac{L}{r}$. Design the column.

Referring to Fig. 653 we note that about 2 is an average value of Y for the larger sections.   A value of 2 for Y would give P' equal to 650,000 pounds, and $P + P'$ equal to 1,800,000 pounds. Using this as a trial value the intersection of the 1,800,000 pound load line with the 15-foot length line indicates the following as a good trial section; one web 14×1, 4 — angles $8 \times 6 \times \frac{5}{8}$, covers on each side $18 \times 2\frac{3}{8}$.   The value of Y for this section is 2.36. $1,300,000 \div 2.36$ equals

$$551,000 = P'.\quad P + P' = 1,701,000.$$

This load and the 15-foot length are used in Fig. 653 and the trial section found to be satisfactory.

20. A wooden column supports a load of 55,000 pounds which has an eccentricity of 3 inches.   The unsupported length is 13 feet.   Design the column using C equal to 1,300 pounds per square inch and the following formula for the allowable stress:

$$S = C\left[1 - .0125\,\frac{L}{D}\right].$$

The actual maximum stress is equal to

$$\frac{P}{A} + \frac{M \cdot c}{I}.\quad M = 55,000 \times 3 = 165,000 \text{ inch-pounds.}$$

Take a 12×12, actual size $11\frac{1}{2} \times 11\frac{1}{2}$, as a trial section.   For this section the allowable unit stress equals

$$1300\left[1 - .0125\,\frac{13\cdot12}{11\frac{1}{2}}\right] = 1080 \text{ pounds per square inch,}$$

the actual stress would be

$$\frac{55,000}{132} + \frac{165,000\cdot5\frac{3}{4}}{\frac{1}{12}\cdot11\frac{1}{2}\cdot(11\frac{1}{2})^3} = 416 + 651 = 1067 \text{ pounds per sq. in.}$$

This is slightly less than the allowable, the trial section will, therefore, be satisfactory. If the trial section is either too large or too small another should be chosen using the first results as a guide, and the stresses recomputed.

21. Design a rectangular reinforced concrete wall column for a direct load of 300,000 pounds, and a moment about the center of 600,000 inch-pounds. Suppose the width of the column is fixed by the sash or some other requirement to 31 inches over all, which means a core width of 28 inches. Use 1 : 1½ : 3 concrete with $n = 12$ and an allowable stress of 560 plus 20 per cent for combined stresses, or 672 pounds per square inch.

The direct load per inch of width equals 300,000 ÷ 28 = 10,700 pounds and the moment per inch equals 21,400 inch-pounds. Find 10,700 on the load scale of Fig. 656 and follow parallel to the horizontal guide lines until the vertical line from $M = 21,400$ is intersected. The diagram indicates that either of the following depths and percentages are strong enough: a 20-inch depth with 1.6 per cent of steel, or, an 18-inch depth and 2.75 per cent of steel. Use the 18-inch core depth which will give an over all depth of 21 inches. Total steel area required equals 28 × 18 × .0275 = 13.85. Use fourteen 1-inch square. This is the amount of steel required if it is all divided equally between the two 28-inch sides, the sides parallel to the axis about which bending occurs.

Refer to the note at the bottom of Fig. 657 for an illustrative problem indicating the additional steel required if some of the bars are placed at the ends.

NOTE: Large blue prints of Figs. or Diagrams may be obtained from the author.

CHAPTER X

MISCELLANEOUS PROBLEMS

A number of miscellaneous problems are considered in this Chapter and solutions presented. It is assumed that the student has studied the other chapters and recalls in a general way the constructions given therein. The solutions will, therefore, not be explained in as much detail as they otherwise would be.

143. Steel Bases with Anchor Bolts.—Fig. 659 shows the elevation of a steel column base which is anchored to its concrete footing by two large anchor bolts. A plan of this base is shown at a larger scale in Fig. 660. The problem is to find the stress produced in the anchor bolt and the maximum bearing on the concrete, when the connection of the column to the concrete footing resists a certain bending moment as well as a direct load. There are a number of approximate methods used for the solution of such a problem but they are more or less in error. An exact solution is given by an article in the *Engineering Record*, Vol. 70, p. 441, but it develops into a rather complicated equation.

The attention of the student is called to the similarity between the connection of the column to its footing and eccentrically loaded reinforced concrete columns. When the moment is large only a part of the concrete under the base plate will take bearing, and there will be tension in one of the anchor bolts. The anchor bolt corresponds to the reinforcing rods in the reinforced concrete columns. A solution similar to that used for eccentrically loaded reinforced concrete columns in Chapter VIII will, therefore, be applied.

Let it be assumed that the base and anchor bolts shown in Figs. 659 and 660 must resist a direct load, P, of 50,000 pounds and a moment, M, of 3,000,000 inch-pounds. This loading is equivalent to a load of 50,000 pounds with an eccentricity of 60 inches. Anchor bolts two inches in diameter will be used as a trial size. The compression area is divided into slices parallel to the axis about which bending occurs, and horizontal lines are

339

drawn from their centroids as well as from the centroid of the tension taking anchor bolt. Attention should be called to the fact that the anchor bolt on the compression side, unlike reinforcing steel, is not able to take compression, because it has only a tension connection at its top. After multiplying the anchor bolt net area at the root of threads by n, the vectors are laid off in Fig. 661, and Fig. 662 drawn, locating the neutral axis by the intersection o. The reader should refer to the constructions given in Chapter VIII.

The line e–f is now drawn in Fig. 663 and b–e is made any convenient stress, in this particular problem equal to 400 pounds per square inch to some scale. The line b–c is drawn from b, passing through the intersection of e–f with the neutral axis, and extending to s. The force polygon of Fig. 664 is drawn and from it the funicular polygon of Fig. 665. This polygon locates the action line of the resultant of the compressive forces for the case when the moment is sufficient to produce a maximum pressure of 400 pounds per square inch on the concrete. The effective depth is found by scaling to be 44.2 inches. This value times 44,245 = 1,956,000 inch pounds (to the nearest thousand), which is the resisting moment when the maximum pressure on the concrete is 400 pounds per square inch.

The moment of the eccentric load P about the neutral axis x–x is 43.4 times 50,000 = 2,170,000. $400 \times \dfrac{2,170,000}{1,956,000} = 444$.

For the eccentric load P we may substitute a direct load of the same magnitude acting at the axis x–x, and a moment which produces a maximum unit pressure on the concrete of 444 pounds per square inch, and a unit tension in the opposite anchor bolt of f–d times 15. The effect of the direct load at c is to move the base line e–f to the position X–Y. Taking into account the effect of this direct load the maximum unit pressure on the concrete is found to be a–X = 611 pounds per square inch, and the maximum unit tension in the anchor bolt is Y–d times

15 = 18,750 pounds per square inch.

Figs. 666 and 667 were used for a check and were not a part of the solution. The student will be interested in noting the similarity between Figs. 659 to 667 and Figs. 574 to 583.

If the stresses are not satisfactory the dimensions may be

revised and the necessary changes made in the figures to give
the new stresses. This can usually be accomplished by redrawing
only a small part of each figure.

The approximate methods are often satisfactory for designing small anchor bolts and bases, but occasionally the designer
desires results from a more correct method.

144. Connection of Wind Bracing Girder to Columns.—A
wind bracing connection similar to that shown in Fig. 668 is
often used in structural steel frames. Fig. 669 shows Section
A–A to a larger scale. Let it be assumed that the moment on
the connection is of direction as indicated by the arrow in Fig.
668, producing compression at the top and tension at the bottom.
Now where there is tension the rivets must, of course, take it,
but when there is compression the rivets are not acting, the compression being taken by direct bearing between the angles and the
column. The area just back of the outstanding legs will be most
efficient in taking this bearing, and the area most remote from
these outstanding legs will be least effective. In Fig. 669 only
the area out to the gauge lines has been considered.

Now this connection to the column is somewhat similar to
a narrow reinforced concrete beam with many layers of steel
uniformly spaced for its entire depth. On the compression side
of the neutral axis x–x the compression is taken by the larger
continuous area, while on the tension side the tension is taken
by the rivets represented in the figure by the black spots. This
neutral axis x–x may be located in the same way that the neutral
axis for concrete beams was located. The compression area is
divided into slices, the dividing being continued until it is evident
that the neutral axis is past. Lines are drawn from the centroids
of these slices parallel to the axis about which bending occurs,
and similar lines are drawn from the rivets on the tension side.
The areas of the slices are computed, also the areas of the rivets
on the tension sides, and these values are used in Fig. 670. The
ratio n does not enter, since only one kind of material is being
used. Starting at W in Fig. 670, the vector for the area of the
top slice is laid off and then the other compression areas in order,
extending to the right. The rivet areas are laid off to the left
of W, starting with the lowest and taking the others in order.
Fig. 671 is drawn from Fig. 670, and the neutral axis x–x located
by the intersection O. The moment of inertia is given by $2 \cdot A \cdot H$,
in which A is the area inclosed by the polygon of Fig. 671. The
area inclosed by this polygon may be found by the use of a planimeter, or it may be divided into parallel slices and the mean lengths
stepped off along a line as shown at the top of the figure. Two
widths of slices were used in Fig. 671, part of them being only
half as wide as the others. To make allowance for this, the mean

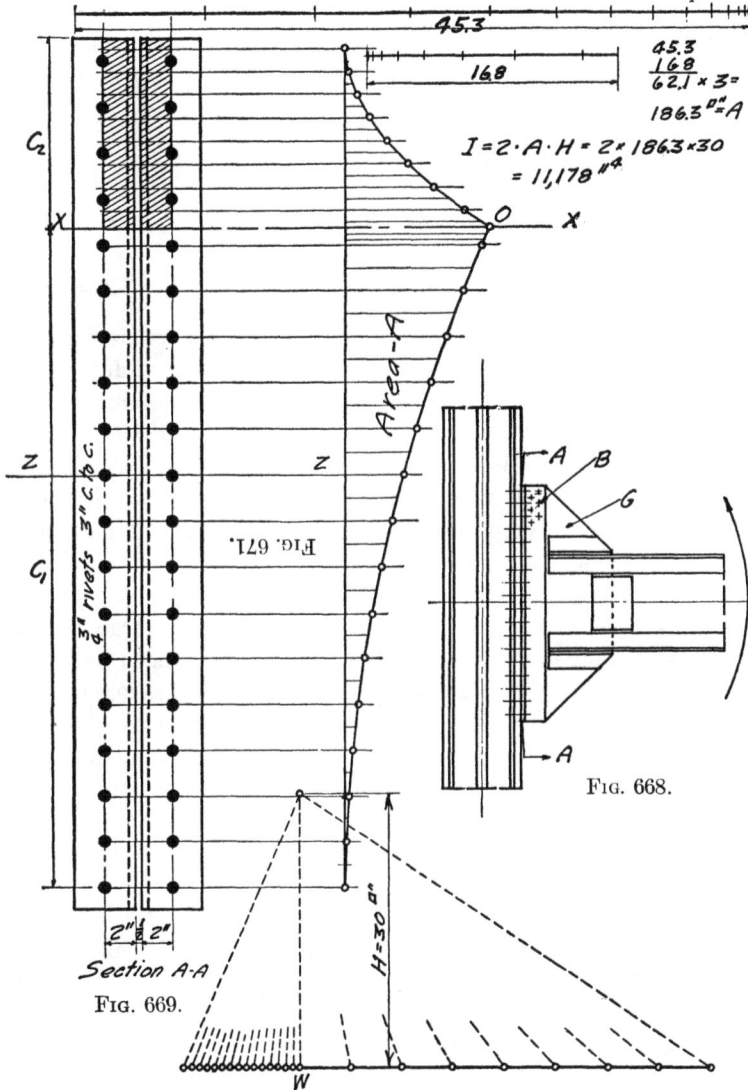

$$\frac{45.3}{\begin{array}{c}16.8\end{array}}$$

$$62.1 \times 3 =$$
$$186.3^{D''} = A$$

$$I = 2 \cdot A \cdot H = 2 \times 186.3 \times 30$$
$$= 11,178^{''^4}$$

FIG. 671.

FIG. 668.

Section A-A
FIG. 669.

FIG. 670.

lengths of the narrow slices were divided by 2. The length on the line or lines along which these mean lengths were stepped off, times the slice width gives the area A, which was here found to be 186.3 square inches, to the scale at which Fig. 669 was drawn. Since the numerical value of H is 30 square inches, $I = 2 \cdot 30 \cdot 186.3 = 11,178$ inches.[4]

The maximum compressive unit stress is given by $\dfrac{M \cdot c_2}{I}$, and the maximum unit rivet stress by $\dfrac{M \cdot c_1}{I}$. Since c_1 is larger than c_2, the governing section modulus is $\dfrac{I}{c_1}$ or $11,178 \div 43.12 = 259$.

Designers often estimate the strength of connections similar to that shown in Fig. 669 by finding the section modulus of the entire group of rivets about the axis z–z. The section modulus of the group of rivets shown in Fig. 669 is $(4 \cdot .442) \cdot (3^2 + 6^2 + 9^2 + 12^2 + 15^2 + 18^2 + 21^2 + 24^2 + 27^2) \div 27 = 168$. The value obtained above is 54 per cent larger, showing that the connection is really very much stronger than the method using the section modulus of the rivets would indicate.

Attention should, however, be called to a possible point of weakness. With the neutral axis at x–x there is a large amount of stress coming out into the angles over a short length near their end. This stress must get into the gusset G by going through the rivets B, and, unless they have received special attention, they may be over-stressed. In case the outer B rivets are over-stressed and give a small amount, the angles will bend a little and thus transmit the stress to rivets farther in. But if the angles bend any appreciable amount the axis x–x will be pulled in towards the axis z–z.

It, therefore, seems desirable not to use the full strength obtained by the solution shown in Figs. 669 to 671, unless the rest of the bracket and girder end is of ample strength and rigidity.

145. Steel Beams Reinforced with Concrete Fireproofing and Slabs.—Steel floor-beams which support concrete slabs are usually fireproofed with concrete. This requires a good deal of concrete, in fact the designer may sometimes think that he has almost enough concrete to make a reinforced concrete beam, which would carry the load, without using structural steel. In such cases we really have concrete and steel acting together whether we figure it that way or not, and the combination is usually a good deal stronger than the steel beam alone. There may be some question about the ability of the flat surfaces of the beam to take high bond stresses, but if the designer figures out the bond stresses for an actual beam he will find that they are very low because of the large surface.

In Fig. 672 a 24-inch BI 84-pound is shown together with fireproofing and a portion of the floor slab. The top of the beam is 2 inches below the top of the slab, and a T projecting 12 inches beyond the edge of the fillet has been considered. The compression area is divided into slices parallel to the axis about which bending occurs, and lines are drawn parallel to this axis from the centroids of the slices. The compression value of each one of these slices is equal to its area of concrete plus 15 times its area of steel. The steel on the tension side is also split into small divisions, and lines drawn from their centroids as shown. Figs. 673 and 674 are now drawn, in the same way that Figs. 670 and 671 were drawn, and the neutral axis x–x located by the intersection O. Fig. 675 is next drawn and f–h made equal to the allowable stress in the steel (16,000 pounds per square inch) divided by 15. The line c–h is drawn and extended to g. The line g–e is found to be less than the allowable stress in the concrete, therefore, the steel governs. The resultant stress on each slice or division is now computed, the compressive stresses and also the tensile stresses are summed up, and by the use of Figs. 676 to 679 the action lines of R_c and R_t are located. The effective depth is found to be 20.8 inches and the mean between R_c and R_t is 168,000 pounds. Therefore the resisting moment of the steel and concrete is

$$\frac{20.8 \times 168,000}{12} = 291,200 \text{ foot-pounds.}$$

The resisting moment for the steel beam alone is 264,660 foot-pounds, when the maximum stress in the steel is 16,000 pounds per square inch. The concrete has, therefore, increased the strength of the beam a little more than 10 per cent. This percentage will, of course, vary with the size of the beam, the thickness of the slab, and the distance the top of the beam is below the top of the slab. It will be relatively large for small beams and will increase very rapidly with the increase in thickness of

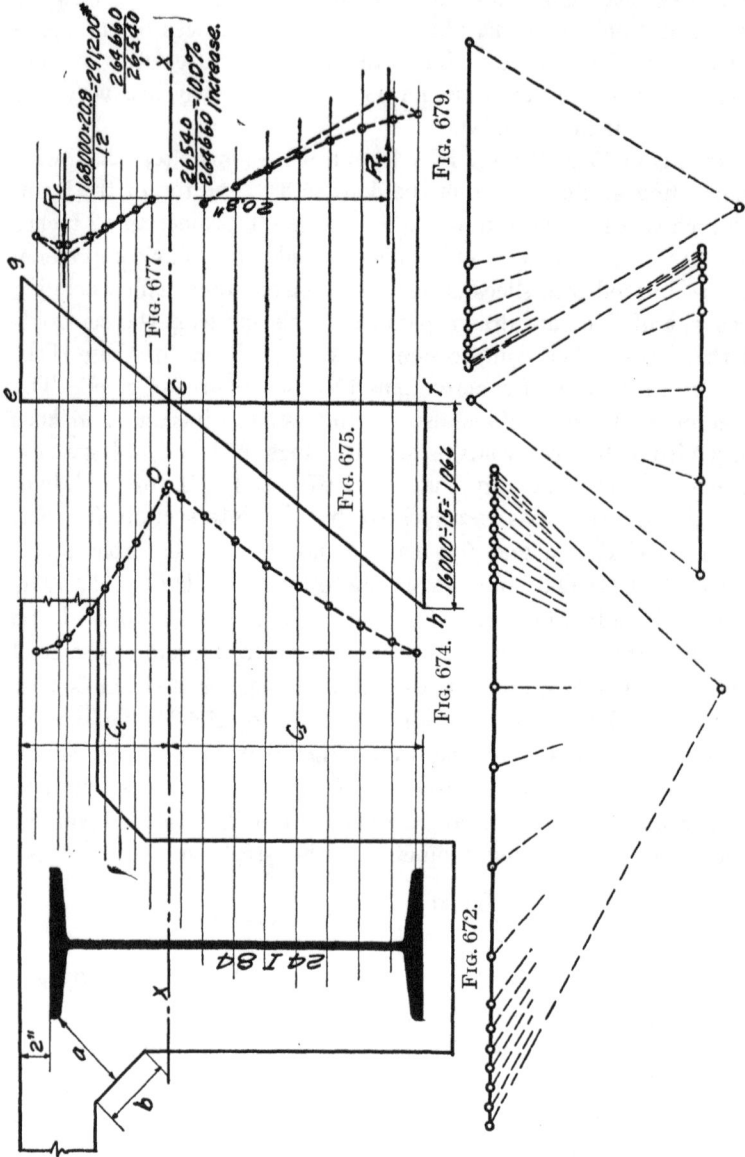

Fig. 672.
Fig. 673.
Fig. 674.
Fig. 675.
Fig. 676.
Fig. 677.
Fig. 678.
Fig. 679.

concrete above the beam.   It is desirable to have a good-sized fillet b, and the dimension a should be at least 3 or 4 inches.

It is interesting to note that the concrete also increases the rigidity of the beam, even for the same stress in the steel.   Without the concrete the deflection would be that for a 24-inch steel beam, with the concrete the deflection is what we would get from a steel beam with a depth equal to 2-C_s or 32.5 inches.

146.   Continuous Frame of Two Fixed Columns and a Girder.— Consider the columns and girder shown in Fig. 680.   The columns are anchored to heavy concrete footings by large anchor bolts in order to fix them at their base, and the girder is rigidly connected to the columns at its ends.   When the girder is loaded it deflects but the ends are somewhat restrained, being partially fixed by the column.   The girder will, therefore, have points of contra-flexure near its ends and the moment diagram will look somewhat like the area e–f–f_1–g–h–Y–X in Fig. 681.   The moment at the top of each column must be just the same as at the corresponding end of the girder.   This moment will then decrease until it passes through zero at a point of contra-flexure some place above the base.   Below this point of contra-flexure the moment will increase again until the base of the column is reached, where it is absorbed by the footing.   The distribution of moment throughout the length of the columns is illustrated by the areas k–c–d–l and b–i–j–a in Fig. 681.   Let these column moment diagrams be revolved around b and c into the positions shown by the dotted lines and then lowered until b and c coincide with e and h respectively.   Note that this has not changed the shape of the diagrams nor their intercepts, but only their position.

With Fig. 681 in mind, consider the frame indicated by center lines in Fig. 683.   This frame is similar to the one given in Fig. 680.   The loading shown will be used.   The uniform load applied to the girder is divided into slices and an equivalent concentrated load substituted for each.   These loads, together with the large concentrated loads, are laid off in order along the load line of Fig. 684, and any convenient pole p chosen with pole distance H.   Then starting at e, any point on the axis of the left column, the funicular polygon for the loading is drawn and the last string extended to h, a point on the axis of the right column.

When the girder is loaded it deflects and, since its ends are not fully fixed, the neutral axis at the ends will have a certain inclination after bending. Since the columns and girder are

Fig. 680.

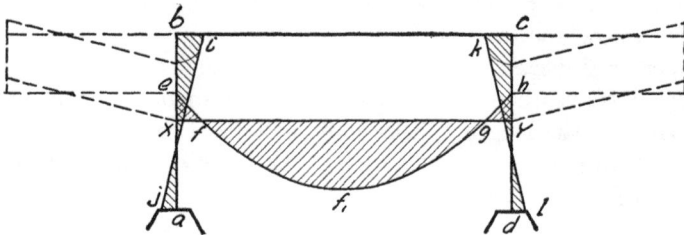

Fig. 681.

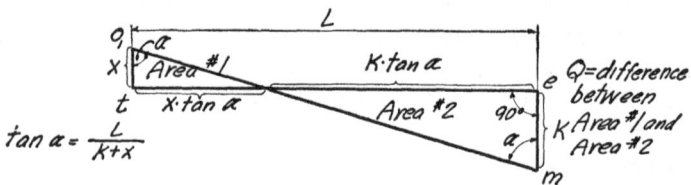

$$\tan \alpha = \frac{L}{K+x}$$

Fig. 682.

rigidly connected, the axis of the girder will intersect the axis of the column at an angle of 90 degrees after bending the same as before. The curvature of the columns will be convex on the outside from their top down as far as the point of contra-flexure, and then convex on the inside on down to the base, where the tangent to the axis is supposed to remain vertical.

If the girder was simply supported and not restrained at the ends, a line connecting e and h would be the base line for its moment diagram. With the girder restrained by the columns the base line for the moment diagram is $X-Y$, and the biggest part of our problem is to locate the points X and Y. The moment diagrams for the colums are shown in the revolved and shifted position explained in connection with Fig. 681. Not only must points X and Y be located but also U and V.

Let the designer assume points m and n which are near where he would expect to find X and Y. Using the trial base line $m-n$, divide the moment diagram for the girder into a number of slices, as shown, and lay off their areas along the load line of Fig. 685, starting at B with the first area at the right end and extending down to C. When the sign of the moment changes, the pole in Fig. 685 is changed to the other side of the load line, and when the moment of inertia changes a corresponding change is made in the length of the pole distance. The student will do well to refer to the constructions for continuous beams and beams having a variable moment of inertia, given in Chapter IV.

When Fig. 685 is completed from B to C, and the poles p'_2 to p'_8 inclusive located, the funicular polygon may be drawn in Fig. 686 from b_1 to c_1. Connect b_1 and c_1 and extend this line in either direction. The straight line b_1-c_1 represents the axis of the beam before bending, and the broken line $b_1-e-f-c_1$ represents it after bending. The line a_1-b_1 represents the axis of the left column before bending, revolved 90 degrees about b_1, also c_1-d_1 represents the axis of the right column before bending.

Since the tangent to the elastic curve at the base of the columns remains vertical, the string at each end of the polygon in Fig. 686 must be parallel to the line a_1-d_1. In order to have these two strings parallel to a_1-d_1 the two rays $A-p'$ and $D-p'_{10}$ must be drawn with this direction. In Fig. 685 the length $A-G$ represents the area of triangle o_1-k-t in the moment diagram of Fig. 683. The length $G-B$ represents the area of triangle $k-e-m$, and in order to have $A-p'$ parallel to a_1-d_1, o_1 must be located so that the area of triangle $m-e-k$ less the area of triangle o_1-k-t is just equal to $B-F$, the line p'_1-F being drawn parallel to a_1-d_1. In other words, o_1 must be located so that the difference between the areas of the two triangles is equal to a known value which is given by the length B–F. In Fig. 682 a diagram similar to the moment diagram of Fig. 683, for the left column, is shown. Let the length o_1–t be called x, and the length e–m, K. Then the area No. 1 $= \frac{1}{2} \cdot x^2 \cdot \tan \alpha$ and area No. 2 $= \frac{1}{2} \cdot K^2 \cdot \tan \alpha$, also $\tan \alpha = \dfrac{L}{x + K}$. Let Q be the difference between area No. 1 and area No. 2, given in magnitude by the length B–F in Fig. 685. Then

$$Q = \tfrac{1}{2} \cdot K^2 \cdot \tan \alpha - \tfrac{1}{2} \cdot x^2 \cdot \tan \alpha = \tfrac{1}{2} \cdot \tan \alpha \, (K^2 - x^2).$$

Substituting the value given above for $\tan \alpha$, we have

$$Q = \frac{L \, (K^2 - x^2)}{2 \, (K + x)},$$

which is a quadratic with x the only unknown. Solving for x, the following expression is obtained:

$$x = K - \frac{2 \, Q}{L},$$

L being the length of the column.

After this discussion it may be well to go back to Fig. 685 and consider the actual construction above B and below C. First the pole distance H'_1 is computed, using the proper value for I, and the pole p'_1 located on the line p'_2–B. In this particular problem the value for H'_1 was rather small, and in order to make this pole distance of convenient length the scale was changed. This change of scale does not produce any error as long as the new scale is used for the vectors from B to A. From p'_1 a line is drawn parallel to a_1–d_1 and the intersection F obtained.

The length B–F measured to the scale used for H'_1 gives the value of Q in the formula $x = K - \dfrac{2 \, Q}{L}$. x is now the only unknown in this formula, K being given by the line m–e of Fig. 683. The value of x is computed and o_1 located at that distance above t. The line m–o_1 is now drawn and the moment area divided into slices. Starting with the slice next to e–m the areas of these slices are laid off above B and extended to A. Then p'_1–G is drawn and extended to the right, p' is next located and p'–A drawn. A check is obtained if A–p' is parallel to a_1–d_1.

In a similar way the lower end of Fig. 685 for the right column is extended from C to D. The funicular polygon representing

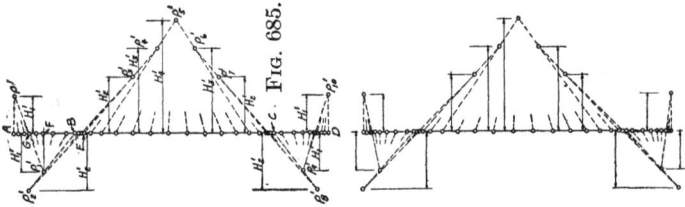

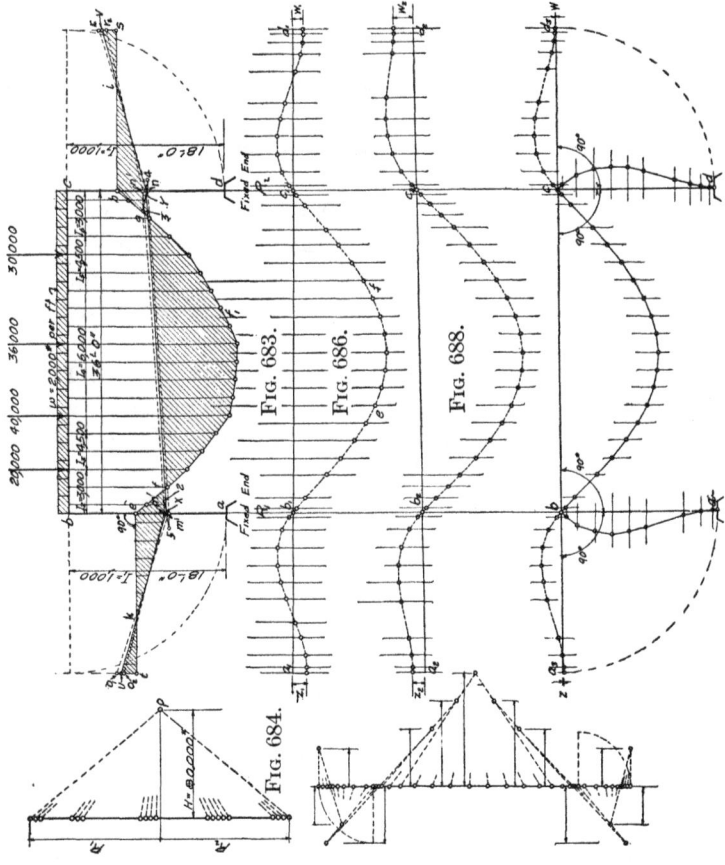

the elastic curve for each of the columns is now drawn and the intercepts Z_1 and W_1 obtained. Since there can be no deflection of the columns at their bases, the Z and W intercepts should be approximately equal to zero.

It is evident that too much moment was assumed at the column tops because the intercepts Z_1 and W_1 are both below the line a_1–d_1. The trial base line for the moment diagram will, therefore, be moved to m'–n', and Figs. 687 and 688 drawn in the same way that Figs. 685 and 686 were drawn. The intercepts Z_2 and W_2 are found to be above the base line, showing that the moments just assumed for the column tops are not large enough.

In Fig. 683 n–4 is made equal to W_1 and n'–3 equal to W_2; 3 and 4 are connected and the intersection Y obtained. In a similar way the intersection X is obtained, U and V are located by the use of the formula $x = K - \dfrac{2\,Q}{L}$, and the base lines U–X, X–Y, and Y–V drawn. Figs. 689 and 690 have been drawn as a check, and the intercepts Z and W are found to be approximately equal to zero. In Fig. 690 the polygons representing the elastic curves for the column have been revolved through 90 degrees into their proper position, as shown by the full lines. The reactions may be found in Fig. 684 by drawing a line from p parallel to X–Y.

A slight approximation was made in locating X and Y from the intercepts Z_1, Z_2, W_1, and W_2. The more exact method would be to draw three elastic curves as was done in connection with three-span continuous beams, and then locate X and Y by the use of a figure similar to Fig. 241. However, the Z and W intercepts are very sensitive to small changes in the moment diagram, and if the trial base lines are carefully chosen so that the Z and W intercepts are relatively small and approximately equal for each curve, good results should be obtained by the use of just two trial elastic curves.

We may, therefore, consider that the problem has been solved. The correct moment diagram is shown by the shaded area in Fig. 683, and the position of the elastic curve after bending is represented by the full line polygon in Fig. 690. The deflection at any part of the beam may be obtained by measuring the vertical intercept between the line b–c and the polygon, and dividing it by the value used for n in the formula $H' = \dfrac{E \cdot I}{H \cdot n \cdot a}$ when computing the pole distances H'_1 to H'_{10}. The deflection of the

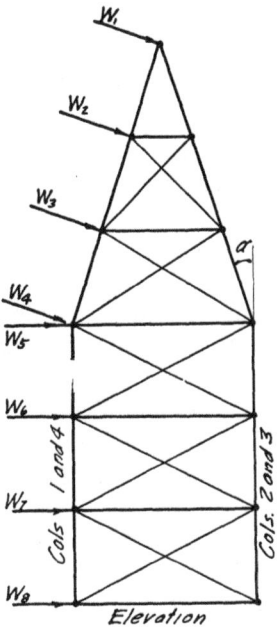

FIG. 691.

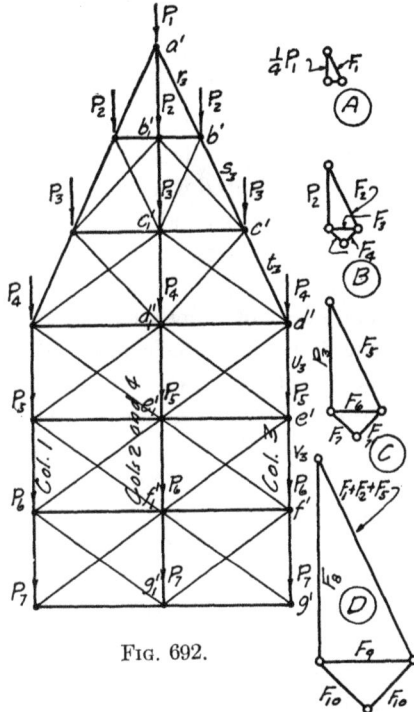

FIG. 692.

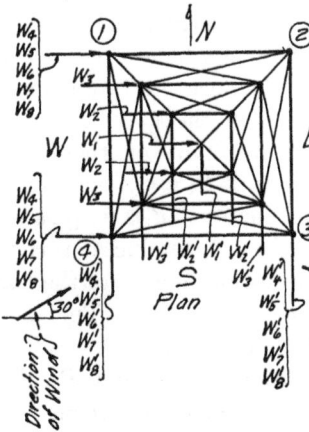

FIG. 693.

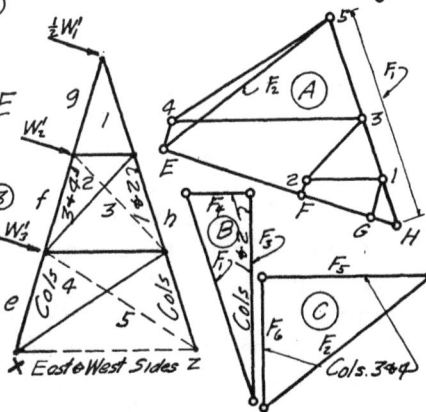

FIG. 694.

left column at any part of its length is given by the horizontal intercept between a–b and the polygon representing its elastic curve, divided by n.

147. Steel Towers.—Towers present a more complicated problem than might at first be expected, especially when the columns change from a vertical to an inclined position for a part of their length, as illustrated by Fig. 691. The problem is still further complicated when the direction of the wind makes an angle with the sides, as indicated in Fig. 693. Fig. 692 shows an elevation looking at right angles to the plane containing columns Nos. 1 and 3.

The effect of a dead load symmetrically placed about the axis of the tower will be considered first, after which the determination of the stresses under wind load will be illustrated. In connection with the dead load refer to Fig. 692 and Figs. 692 Ⓐ to 692 Ⓓ. Under this loading the stresses in each of the four columns will be the same, also the stress in each of the four horizontal members at any one level will be the same.

At the top the load P_1 is supported by the four upper column sections, and we may consider that each one of them has a stress such that its vertical component is just equal to $\frac{1}{4} P_1$. In Fig. 692 Ⓐ a force triangle is drawn and F_1, the stress in these upper column sections, determined. At the panel point b' the load P_2 is applied. The vertical component of the stress which P_2 produces in b'–c' must be equal to P_2. The magnitude of this stress is found in Fig. 692 Ⓑ by scaling F_2. The horizontal component of F_2 must be balanced by compression in the two horizontal members which intersect at b'. The resultant of these two compressive stresses is F_3, and it follows that F_4 gives the magnitude of each one of them. The compression in b'–c' is F_1 plus F_2.

By the use of Fig. 692 Ⓒ we find that P_3 produces a compression in c'–d' equal to F_5 and a compression in each of the two horizontal members that intersect at c' equal to F_7. The total compression in c'–d' is equal to F_1 plus F_2 plus F_5.

Below d' the column is vertical, which means that P_4, P_5, etc., will go directly into the column, increasing its stress an amount just equal to their magnitude, but not producing stress in any other member. The stress in c'–d' is not a vertical force, but in Figs. 692 Ⓓ it is broken into its vertical and horizontal components. The vertical component F_8, plus P_4 gives the com-

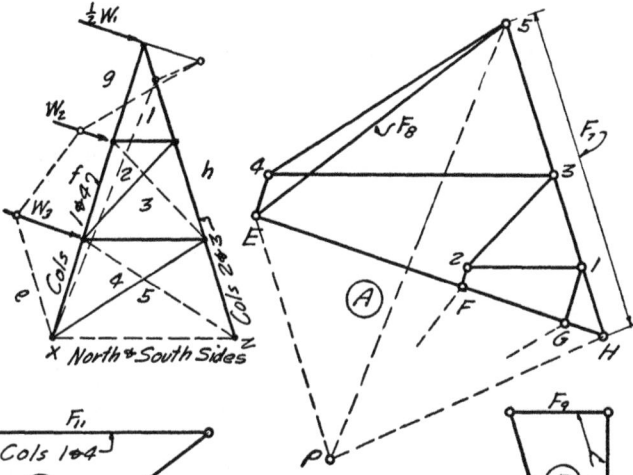

FIG. 695.

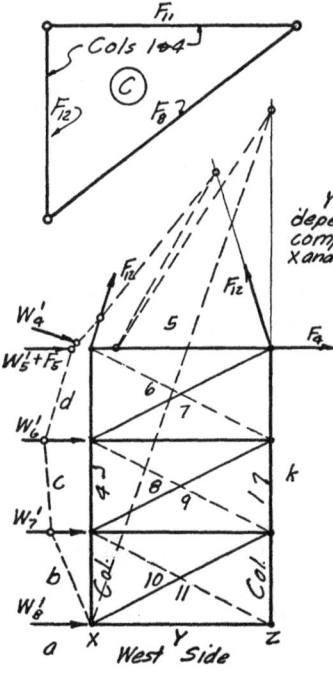

FIG. 696.

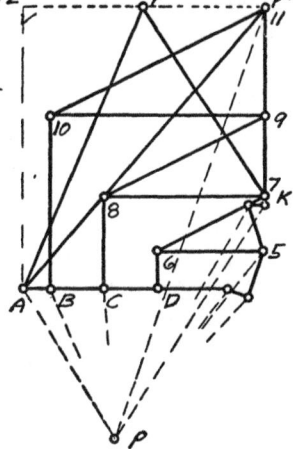

Y is some place on the line V-11 depending upon how the horizontal component is divided between X and Z

FIG. 697.

pression in $d'-e'$, and the horizontal component produces a tension in each of the horizontal members that intersect at d' equal to F_{10}. The stresses in the corresponding members of the other three columns are the same as those just obtained for the various members of column No. 3. It is of interest to note that the dead load produces no stress in the diagonals and no direct stress in the horizontals below the level at which the columns become vertical.

For the purpose of illustrating a method of solution under wind loads, a wind, the direction of which makes an angle of 30 degrees with the south side, will be considered. Let it be assumed that the loads applied to the various panel points are as indicated in Figs. 691 and 693. In case any bending stresses are produced in the members of the frame in order to get the loads over to the panel points, these stresses should be combined with the stresses obtained below. The unit pressure on the west and also on the south side is estimated, and the magnitudes of W_1 to W_8 inclusive, and of W'_1 to W'_8 inclusive are computed. The wind pressure on the south side is, of course, less than that on the west side, because the wind strikes the west side more directly.

The four sides of the tower above the level at which the columns become inclined will be considered first. The stresses in the members of the east side, produced by the wind load on the south side, will be the same as those produced in the members of the west side by this wind load. Figs. 694 to 694 ⓒ will, therefore, serve for the west side as well as for the east side. When drawing the space diagram in Fig. 694, it was necessary to revolve the plane of the side into the plane of the paper in order to get the proper directions for use in the construction of the stress diagram. The stress diagram is shown in Fig. 694 Ⓐ. The reactions are F_1 at Z and F_2 at X. Or we may say that the portion above delivers a thrust at Z equal in magnitude and direction to F_1, and a pull at X equal in magnitude and direction to F_2. Remember that these forces have their action lines in the plane of the tower side which makes an angle α with the vertical, as shown in Fig. 691. In Fig. 694 Ⓑ the thrust F_1 is broken into two components, one horizontal, and the other falling at the intersection of the vertical plane containing the lower part of columns 1 and 2, and the inclined plane containing the upper part of col-

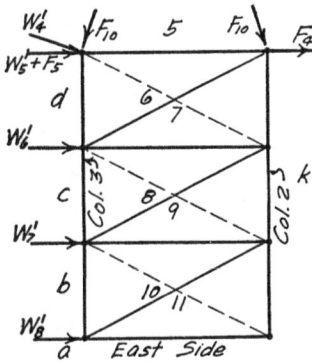

FIG. 698.

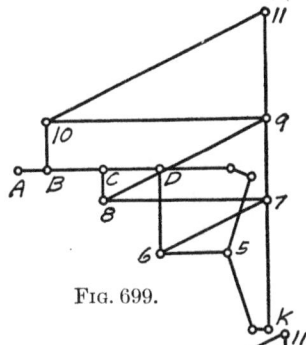

FIG. 699.

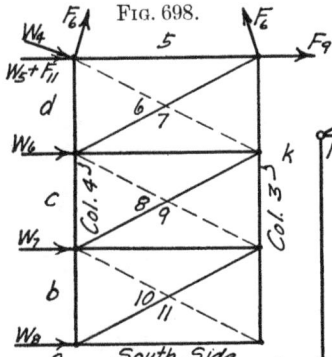

FIG. 700.

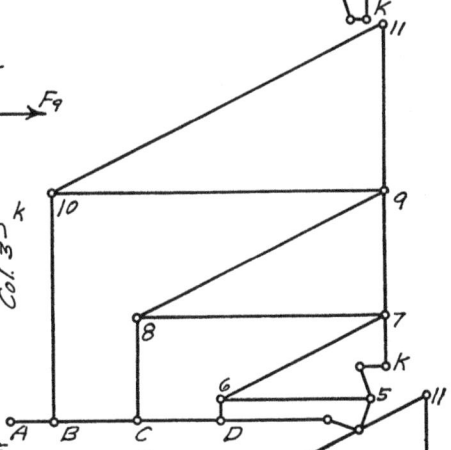

FIG. 701.

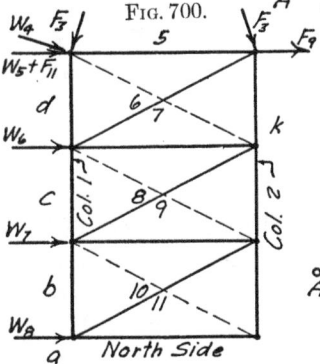

FIG. 702.

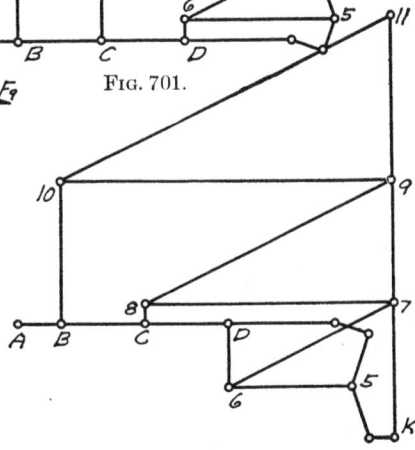

FIG. 703.

umns 2 and 3. In Fig. 694 Ⓒ, F_2 is broken into similar components.

Figs. 695 to 695 Ⓒ correspond to Figs. 694 to 694 Ⓒ. By their use, the stresses in the north and south side members produced by the wind load on the west side are obtained. The reactions at the tops of the vertical columns are also obtained. It should be noted that the stress diagram of Fig. 694 Ⓐ gives a stress for each inclined member of each of the four columns since it serves as the diagram for the west side as well as for the east side. Also the stress diagram of Fig. 695 Ⓐ gives a stress for each of these column members. The resultant stress in each member is the algebraic sum of the two stresses. This resultant stress is found to be large for columns 2 and 4, but comparatively small for columns 1 and 3.

The wind load on the south side produces a reaction at the top of each vertical column. These reactions are broken into components by Figs. 694 Ⓑ and Ⓒ. The wind load from the west also produces a reaction at the top of each of these columns, see Figs. 695 Ⓑ and Ⓒ. These reactions should be considered as loads when the stress diagrams are drawn for the straight part of the tower. The loads applied at the top of any column may be divided between the two stress diagrams for the adjoining sides in any desired way, provided all the forces considered in connection with any one side are in the plane of that side. The loads should be broken into components such that each component falls within the plane of one of the two sides adjacent to the column on which the load is applied. It is, of course, important that all of the loads be considered, but that none of them be considered twice.

Figs. 696, 698, 700 and 702 show the center line diagrams and loads for the four sides of the straight part of the tower. The corresponding stress diagrams are drawn in Figs. 697, 699, 701, and 703. It should be noted that there are two stresses found for each member of each of the four vertical columns. The algebraic sum of the two stresses found for each column member gives the resultant stress for that member. The diagonals which are stressed under the given loading are shown by the full lines.

The reactions may be checked by drawing funicular polygons as illustrated in Figs. 695 and 695 Ⓐ, also in Figs. 696 and 697.

148. Longitudinal Strength Calculations for a Ship.—Ships present a great many interesting problems for the structural engineer, especially for the engineer who is interested in graphical constructions and solutions.   The distribution of weight along the length of the hull differs from the distribution of the buoyancy forces.   This variation between the distribution of weight and buoyancy produces longitudinal bending stresses along the length of the hull.

This article will be devoted to the explanation of a method which may be used in determining the distribution of such moments, as well as their numerical value.   The hull of a ship may be thought of as a large box girder of variable cross-section, which is loaded with a distributed load consisting of its own weight, cargo, equipment, etc., and is supported by a distributed support which is the buoyancy of the water.

Let the base line *A–B* in Fig. 704 represent the length of the ship with the forward perpendicular at *A* and the aft perpendicular near *B*.   Above this base line the weight curve is first drawn with ordinates which represent the distribution of total weight along the length of the hull.   An inch of ordinate is made equal to some convenient number of tons per lineal foot of hull.

The total area under this curve, measured to the proper scale, is equal to the total weight of the ship including the cargo, equipment, etc.   The number of feet which one inch of base line represents, times the number of tons per lineal foot which one inch vertical represents, gives the number of tons which one square inch of area under the weight curve represents.

The next step is to find the longitudinal position of the centroid of the area under the weight curve, which locates a line containing the center of gravity of the ship.   This centroid is located by a construction similar to that used for locating the neutral axis of a reinforced concrete beam.   The area under the weight curve is divided into slices, all of equal width, most of them being either trapezoids or rectangles.   The areas of these slices are found, usually by scaling the mean ordinate, and recorded in the top line of figures just below the station numbers.   Starting with the slice at the right end, and taking the others in order, the areas are laid off along a load line beginning at *G* and extending down towards *H*.   Lay off enough of these areas so that it is evident the centroid has been passed.   In a similar way, areas are laid off along another load line starting at E with the extreme left area and extending down, with the other areas taken in order. The load line should be extended until it is evident that the centroid has been past. The poles p_1 and p_2 are chosen with pole distance H_1 and H_2. These two pole distances are made the same length, the same scale is used for each load line, and both p_2–G and p_1–E are drawn horizontal. The areas might have been laid off with E–F extending on up above G, and E–p_1 coinciding with G–p_2, but in this particular problem it would not have been quite so convenient because of the greater distance the direction of lines would have to be transferred.

The base line C–D is drawn parallel to p_2–G and p_1–E, and the funicular polygons drawn from D and C and continued until they intersect at V, thus locating the vertical line z–z which contains the centroid.

The most severe longitudinal bending stresses are produced when the ship is centered on the crest of a wave or over the hollow of a wave. When the ship is centered on the crest it is said to be in the hogging position, and when centered over the hollow it is in the sagging position. In all cases the longitudinal center of buoyancy should be in approximately the same vertical line as the L.C.G. (longitudinal center of gravity). The buoyancy curve for the hogging position is shown by the full line smooth curve in the figure. It has been drawn upon the same base line as was used for the weight curve and at the same scale. Points on this curve were located by the use of the Bonjean curves, a standard wave curve, and Sympson's Rule. See " Naval Architecture " by Peabody. The area under this buoyancy curve should be just the same as the area under the weight curve, and the same vertical line should contain their centroids. If these conditions are not met, an error will be produced which will show up later. The areas between the weight and buoyancy curves represent the load which is effective in producing moment. The area under the buoyancy curve is divided into slices of equal width, the same as the area under the weight curve, and the areas found by scaling the mean ordinates. These areas are recorded in the line just below the line for the weight curve areas. If these two lines of slice areas are subtracted one from the other, areas representing the effective load are obtained. The effective load areas are shown in the lower line, some of them being negative and others

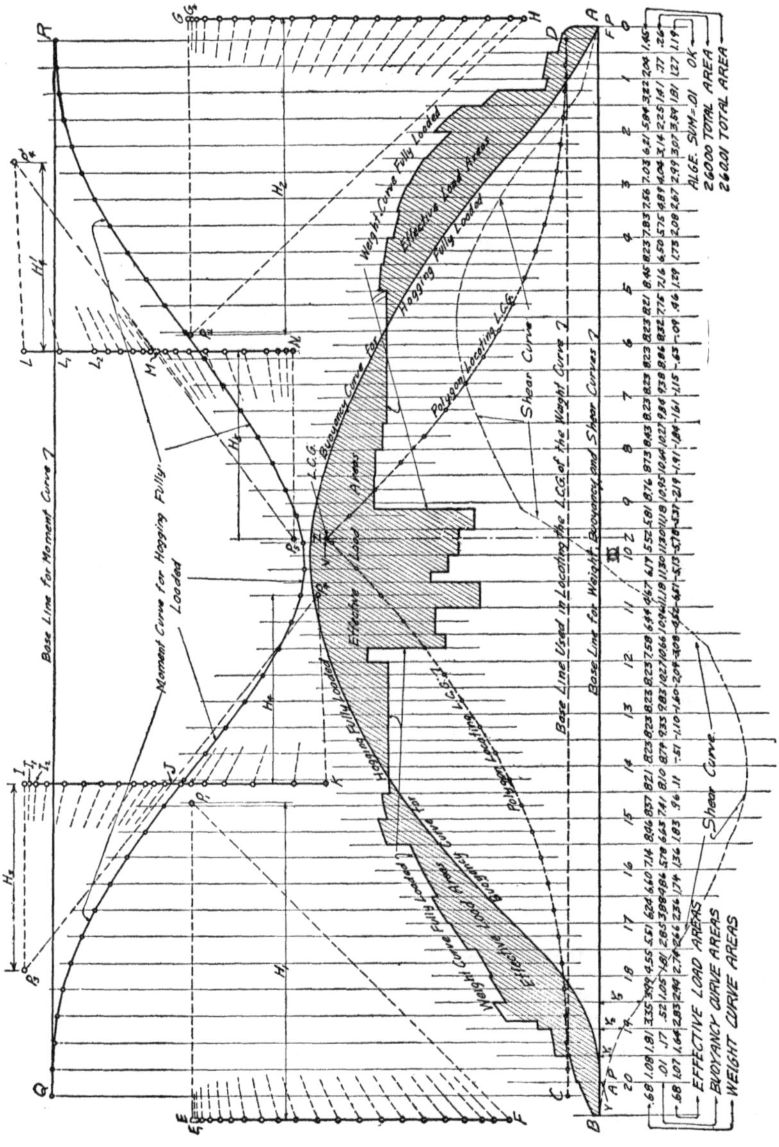
Base Line for Moment Curve
Moment Curve for Hogging Fully Loaded
Weight Curve Fully Loaded
Effective Load Areas
Hogging Fully Loaded
Buoyancy Curve for Hogging Fully Loaded
Effective Load Areas
Polygon Locating L.C.G.
Shear Curve
Base Line Used in Locating the L.C.G. of the Weight Curve
Base Line for Weight, Buoyancy and Shear Curves
EFFECTIVE LOAD AREAS
BUOYANCY CURVE AREAS
WEIGHT CURVE AREAS
ALGE. SUM = .01
OK
26.000 TOTAL AREA
26.01 TOTAL AREA

FIG. 704.

positive.   Their algebraic sum should, of course, be equal approximately to zero.   In this problem its value is found to be .01, which will be considered satisfactory.

Beginning with the one at the left, these effective load areas are laid off in order to some convenient scale along a load line, starting at I and extending down.   When the areas change sign the pole is changed from p_3 to p_4 on the other side of the load line, but p_3–J–p_4 must be a straight line.   When the point K is reached it seems convenient to break the load line and continue its construction from L.   The pole distance H'_4 must be the same as H_4, and L–p'_4 must be parallel to K–p_4.   For convenience, the ray I–p_3 was made horizontal, and if the loads have been laid off correctly, the ray p_5–N will come out horizontal.

Now starting at any point Q on the vertical line from the centroid of the first area at the left, the large full-line polygon is drawn which extends around to R, on the vertical line from the centroid of the first area at the right.   Connect Q and R by a straight line which should be parallel to the ray I–p_3.   The area between the closing string and the full line funicular polygon represents the moment diagram for the ship fully loaded in the hogging position.   Vertical intercepts measured in this diagram will give the numerical value for the moment, at any desired section. The number of feet which one inch of base line represents, times the number of tons which the pole distance H_3 represents gives the number of foot-tons which each inch of intercept in the moment diagram represents.   The number of tons which the pole distance H_3 represents is the number of tons which one square inch of weight curve area represents times H_3 measured in square inches to the scale that was used in laying off the areas along the load line I–K.

The vertical lines on which the polygon, used for locating the L.C.G., changes direction were drawn from the approximate centroids of the weight curve area slices.   For the moment diagram polygon these vertical lines should be drawn from the approximate centroids of the effective load area slices.   For most of the large areas these centroids are very close to the vertical lines already drawn.   In Fig. 704 the same vertical lines that were used for locating the L.C.G. are therefore used for the moment diagram.

Other moment diagrams for other positions of the ship on the wave or for other loadings may be drawn in the same way as the one shown. The base line *A–B* may be used for several different buoyancy curves if desired.

The shear is often of importance. It is given for any section by the algebraic sum of the effective load areas to the left or to the right of the section. It is sometimes desirable to draw a shear diagram, such as is shown bounded by the dotted line partly above and partly below the base line *A–B*. The intercept *Y* is equal to the effective load area to the left or .68. The intercept Y_1 is equal to the effective load area to the left or .68 + 1.07 = 1.75. The intercept Y_2 is equal to .68 + 1.07 + 1.64 = 3.39 etc. When the negative areas are reached, the shear curve turns back towards the base line and crosses it near the center. The shear at any section may be obtained in tons by measuring the corresponding intercept in the shear diagram to the proper scale. If the number of tons which one square inch of area under the load curve represents, be multiplied by the number of square inches which one inch vertical in the shear diagram was made to represent, we have the number of tons which one inch of intercept in the shear diagram represents.

149. Moment of Inertia of a Large Reinforced Concrete Section.—Fig. 705 shows a half section of a reinforced concrete ship hull. Let it be required to locate the neutral axis and find the moment of inertia for both positive and negative moments about a horizontal axis.

First consider the case when the ship is in the hogging position, the moment producing tension in the top and compression in the bottom. The rods are grouped together, and horizontal lines drawn from the approximate centroids of the groups are extended some little distance to the right. It will be found much more convenient to group the rods, and have each group represented by one line, than to have a line for each rod. However, the vertical dimension of the groups should not be too large or an error will be produced. The concrete on the compression side is divided into slices of small vertical dimensions, and horizontal line drawn from their centroids are extended to the right. Starting at the top, the steel area for each group of bars times the ratio *n* is computed, and the values scaled off in order along the load line of Fig. 706, extending from *z* towards the left. Then starting at the bottom, the area for each slice is computed; also the steel area of each group of bars times *n*–1 is computed.

The value $n-1$ is used in place of n for the steel on the compression side because the steel displaces one times its area of compression

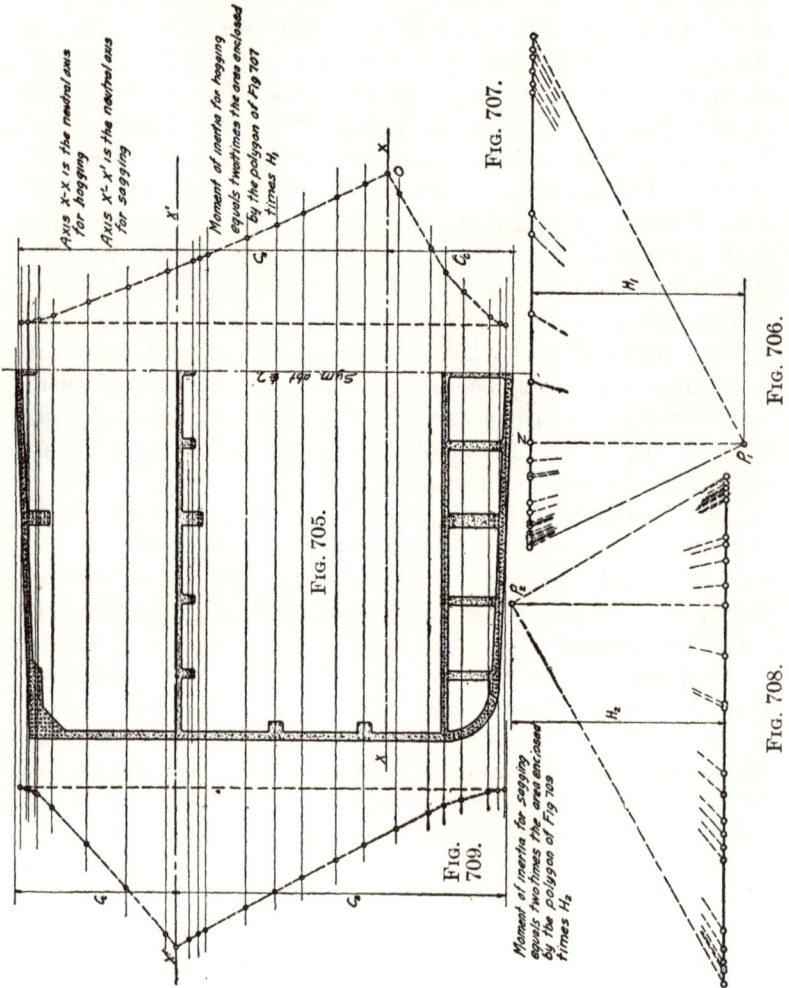

Axis X–X is the neutral axis for hogging

Axis X'–X' is the neutral axis for sagging

Moment of inertia for hogging equals two times the area enclosed by the polygon of Fig 707 times H_1

Sum 190 tons

Moment of inertia for sagging equals two times the area enclosed by the polygon of Fig 709 times H_2

FIG. 705.

FIG. 707.

FIG. 706.

FIG. 708.

FIG. 709.

concrete. These area values for the slices and the compression steel groups are laid off from z towards the right, taking them in order. When it is evident that we are above the neutral axis there is no use of laying off any more compression area, neither is there any use laying off more tension values when it is evident that the neutral axis has been passed.

From Fig. 706 the funicular polygon of Fig. 707 is drawn, and the neutral axis located by the intersection o. The moment of inertia is equal to 2 times the pole distance H, times the area inclosed by the polygon of Fig. 707. Then using the common formula $\dfrac{M}{S} = \dfrac{I}{c}$, the stress in the concrete and steel for any moment may be easily determined. When finding the maximum stress in the concrete, c_c should be used for c, and for finding the maximum steel stress use c_s and multiply the result by n. Formulas expressing this may be written thus

$$S_c = \frac{M \cdot c_c}{I} \quad \text{and} \quad S_s = \frac{M \cdot c_s}{I} \cdot n.$$

In a similar way, by the use of Figs. 708 and 709, the neutral axis $x'-x'$ is located and the moment of inertia found for the case when the ship is in the sagging position. If the stresses are desired, they may be found by the method just explained.

150. Stresses in a Reinforced Concrete Beam Foundation.— Let it be required to find the distribution of moment and shear along the length of the continuous footing shown in Fig. 710. The column loads are given, also the allowable soil pressure. The footing has a rectangular cross-section, is of length such that the center of gravity of the loads is directly above the centroid of the footing area, and has a width such that the soil pressure is not excessive. It resists two sets of forces, the four column loads acting down, and the distributed soil pressure acting up.

The area representing the distributed pressure from the soil is divided into small vertical slices, and an equivalent concentrated load substituted for each slice. The loads acting upon the footing are now laid off in order along the load line of Fig. 710. Any convenient pole distance H is chosen, and the pole changed from side to side as required by the different sense of the forces. This is more convenient than using just one pole and doubling up the load line. After Fig. 711 is completed it should be checked by seeing if the last ray is parallel to the first. Starting at any point S on the vertical line from the first pressure area, a funicular polygon is drawn for the loads and pressures and the point T located. The line $S-T$ is the closing string and it should be parallel

to the first and last rays in Fig. 711. The intercept between
the polygon and the base line S–T times the pole distance H gives
the moment at any section. The intercepts in Fig. 712 are

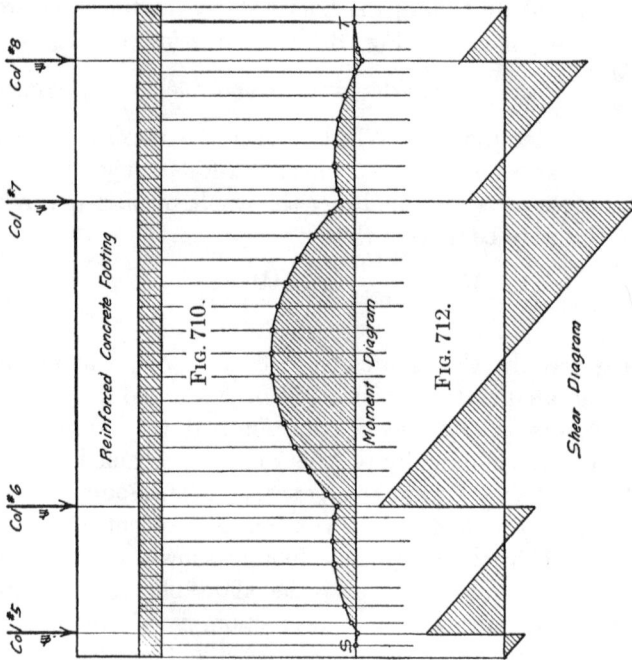

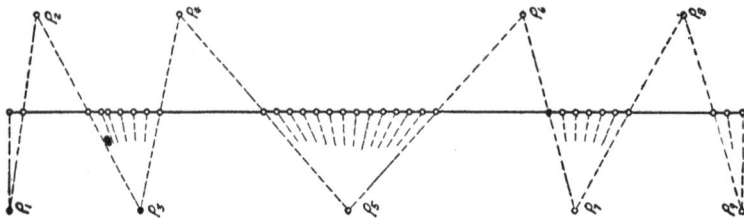

to be measured to the scale used for Fig. 710, and H should be
measured to the scale used for the load line of Fig. 711.

The shear diagram is shown in Fig. 713.

The footing just studied is really an inverted continuous
beam with the column loads corresponding to the reactions and
the soil pressure corresponding to a distributed load. There is,

however, an important feature which it may be well to discuss. Ordinary continuous beams have their loading given and the reactions are determined from the loads and the spans. In other words, given the beam, the loading, and the location of the supports, the magnitudes of the reactions are fixed, as long as the supports remain at the same elevation. But by varying the elevation of the support, almost any desired moment may be obtained at any support. In the footing problem just considered the column loads which correspond to the reactions of a continuous beam, are given, and a uniform distribution for the soil pressure assumed. It would, therefore, seem that we have assumed something which we had no right to assume, and to a certain extent this may be true. The moments given by the moment diagram of Fig. 712 exist only after certain deflections have been produced in the footing. These deflections will usually be such as to leave the column bases at slightly different elevations. Deflections in the footing mean that the soil will be compressed more in some places than in others and we would, therefore, expect a slight variation in the pressure which the soil exerts against the footing. However, beam footings are in most cases rather deep, therefore the deflection and also the variation in soil pressure will be small.

The redistribution of the soil pressure due to deflection occurs in all spread or beam footings. The effect is usually small, and in most cases tends to reduce the maximum moments. It is customary to neglect it, but the designer should keep in mind the fact that there is such an action.

INDEX

Action line of a force, 2, 5
Advantage of graphical method, 1
Analytical moments, stresses in, trusses by, 117
Anchor bolts, steel bases with, 339
Angle, graphical construction of, 3
    graphical measurement of, 3
Anti-resultant, 3, 7
Application, point of, 2, 5
Arc, centroid of an, 42
Arches, hingeless, 242
    line of pressure in, 228
    three-hinged, 231
    three-hinged symmetrically loaded, 237
    two-hinged, 238
    two-hinged, method of least work, 238
Areas, centroids of, 45
    centroids of irregular, 51
Area curves, centroid located by, 59
Area, effective, 283
    first moment of, 67
    moment of inertia of, 72

Bases, steel, with anchor bolts, 339
Beams, deflection of reinforced concrete, 278
    design of, 281
Beams, design of reinforced concrete, 287
    design of steel, 286
    design of wooden, 285
    effective area of, 283
    effective depth of, 282
    fixed at both ends, deflection of, 99

Beams, fixed at both ends, moment diagram for, 100
    horizontal shear in, 285
    moment diagram for, 68
    reactions for, 23, 25, 26
    reactions from distributed load, 25
    reactions from inclined loads, 26
    steel, reinforced with concrete fireproofing and slabs, 344
    trussed, 319
    with an overhanging end, deflection of, 91
    with an overhanging end, moment diagram for, 91
    with an overhanging end, shear diagram for, 91
    with a variable moment of inertia, deflection of, 94
    with one end fixed, deflection of, 96
    with one end fixed, moment diagram for, 98
    with one end fixed, shear diagram for, 98
Bending stresses in complex reinforced concrete sections, 267
Bicycle frame, stresses in, 35
Bridge truss of the K type, stress diagram for, 156
Broken line, centroid of, 40

Cantilever beams, deflection of, 89
Cantilever beams, moment diagram for, 90
    shear diagrams for, 90
Cantilever trusses, stress diagrams for, 135

Cast-iron O-gee washers, diagram for designing, 316
Central circle of inertia, 79
Central ellipse of inertia, 79
Centroid of, an arc, 42
 a broken line, 40
 a circular segment, 50
 a curve, 44
 a quadrilateral, 46
 a sector, 49
 a trapezoid, 47
 a triangle, 45
 triangular pyramid, 53
Centroid located by area curves, 59
Centroids of, areas, 45
 complex volumes, 54
 irregular areas, 51
 irregular volumes, 57
 volumes, 53
Chimneys, masonry, 224
 reinforced concrete, 275
Columns, eccentrically loaded reinforced concrete, 271
 two fixed and a girder, continuous frame of, 347
Combined loads, stress diagram for, 127
Combined stresses, 317
Complex reinforced concrete sections, bending stresses in, 267
Complex volumes, centroids of, 54
Complicated trusses, stress diagrams for, 153, 155
Component, 3, 8
Composition of forces, 2, 6
Compression, diagonal, 284
Compression members, wooden, diagram for designing, 315
Concurrent-coplanar forces, 2, 10
Concurrent forces, 2
Conditions for equilibrium, 10, 19
Connection, wind bracing, 342
Construction of the leastic curve, 81
Continuous beams of three spans, deflection of, 107
 moment diagram for, 109
 shear diagram for, 109

Continuous beams of two spans, deflection of, 102
 moment diagrams for, 104
Continuous frame of two fixed columns and a girder, 347
Coplanar forces, 2
Couple, resultant a, 18
Culmann's method, 73
Curve, centroid of, 44

Deflection of beams, fixed at both ends, 99
 with an overhanging end, 91
 with a variable moment of inertia, 94
 with one end fixed, 96
Deflection of, cantilever beams, 89
 continuous beams of three spans, 107
 continuous beams of two spans, 102
Deflection of, reinforced concrete beams, 278
 simple beams, 85
Depth, effective, 282
Design of, beams, 281
 plate girders, 297
 reinforced concrete beams, 287
 reinforced concrete slabs, 289
 steel beams, 286
 steel trusses, 300
 wooden beams, 285
 wooden trusses, 305
Details of, steel trusses, 300
 wooden trusses, 305
Diagonal, compression, 284
 tension, 284
Diagram, stress, 120
Diagram for designing, cast-iron O-gee washers, 316
 eccentrically loaded reinforced concrete columns, 328
 eccentrically loaded steel columns, 323
 steel angle struts, 317
 wooden compression members, 315
Direction, of a force, 2, 5
Domes, 250
Double, reinforced concrete beams, 264

Eccentrically loaded columns, reinforced concrete, 271
reinforced concrete columns, diagram for designing, 328
steel columns, diagram for designing, 323
Effective, area, 283
depth, 282
Elastic curve, construction of, 81
Equilibrium, 2
conditions for, 10, 19
polygon, 3, 14
Exact method, moment of inertia by, 76

First moment of, an area, 67
a force, 65
Force, direction of, 2, 5
first moment of, 65
magnitude of, 2, 5
polygon, 3, 8
Forces, composition of, 2, 6
concurrent, 2
concurrent-coplanar, 2, 10
coplanar, 2
non-concurrent-coplanar, 13
representation of, 4
resolution of, 3, 8
Foundation, stresses in a reinforced concrete beam, 365
Frames, stresses in the members of, 33
Funicular polygon, 3, 14
passed through three points, 37

Girders, plate, design of, 297
Gothic vault, investigation of, 245
Graphical method, advantage of, 1
Graphical moments, stresses in, trusses by, 117
Gyration, radius of, 74

Hingeless arches, 242
Higher moments, 77
Horizontal shear, 285

Inertia, central circle of, 79
central ellipse of, 79
Investigation of a gothic vault, 245

Irregular, piers, stresses in, 206
volumes, centroids of, 57

Kerns, 209
location of points on edge of analytically, 214
Ketchum, Milo S., 113, 115

Least crown pressure, theory of, 245
Line of pressure in, an arch, 228
a pier, 222
Loads (also see moving loads)
snow, 115
winds, 115
Location of points on edge of Kern, analytically, 214
Longitudinal strength calculations for a ship, 359
Lower chord loads, stress diagram for, 123

Magnitude, of a force, 2, 5
Masonry arch, true pressure line in, 238
Masonry chimneys, 224
Maximum and minimum stresses, 128
Maximum shears and moments in a turntable, 193
Method, Culmann's, 73
Mohr's, 73
Method of least work, hingeless arches, 242
two-hinged arches, 238
Mill bent, stress diagram for, 137, 150
Mohr's method, 73
Moment diagram for a beam, 68
Moment diagram for beams with, an overhanging end, 91
both ends fixed, 100
one end fixed, 98
Moment diagram for cantilever beams, 90
Moment diagram for, continuous beams of three spans, 109
two spans, 104
Moment diagram for simple beams, 86
Moments, higher, 77

Moments, maximum in a turntable, 193

second, 69

Moment of, an area, first, 67

a force, first, 65

Moment of inertia, by exact method, 76

of an area, 72

of a large reinforced concrete section, 363

Moving loads, a large number of concentrated, 190

four concentrated, with uniform dead load, 187

on trusses, 196

single concentrated, 168

single concentrated and uniform dead, 171

three concentrated, 184

two concentrated, 179

Moving loads, uniform longer than the span, 175

uniform shorter than the span, 177

uniform shorter than the span with uniform dead load, 179

Non-concurrent coplanar forces, 13

Notation, xiii

for trusses, 116

O-gee washers, cast iron, diagrams for designing, 316

Parallel forces, resultant of, 19, 21

Parallelograms, radius of gyration of, 75

Pier, line of pressure in, 222

Piers, stresses in irregular, 206

stresses in rectangular, 201

Plate girders, design of, 297

Point of application, 2, 5

Polygon, equilibrium, 3, 14

force, 3, 8

funicular, 3, 14

Pressure, line of, in an arch, 228

on wall footings, 216

Problems, design, 330

Purlins, steel, diagram for use in designing, 311

wooden, diagram for use in designing, 310

Pyramid, centroid of a triangular, 53

Quadrilateral, centroid of, 46

Radius of gyration, 74

of parallelograms, 75

of rectangles, 75

of triangles, 75

Rafter, reactions for, 33

Reactions for, beams, 23, 25, 26

a rafter, 33

trusses, 27

Rectangles, radius of gyration of, 75

Rectangular piers, stresses in, 201

Reinforced concrete beams, deflection of, 278

design of, 287

Reinforced concrete, beam foundation, stresses in, 365

bending stresses in complex sections, 267

chimneys, 275

columns, combined stresses in, 270

columns, eccentrically loaded, diagram for designing, 328

double reinforced beams, 264

eccentrically loaded columns, 271

moment of inertia of a large section, 363

moment of inertia for rectangular beams, 259

moment of inertia of T-beams, 264

simple rectangular beams, 254

slabs, design of, 289

T-beams, 260

Representation of forces, 4

Resolution of forces, 3, 8

Resultant, 3, 7

a couple, 18

of parallel forces, 19, 21

Retaining walls, 218

Reversal of stress in truss members, 128

Ricker, Dr. N. C., 113

Ring dome, stress diagrams for, 159, 163
Roof coverings, weights of, 114

Second moments, 69
Sector, centroid of, 49
Segment, centroid of a circular, 50
Sense, 2, 5
Shear diagram for beams with, an overhanging end, 91
one end fixed, 98
Shear diagram for, cantilever beams, 90
continuous beams of three spans, 109
simple beams, 86
Shear, horizontal, 285
Shears, maximum in a turntable, 193
Ships, longitudinal strength calculations for, 359
Simple beams, deflection of, 85
moment diagram for, 86
shear diagram for, 86
Slabs, design of reinforced concrete, 289
Smith, W. M., 238
Snow loads, 115
Steel angle struts, diagram for designing, 317
Steel bases with anchor bolts, 339
Steel beams, design of, 286
reinforced with concrete fireproofing and slabs, 344
Steel columns, eccentrically loaded, diagram for designing, 323
Steel purlins, diagrams for use in designing, 311
Steel towers, 354
Steel trusses, design and details of, 300
Stress diagram, 120
for bridge truss of K type, 156
for cantilever trusses, 135
for combined loads, 127
for complicated trusses, 153, 155
for lower chord loads, 123
for mill bent, 137, 150
for ring dome, 159, 163

Stress diagram, for three-hinge arch truss, 141, 146
for trussed dome, 158
for upper and lower chord loads, 122
for wind loads, 124
Stress volumes, 61, 204
Stresses, combined, 317
combined in reinforced concrete columns, 270
in a bicycle frame, 35
in a reinforced concrete beam foundation, 365
in irregular piers, 206
in rectangular piers, 201
in the members of a frame, 33
maximum and minimum, 128
obtained analytically in trusses, 116
in trusses by analytical moments 117
in trusses by graphical moments, 117
in trusses obtained by stress diagram, 120
in trusses obtained graphically by joints, 118
Struts, steel angle, diagram for designing, 317

T-beams, moment of inertia of reinforced concrete, 264
reinforced concrete, 260
Tension, diagonal, 284
Theory of least crown pressure, 245
Three-hinged arches, 231
Three-hinge, a r c h symmetrically loaded, 237
truss, stress diagram for, 141, 146
Towers, steel, 354
Trapezoid, centroid of, 47
Triangle, centroid of, 45
radius of gyration of, 75
True pressure line in a masonry arch, 238
Trussed beams, 319
Trussed dome, stress diagram for, 158
Trusses, maximum and minimum stresses in, 128
moving loads on, 196

Trusses, notation for, 116
  reactions for, 27
  reversals in members of, 128
  steel, design and details of, 300
  stresses obtained analytically in, 116
  stresses in, obtained by stress diagram, 120
  stresses in, obtained graphically by joints, 118
  weight of, 113
  wooden, design and details of, 305
Turntable, maximum shears and moments in, 193
Two-hinged arches, 238
  method of least work, 238

Upper and lower chord loads, stress diagram for, 122

Vault, investigation of a gothic, 245

Vector, 2–5
Volumes, centroids of, 53
  centroid of complex, 54
  centroids of irregular, 57
  stress, 61, 204

Wall footings, pressure on, 216
Walls, retaining, 218
Weight of trusses, 113
Weights of roof coverings, 114
Wind-bracing connection, 342
Wind loads, 115
  stress diagram for, 124
Wooden beams, design of, 285
Wooden compression members, diagram for designing, 315
Wooden purlins, diagram for use in designing, 310
Wooden trusses, design and details of, 305

www.ingramcontent.com/pod-product-compliance
Lightning Source LLC
Chambersburg PA
CBHW030935150426

42812CB00064B/2888/J